PARTIAL DIFFERENTIAL EQUATIONS
OF MATHEMATICAL PHYSICS

S. L. SOBOLEV

PARTIAL DIFFERENTIAL EQUATIONS OF MATHEMATICAL PHYSICS

Translated from the third Russian edition by

E. R. DAWSON
Lecturer in Mathematics at Queen's College, Dundee
University of St. Andrews

English translation edited by

T. A. A. BROADBENT
Professor of Mathematics
Royal Naval College, Greenwich

DOVER PUBLICATIONS, INC.
New York

Published in Canada by General Publishing Company, Ltd., 30 Lesmill Road, Don Mills, Toronto, Ontario.

This Dover edition, first published in 1989, is an unabridged and unaltered republication of the work first published by Pergamon Press in 1964. The original Russian edition, *Uravneniya matematicheskoi fiziki,* was published by Gostekhizdat, Moscow. This edition is reprinted by special arrangement with Pergamon Books Ltd., Headington Hill Hall, Oxford, OX3 OBW, England.

Manufactured in the United States of America
Dover Publications, Inc., 31 East 2nd Street, Mineola, N.Y. 11501

Library of Congress Cataloging-in-Publication Data

Sobolev, S. L. (Sergeĭ L'vovich), 1908–
    [Uravneniiã matematicheskoĭ fiziki. English]
    Partial differential equations of mathematical physics / S.L. Sobolev ; translated from the third Russian edition by E.R. Dawson ; English translation edited by T.A.A. Broadbent.
        p.   cm.
    Translation of: Uravneniiã matematicheskoĭ fiziki.
    Reprint. Originally published: London : Pergamon Press, 1964.
    Includes index.
    ISBN 0-486-65964-X
    1. Mathematical physics.   2. Differential equations, Partial.   I. Title.
QA401.S613   1989   89-1084
530.1'55353—dc19   CIP

CONTENTS

v

TRANSLATION EDITOR'S PREFACE

THE classical partial differential equations of mathematical physics, formulated and intensively studied by the great mathematicians of the nineteenth century, remain the foundation of investigations into waves, heat conduction, hydrodynamics, and other physical problems. These equations, in the early twentieth century, prompted further mathematical researches, and in turn themselves benefited by the application of new methods in pure mathematics. The theories of sets and of Lebesgue integration enable us to state conditions and to characterize solutions in a much more precise fashion; a differential equation with the boundary conditions to be imposed on its solution can be absorbed into a single formulation as an integral equation; Green's function permits a formal explicit solution; eigenvalues and eigenfunctions generalize Fourier's analysis to a wide variety of problems.

All these matters are dealt with in Sobolev's book, without assumption of previous acquaintance. The reader has only to be familiar with elementary analysis; from there he is introduced to these more advanced concepts, which are developed in detail and with great precision as far as they are required for the main purposes of the book. Care has been taken to render the exposition suitable for a novice in this field: theorems are often approached through the study of special simpler cases, before being proved in their full generality, and are applied to many particular physical problems.

Commander Dawson has taken pains to render his translation idiomatic as well as accurate, thus assisting the English reader to avail himself readily of the vast amount of information contained in this volume.

T.A.A.BROADBENT

AUTHOR'S PREFACE TO THE FIRST EDITION

THIS book is based on a course of lectures given in the "Lomonosov" State University in Moscow. The author has therefore retained the name "lectures" for the various sections. The same circumstance also explains the selection of material, the extent of which was limited by the number of lecture periods.

The author expresses his deep gratitude to Academician V.I.Smirnov, who read through the manuscript, and also to Professor V.V.Stepanov for his useful comments.

S.SOBOLEV

AUTHOR'S PREFACE TO THE THIRD EDITION

THE third edition of this course on "Equations of Mathematical Physics" differs little from the second edition, which underwent extensive revision. In the second edition the lecture on the Ritz method was omitted since that subject lies somewhat apart from the rest of the course. The theory of multiple Lebesgue integrals and of integral equations has been simplified somewhat, and the proof of the Fourier method has been made more precise.

As in the second edition, various improvements in style and clarifications of the presentation have been made. Moreover, in this third edition the lecture on the dependence of the solutions of equations of mathematical physics on the boundary conditions has been developed in greater detail by the editor, V.S.Ryaben'kii.

The author expresses his gratitude for valuable comments made by various people when the second and third editions were being prepared. Particularly valuable comments were made by Academician V.I.Smirnov and the editor of the third edition, V.S.Ryaben'kii.

S.SOBOLEV

PARTIAL DIFFERENTIAL EQUATIONS
OF MATHEMATICAL PHYSICS

DERIVATION OF THE FUNDAMENTAL EQUATIONS

THE theory of the equations of mathematical physics has as its object the study of the differential, integral, and functional equations which describe various natural phenomena. It is somewhat difficult to define the precise limits of the subject as it is usually understood. Moreover, the great variety of problems relating to the equations of mathematical physics does not allow them to be dealt with at all fully in a university course. The present book contains only a fraction of the whole theory of the equations of mathematical physics: it includes only what seemed to be most important for an introduction to the subject.

The course is devoted for the most part to the study of second-order partial differential equations with one unknown function; in particular, we shall deal with what are usually called the *classical equations of mathematical physics*, namely, the wave equation, Laplace's equation, and the equation of heat conduction. We shall develop the necessary theory of related problems as we go along.

§ 1. Ostrogradski's Formula †

Before we undertake the derivation of those equations of mathematical physics with which we shall be concerned, we recall a formula of integral calculus dealing with the transformation of surface integrals into volume integrals.

Let $P(x, y, z)$, $Q(x, y, z)$, $R(x, y, z)$ be three functions which:

(i) are specified in a certain domain D of the variables x, y, z;

(ii) are continuous right up to the boundary of D; and

(iii) have continuous first-order partial derivatives with respect to x, y, z throughout D.

Consider within D some closed surface S consisting of a finite number of pieces for each of which the tangent plane varies continuously. Such a surface

---

† Mikhail Vassilievich Ostrogradski (1801–1862). See *Mem. Acad. Imp. Sci.*, St. Petersburg (6), **1**, 130 (1831). The result is otherwise known as Green's lemma, Gauss's theorem, or the Divergence theorem.—*Translator*.

is said to be *piecewise smooth*. We shall further suppose that any straight line parallel to any of the coordinate axes either intersects S in a finite number of points or has a whole interval in common with it.

Consider the integral

$$\iint_S [P \cos (n, x) + Q \cos (n, y) + R \cos (n, z)] \, dS. \tag{1.1}$$

where $\cos (n, x)$, $\cos (n, y)$, $\cos (n, z)$ denote the cosines of the angles formed by the *inward-directed normal* to the surface S at the point (x, y, z), and dS is an element of the surface.

Using vector notation, we can regard P, Q, R as the components of a certain vector T.

Then

$$P \cos (n, x) + Q \cos (n, y) + R \cos (n, z) = T_n,$$

where T_n is the projection of the vector T in the direction of the inward normal.

A classical theorem of integral calculus enables us to transform the surface integral (1.1) into a volume integral over the region D bounded by the surface S.

This theorem asserts that:

$$\iint_S [P \cos (n, x) + Q \cos (n, y) + R \cos (n, z)] \, dS$$

$$= - \iiint_D \left[\frac{\partial P}{\partial x} + \frac{\partial Q}{\partial y} + \frac{\partial R}{\partial z} \right] dx \, dy \, dz$$

or in vector notation

$$\iint_S T_n \, dS = - \iiint_D \operatorname{div} T \, dv \tag{1.2}$$

where dv denotes an infinitesimal volume-element and

$$\operatorname{div} T = \frac{\partial P}{\partial x} + \frac{\partial Q}{\partial y} + \frac{\partial R}{\partial z}. \tag{1.3}$$

The formula just obtained is valid under rather more general assumptions with regard to S. In particular formula (1.2) holds for any piecewise smooth surface bounding a certain region D.

We shall in future take the word "surface" to mean a piecewise smooth surface unless a further restriction on its meaning is made.

An important result follows from formula (1.2):

LEMMA 1. *Let F be a continuous function defined in some domain in three-dimensional Euclidean space. Let S be any closed surface within the domain over which a vector function T is specified, and let S bound the region Ω. Then the necessary and sufficient condition for the equality*

$$\iint_S T_n \, dS - \iiint_\Omega F \, dv = 0 \qquad (1.4)$$

to hold good is that

$$\operatorname{div} \boldsymbol{T} + F = 0.$$

For, using formula (1.2), we can put the equality (1.4) into the form

$$\iiint_\Omega (\operatorname{div} \boldsymbol{T} + F) \, dv = 0,$$

and then the sufficiency of the condition in the lemma becomes obvious. It is also necessary. For, suppose, if possible, that the function $\operatorname{div} \boldsymbol{T} + F$ is different from zero, say positive, at some point A; then because of continuity it would also be positive in the neighbourhood of A, and the integral

$$\iiint_\omega (\operatorname{div} \boldsymbol{T} + F) \, dv$$

taken over a small region ω round A would be non-zero, and so the left-hand member of (1.4) would also be different from zero. Hence our supposition contradicts (1.4), and the necessity of the equality

$$\operatorname{div} \boldsymbol{T} + F = 0$$

is proved.

The analogous lemma for a two-dimensional region lying in a plane can be proved in a similar way.

§ 2. Equation for Vibrations of a String

Consider a string stretched between two points. By *string* we mean a rigid body whose other dimensions are small compared with its length; and we also suppose that the tension in it is considerable, so that its resistance to flexure can be neglected in comparison with the tension.

We take the x-axis to be along the string when it is in equilibrium under the action of the tension only. When transverse forces act on the string it will assume some other form, in general non-rectilinear.

Imagine the string cut into two pieces at some point x and consider the interaction between the two parts. The force which the right-hand part exerts on the left-hand part is directed along the tangent at x to the curve representing the string and is denoted by $T(x)$ (see Fig. 1).

To simplify the discussion, let us suppose that the motion of the string takes place in one plane, and let u denote the displacement of the string from its rest position. Let $u = u(x, t)$ be the equation of the curve assumed

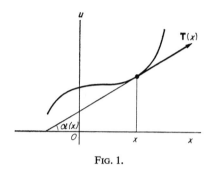

FIG. 1.

by the string in the plane xOu. Let $\varrho(x)$ denote the linear density of the string at the point x, *i.e.*, the limit of the ratio of mass to length for a small part of the string.

We consider first the equilibrium position of the string under the influence of a transverse loading $p(x)$: by which we mean that the part of the string defined by $x_1 \leq x \leq x_2$ is acted on by a force, directed along the u-axis, of magnitude $\int_{x_1}^{x_2} p(x)\,dx$. Let $\alpha(x)$ be the angle formed with the x-axis by the tangent to the string at the point x; then the component along the u-axis of the tension acting at the point x_2 is given by

$$\left|T(x_2)\right| \sin \alpha(x_2) = T(x_2) \sin \alpha(x_2)$$

where $T(x)$ is the absolute magnitude (the length) of the vector $T(x)$.

Similarly, the component along the u-axis of the tension at x_1 is given by

$$-\left|T(x_1)\right| \sin \alpha(x_1) = -T(x_1) \sin \alpha(x_1).$$

Now

$$\sin \alpha = \frac{\dfrac{\partial u}{\partial x}}{\sqrt{1 + \left(\dfrac{\partial u}{\partial x}\right)^2}}$$

and if we take $\partial u/\partial x$ to be so small that its square can be neglected, we obtain as the equilibrium condition for the string

$$\left[T\frac{\partial u}{\partial x}\right]_{x_2} - \left[T\frac{\partial u}{\partial x}\right]_{x_1} + \int_{x_1}^{x_2} p(x)\,\mathrm{d}x = 0 \tag{1.5}$$

But, clearly,

$$\left[T\frac{\partial u}{\partial x}\right]_{x_2} - \left[T\frac{\partial u}{\partial x}\right]_{x_1} = \int_{x_1}^{x_2} \frac{\partial}{\partial x}\left(T\frac{\partial u}{\partial x}\right)\mathrm{d}x$$

so that the condition (1.5) can be written as

$$\int_{x_1}^{x_2}\left[\frac{\partial}{\partial x}\left(T\frac{\partial u}{\partial x}\right) + p(x)\right]\mathrm{d}x = 0. \tag{1.6}$$

Since (1.6) holds for *any* values of x_1 and x_2, the integrand must be identically zero, *i.e.*,

$$\frac{\partial}{\partial x}\left(T\frac{\partial u}{\partial x}\right) + p(x) = 0 \tag{1.7}$$

and this is the required *equation for equilibrium of the string under the transverse loading* $p(x)$.

Next we pass from statics to dynamics and consider vibrations of the string. To do this we apply d'Alembert's principle and include in the equation for equilibrium the inertial forces for the string as well; these take the form

$$\int_{x_1}^{x_2}\left(-\varrho(x)\frac{\partial^2 u}{\partial t^2}\right)\mathrm{d}x.$$

The condition for equilibrium becomes

$$\int_{x_1}^{x_2}\left[\frac{\partial}{\partial x}\left(T\frac{\partial u}{\partial x}\right) - \varrho(x)\frac{\partial^2 u}{\partial t^2} + p(x)\right]\mathrm{d}x = 0,$$

and the equation for vibrations of the string will be

$$\frac{\partial}{\partial x}\left(T\frac{\partial u}{\partial x}\right) - \varrho(x)\frac{\partial^2 u}{\partial t^2} + p(x) = 0. \tag{1.8}$$

We suppose that the vibrations of the string are transverse. The components along the x-axis of all the forces acting on any element of the string must therefore add up to zero. Since the loading $p(x)$ is also transverse, we have

$$[T\cos\alpha]_{x_2} - [T\cos\alpha]_{x_1} = 0$$

for any values of x_1, x_2.

Hence, neglecting α^2 and higher powers, we get

$$[T]_{x_2} = [T]_{x_1}$$

i.e., T is independent of x. Hence T can be taken outside the differential co-efficient with respect to x in (1.8). If we also assume that T is independent of the time, *i.e.*, that it is constant, and also that the density ϱ is constant and the loading $p(x)$ zero, then equation (1.8) takes the form

$$\frac{\partial^2 u}{\partial t^2} = a^2 \frac{\partial^2 u}{\partial x^2} \tag{1.9}$$

where $a^2 = T/\varrho = \text{constant}$.

The equation (1.9) was discussed by Daniel Bernoulli, d'Alembert, and Euler as early as the 18th century.

§ 3. Equation for Vibrations of a Membrane

Consider a film, *i.e.*, a very thin rigid body stretched uniformly in all directions. We suppose that the film is so thin that it offers no resistance to flexure. Such a film is called a *membrane*.

Let its rest position be in the plane xOy and let

$$u = u(t, x, y)$$

be its equation when displaced.

Considering any piece S of the membrane, we suppose that the rest of the membrane exerts on it a uniformly distributed tension T which, at any

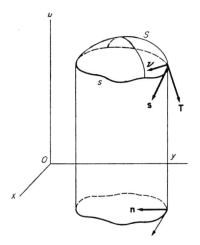

FIG. 2.

point on the boundary of S, is directed along the normal to the boundary and lies in the tangent plane to the membrane (see Fig. 2).

We shall establish the equation for the equilibrium of the part S of the membrane which is bounded by the curve s under the action of transverse forces. The component along the u-axis due to the tension is given by the integral

$$\int T \cos (l, u) \, \mathrm{d}s \qquad (1.10)$$

where T is the length of the vector $\boldsymbol{T}$

and l is a vector directed along the line of action of the tension.

We next evaluate $\cos (l, u)$.

By hypothesis, the vector l is perpendicular both to the boundary s and to a vector $\boldsymbol{v}$ along the inward-directed normal $\boldsymbol{v}$ to the surface $u = u(t, x, y)$. Again, any vector s directed along a tangent to the boundary s is perpendicular to $\boldsymbol{v}$ and to a unit vector $\boldsymbol{n}$ along the inward normal (at the corresponding point) to the projection of the boundary s on to the plane xOy (since the tangent vector s and the tangent to the projection of s on to the plane xOy lie in a plane touching the projecting cylinder). Hence as the vector s we may take the vector product $\boldsymbol{n}_\wedge\boldsymbol{v}$, and the vector l_1 defined by $l_1 = s_\wedge\boldsymbol{v} = (\boldsymbol{n}_\wedge\boldsymbol{v})_\wedge\boldsymbol{v}$ can also be written as $l_1 = -\boldsymbol{n}\boldsymbol{v}^2 + \boldsymbol{v}(\boldsymbol{n}\boldsymbol{v})$.

Since $\boldsymbol{n}$ has components $\{\cos (\boldsymbol{n}, x), \cos (\boldsymbol{n}, y), \cos (\boldsymbol{n}, z)\}$, and $\boldsymbol{v}$ has components $\left\{ -\dfrac{\partial u}{\partial x}, -\dfrac{\partial u}{\partial y}, 1 \right\}$, we find for the components of l_1 the expressions $\left\{ -\cos (\boldsymbol{n}, x), -\cos (\boldsymbol{n}, y), -\dfrac{\partial u}{\partial x}\cos (\boldsymbol{n}, x) - \dfrac{\partial u}{\partial y}\cos (\boldsymbol{n}, y) \right\}$, from which we have dropped $\left(\dfrac{\partial u}{\partial x}\right)^2, \left(\dfrac{\partial u}{\partial y}\right)^2, \dfrac{\partial u}{\partial x}\dfrac{\partial u}{\partial y}$ as being quantities of the second order of smallness.

To the same order, the length of vector l_1 is unity, and so we may now regard l_1 as the unit vector l directed along the line of action of the tension, and thus

$$\cos (l, u) = -\frac{\partial u}{\partial x}\cos (\boldsymbol{n}, x) - \frac{\partial u}{\partial y}\cos (\boldsymbol{n}, y).$$

The equation for equilibrium of the membrane has the form

$$\iint_\omega p(x, y) \, \mathrm{d}x \, \mathrm{d}y + \int_s T \cos (l, u) \, \mathrm{d}s = 0$$

where $p(x, y)$ is the magnitude of the transverse loading per unit area, and ω is the projection of S on to the plane xOy.

Substituting for cos (l, u) we get

$$\iint_\omega p(x, y)\, dx\, dy - \int_s \left(\frac{\partial u}{\partial x} \cos (n, x) + \frac{\partial u}{\partial y} \cos (n, y)\right) T\, ds = 0,$$

or by virtue of Lemma 1,

$$\frac{\partial}{\partial x}\left(T\frac{\partial u}{\partial x}\right) + \frac{\partial}{\partial y}\left(T\frac{\partial u}{\partial y}\right) + p(x, y) = 0. \tag{1.11}$$

The equation for vibrations of the membrane has the form

$$\iint_\omega \left(p(x, y) - \varrho\frac{\partial^2 u}{\partial t^2}\right) dx\, dy - \int_s T\left(\frac{\partial u}{\partial x} \cos (n, x) + \frac{\partial u}{\partial y} \cos (n, y)\right) ds = 0$$

where $\varrho = \varrho(x, y)$ is the density per unit area of the membrane,

or, by Lemma 1,

$$\frac{\partial}{\partial x}\left(T\frac{\partial u}{\partial x}\right) + \frac{\partial}{\partial y}\left(T\frac{\partial u}{\partial y}\right) + p(x, y) - \varrho(x, y)\frac{\partial^2 u}{\partial t^2} = 0. \tag{1.12}$$

If T and ϱ are constant, we get from (1.12)

$$T\left(\frac{\partial^2 u}{\partial x^2} + \frac{\partial^2 u}{\partial y^2}\right) + p(x, y) = \varrho\frac{\partial^2 u}{\partial t^2}. \tag{1.13}$$

The sum $\dfrac{\partial^2 u}{\partial x^2} + \dfrac{\partial^2 u}{\partial y^2}$ or, in three-dimensional space, $\dfrac{\partial^2 u}{\partial x^2} + \dfrac{\partial^2 u}{\partial y^2} + \dfrac{\partial^2 u}{\partial z^2}$, is often called *Laplace's operator* and is denoted by $\nabla^2 u$. Using this notation we can write (1.13) as

$$\frac{\partial^2 u}{\partial t^2} = a^2\, \nabla^2 u + \frac{p(x, y)}{\varrho} \tag{1.14}$$

where

$$a^2 = \frac{T}{\varrho} = \text{a constant.}$$

§ 4. Equation of Continuity for Motion of a Fluid. Laplace's Equation

Before deriving the equation of continuity, we establish an important formula.

Consider a closed, piecewise smooth, time-dependent surface $S(t)$ enclosing a variable volume $\Omega(t)$. Let $\varrho(x, y, z, t)$ be some function of the coordinates and the time t, and consider the integral

$$Q(t) = \iiint_{\Omega(t)} \varrho \, dx \, dy \, dz:$$

our aim is to calculate the time-derivative of Q.

Consider first the special case when the volume $\Omega(t)$ is bounded by a cylindrical surface with generators parallel to Oz and having a fixed base Ω_1 in the plane $z = 0$ and with an upper surface $z = \varphi(x, y, t)$. Suppose that $z = \varphi(x, y, t)$ is the equation of a piecewise smooth surface $S(t)$. Suppose also that the derivative $\dfrac{\partial \varphi}{\partial t}$ is bounded: $\left| \dfrac{\partial \varphi}{\partial t} \right| \leqq M$.

Then

$$Q(t) = \iint_{\Omega_1} \left\{ \int_0^\phi \varrho(x, y, z, t) \, dz \right\} dx \, dy,$$

where the plane region Ω_1 is the part of the boundary of the volume Ω which lies in the plane xOy.

To calculate $\partial Q/\partial t$ we first set up an equation for the ratio of increments. We have:

$$\Delta Q = \iiint_0^{\phi(x,y,t+\Delta t)} \varrho(x, y, z, t + \Delta t) \, dz \, dx \, dy$$

$$- \iiint_0^{\phi(x,y,t)} \varrho(x, y, z, t) \, dz \, dx \, dy$$

$$\frac{\Delta Q}{\Delta t} = \frac{1}{\Delta t} \iint \left\{ \int_{\phi(x,y,t)}^{\phi(x,y,t+\Delta t)} \varrho(x, y, z, t + \Delta t) \, dz \right\} dx \, dy$$

$$+ \frac{1}{\Delta t} \iint \left\{ \int_0^{\phi(x,y,t)} [\varrho(x, y, z, t + \Delta t) - \varrho(x, y, z, t)] \, dz \right\} dx \, dy$$

$$= \iint_{\Omega_1} \frac{\Delta \varphi}{\Delta t} \cdot \frac{1}{\Delta \varphi} \left\{ \int_\phi^{\phi + \Delta \phi} \varrho(t + \Delta t) \, dx \right\} dx \, dy$$

$$+ \iint_{\Omega_1} \left\{ \int_0^\phi \frac{\varrho(t + \Delta t) - \varrho(t)}{\Delta t} \, dz \right\} dx \, dy.$$

Passing to the limit as $\Delta t \to 0$, we find

$$\lim_{\Delta t \to 0} \frac{\Delta Q}{\Delta t} = \lim_{\Delta t \to 0} \iint_{\Omega_1} \frac{\Delta \varphi}{\Delta t} \cdot \frac{1}{\Delta \varphi} \left\{ \int_\phi^{\phi + \Delta \phi} \varrho(t + \Delta t) \, dz \right\} dx \, dy$$

$$+ \lim_{\Delta t \to 0} \iint_{\Omega_1} \left\{ \int_0^\phi \frac{\varrho(t + \Delta t) - \varrho(t)}{\Delta t} \, dz \right\} dx \, dy$$

or

$$\lim_{\Delta t = 0} \frac{\Delta Q}{\Delta t} = \iint_{\Omega_1} \lim_{\Delta t \to 0} \frac{\Delta \varphi}{\Delta t} \cdot \frac{1}{\Delta \varphi} \left\{ \int_{\phi}^{\phi + \Delta \phi} \varrho(t + \Delta t) \, dz \right\} dx \, dy$$

$$+ \iint_{\Omega_1} \left\{ \lim_{\Delta t \to 0} \int_0^{\phi} \frac{\varrho(t + \Delta t) - \varrho(t)}{\Delta t} \, dz \right\} dx \, dy,$$

[this change in the order of the limiting processes can be justified],

$$= \iint_{\Omega_1} \frac{\partial \varphi}{\partial t} \varrho(x, y, \varphi) \, dx \, dy + \iiint_{\Omega} \frac{\partial \varrho}{\partial t} \, dx \, dy \, dz$$

$$\text{or} \quad \frac{dQ}{dt} = \lim_{\Delta t \to 0} \frac{\Delta Q}{\Delta t} = \iiint_{\Omega} \frac{\partial \varrho}{\partial t} \, dx \, dy \, dz + \iint_{z = \phi} \varrho \frac{\partial \varphi}{\partial t} \, dx \, dy$$

$$= \iiint_{\Omega} \frac{\partial \varrho}{\partial t} \, dx \, dy \, dz - \iint_S \varrho \frac{\partial \varphi}{\partial t} \cos(\boldsymbol{n}, z) \, dS$$

where $\boldsymbol{n}$ is the direction of the inward normal to the surface S.

$\partial \varphi / \partial t$ is called the *apparent velocity of movement of the surface $S(t)$ in the direction Oz*. It has very obvious interpretation: $\partial \varphi / \partial t$ is the velocity of movement along the straight line given by $x = \text{const.}$, $y = \text{const.}$, of the point of intersection of the surface $z = \varphi$ with this straight line.

The apparent velocity can be expressed in another form. To do this we represent the family of surfaces $z = \varphi(x, y, t)$ in the form of the equation solved for t:

$$t = \psi(x, y, z).$$

Then

$$\frac{\partial z}{\partial t} = \frac{1}{\dfrac{\partial t}{\partial z}} . \dagger$$

But $\partial t / \partial z$ is the component along Oz of the vector grad t, which is directed along the normal to the surface S and has the components

$$\frac{\partial t}{\partial n} \cos(n, x), \qquad \frac{\partial t}{\partial n} \cos(n, y), \qquad \frac{\partial t}{\partial n} \cos(n, z).$$

Hence

$$\frac{\partial \varphi}{\partial t} = \frac{\partial z}{\partial t} = \frac{1}{\dfrac{\partial t}{\partial z}} = \frac{1}{\dfrac{\partial t}{\partial n} \cos(n, z)}.$$

† *Editor's note.* We spend so much time proving that, in general $\partial y / \partial x \neq \dfrac{1}{\partial y / \partial x}$, that it is a little of a shock to come across an instance where the relation is in fact true, simply because x and y are in each case the second and third variables.

The expression $\dfrac{1}{\partial t/\partial n}$ is called *the apparent velocity of movement of the surface along the normal* and we shall denote it by v_n. For the apparent velocity of movement of the surface along the axis O_z (which we now denote by v_z) we have

$$v_z = \frac{\partial \varphi}{\partial t} = \frac{v_n}{\cos (n, z)}.$$

If the surface S consists of material particles moving with velocity v, then the velocity along the normal will be

$$v_n = v_x \cos (n, x) + v_y \cos (n, y) + v_z \cos (n, z),$$

and we can write our formula for dQ/dt on p. 10 as

$$\frac{d}{dt} \iiint_{\Omega(t)} \varrho \, dx \, dy \, dz = \iiint_{\Omega(t)} \frac{\partial \varrho}{\partial t} \, dx \, dy \, dz - \iint_{S(t)} \varrho v_n \, dS. \quad (1.15)$$

Exactly the same formula can be obtained in the general case, by dividing the surface $S(t)$ into a finite number of suitable pieces.

It may be helpful to give an easily visualized physical illustration of formula (1.15). At a certain instant of time, t, we pick out on the surface S some element dS and draw the trajectories of all points of dS during the time-interval $(t, t + \Delta t)$. The element dS traces out a volume which coincides approximately with an oblique cylinder the length of whose generators is given by $v \, dt$ where the vector v is the velocity of movement of the surface. The volume of this cylinder is equal to the product of the area of its base by the height, *i.e.*, $v_n \, dt \, dS$. The total contribution of the integral $\iiint \varrho \, d\Omega$ arising from the movement of the surface S is approximately equal to the integral

$$\iint_S \varrho v_n \, dt \, dS.$$

On dividing this expression by dt, and after calculating separately the contribution to Q arising from the change in ϱ, we obtain (1.15).

The formula (1.15) may be used to obtain an equation which expresses the conservation of mass of fluid during the motion. Suppose that fluid motion is taking place in some region of space and that the velocity components $v_x(x, y, z, t)$, $v_y(x, y, z, t)$, $v_z(x, y, z, t)$ are given functions of the coordinates and time. Imagine a surface $S(t)$ which consists always of the same moving material particles and which encloses a varying volume $\Omega(t)$. The mass of fluid enclosed within $\Omega(t)$ is given by

$$Q = \iiint_{\Omega(t)} \varrho(x, y, z, t) \, dx \, dy \, dz$$

where $\varrho(x, y, z, t)$ is the density of the fluid.

The mass of fluid within such a volume must remain constant because fluid neither enters it from outside nor disappears within it. So that, differentiating with respect to time,

$$\frac{\partial Q}{\partial t} = 0,$$

or, using (1.15),

$$\iiint_{\Omega(t)} \frac{\partial \varrho}{\partial t} \, d\Omega - \iint_{S(t)} \varrho v_n \, dS = 0$$

This equality must hold good for any surface S and for any time t. Hence, using Lemma 1, we get

$$\frac{\partial \varrho}{\partial t} + \operatorname{div}(\varrho v) = 0 \tag{1.16}$$

or in explicit form

$$\frac{\partial \varrho}{\partial t} + \frac{\partial}{\partial x}(\varrho v_x) + \frac{\partial}{\partial y}(\varrho v_y) + \frac{\partial}{\partial z}(\varrho v_z) = 0.$$

This is known as the *equation of continuity*.

Let us by way of illustration apply this equation to the motion of an incompressible, homogeneous fluid — for which the density is constant. The equation then reduces to

$$\varrho \left(\frac{\partial v_x}{\partial x} + \frac{\partial v_y}{\partial y} + \frac{\partial v_z}{\partial z} \right) = 0.$$

The problem of the motion of an incompressible fluid is equivalent to that of finding an unknown function V (the velocity-potential) such that

$$v = \operatorname{grad} V, \quad v_x = \frac{\partial V}{\partial x}, \quad v_y = \frac{\partial V}{\partial y}, \quad v_z = \frac{\partial V}{\partial z}.$$

Substituting these expressions for the velocity components in the continuity equation, we get

$$\varrho \left(\frac{\partial^2 V}{\partial x^2} + \frac{\partial^2 V}{\partial y^2} + \frac{\partial^2 V}{\partial z^2} \right) = 0$$

or

$$\nabla^2 V = 0 \tag{1.17}$$

where $\nabla^2 V$ stands for the previously introduced Laplacian operator

$$\frac{\partial^2}{\partial x^2} + \frac{\partial^2}{\partial y^2} + \frac{\partial^2}{\partial z^2}.$$

(1.17) is known as *Laplace's equation*.

Later we shall write down the complete set of equations of motion for a fluid and we shall show that any function V which satisfies (1.17) does indeed describe a possible motion of the fluid. Thus to solve a problem of fluid motion it suffices to know how to find the requisite solutions of equation (1.17).

In some circumstances, the velocity v and so also the function V do not depend on the time t; the motion is then one of *steady flow*.

§ 5. Equation of Heat Conduction

We know from physics that heat may be regarded as the result of random motions of material particles. The thermal state of a body is defined by its temperature. There is a simple relation between the energy of thermal motion of a body and its temperature, namely

$$Q = \iiint_D c\varrho T \, \mathrm{d}v$$

where D is the volume occupied by the body,

   Q is the energy of thermal motion, or, what comes to the same thing, the quantity of heat in calories,

   ϱ is the density of the substance,

   T is the absolute temperature,

and c is the heat-capacity of the body.

The transmission of heat from one body to another can take place in several ways. We shall pay no attention here to radiation, chemical processes and so on, but concentrate on the transfer of heat by direct transmission of kinetic energy from one particle to another.

Imagine within the medium under consideration a region bounded by a smooth surface S, and let n be a unit vector normal to S. The thermal energy due to the motion of the particles situated on either side of this surface may change in course of time either because of their mutual collisions or because particles move across the surface.

A particle whose mass-centre lies on one side of the surface and which has a certain energy can transmit energy either by itself passing to the other side of the surface or by colliding with another particle whose mass-centre lies on the other side of the surface. Let $\Delta_s Q$ denote the amount of energy per unit time which at the time t is being transmitted across the surface S to the particles situated on the side to which the normal is directed from the particles situated on the other side.

We shall suppose that the quantity $\Delta_s Q$ can be expressed in the form

$$\Delta_s Q = \iint_S \chi(S, t) \, \mathrm{d}S$$

where

$$\chi(S, t) = f(x, y, z, \boldsymbol{n}, t).$$

This formula is equivalent to supposing that the amount of thermal energy passing across an element of area depends only on the position of the centre of the element and the direction of the vector $\boldsymbol{n}$. The heat flow is regarded as positive if it is in the direction of $\boldsymbol{n}$. We shall further suppose that f, ϱ, c, and T are everywhere differentiable functions of their arguments.

In considering the diffusion of heat within a certain body, we shall suppose for the sake of generality that there is within the body a continuous distribution of heat sources with intensity $q(x, y, z, t)$. Then striking a heat balance for the volume D, we obtain

$$\frac{dQ}{dt} = \frac{d}{dt} \iiint_D c\varrho T \, dv = \iiint_D \frac{\partial(c\varrho T)}{\partial t} \, dv$$

$$= -\iint_S f(x, y, z, \boldsymbol{n}, t) \, dS + \iiint_D q \, dv \qquad (1.18)$$

where the vector $\boldsymbol{n}$ is directed along the outward normal.

Equation (1.18) holds good for any volume D. We apply it now, taking as D the tetrahedron Ω shown in Fig. 3, having a vertex at the point

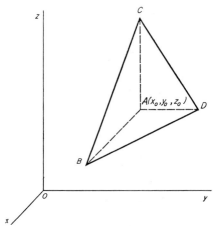

Fig. 3.

$A(x_0, y_0, z_0)$ and the three faces through A parallel to the coordinates planes. Let S_x, S_y, S_z denote the faces perpendicular to Ox, Oy, Oz respectively, and S_0 be the inclined face; let $\sigma_x, \sigma_y, \sigma_z, \sigma_0$ be their areas. We then have

$$\sigma_x = \sigma_0 \cos(n_0, x), \qquad \sigma_y = \sigma_0 \cos(n_0, y), \qquad \sigma_z = \sigma_0 \cos(n_0, z),$$

where $\cos(n_0, x)$, $\cos(n_0, y)$, $\cos(n_0, z)$ are the direction-cosines of the outward normal to the face S_0.

Applying (1.18) to the tetrahedron Ω we get

$$\iint_\Omega \frac{\partial}{\partial t}(c\varrho T)\,\mathrm{d}v - \iiint_\Omega q\,\mathrm{d}v = \iint_{S_x} f(x, y, z, \boldsymbol{i}, t)\,\mathrm{d}\sigma_x$$

$$+ \iint_{S_y} f(x, y, z, \boldsymbol{j}, t)\,\mathrm{d}\sigma_y + \iint_{S_z} f(x, y, z, \boldsymbol{k}, t)\,\mathrm{d}\sigma_z$$

$$+ \iint_{S_0} f(x, y, z, -\boldsymbol{n}_0, t)\,\mathrm{d}\sigma_0$$

where $\boldsymbol{i}, \boldsymbol{j}, \boldsymbol{k}$ are unit vectors parallel to Ox, Oy, Oz. If the volume of Ω is ω, then by a mean-value theorem,

$$\omega\,\overline{\frac{\partial}{\partial t}(c\varrho T)} - \omega\overline{q} = \sigma_x\overline{v_x} + \sigma_y\overline{v_y} + \sigma_z\overline{v_z}$$
$$+ \sigma_0\overline{f(x, y, z, -\boldsymbol{n}_0, t)}$$

where the bars denote mean values, and we have written for brevity

$$v_x = f(x, y, z, \boldsymbol{i}, t), \quad v_y = f(x, y, z, \boldsymbol{j}, t), \quad v_z = f(x, y, z, \boldsymbol{k}, t).$$

If h is the perpendicular distance of A from S_0, the above equation can be written

$$\tfrac{1}{3}h\sigma_0\,\overline{\frac{\partial}{\partial t}(c\varrho T)} - \tfrac{1}{3}h\sigma_0\overline{q} = \sigma_0\{\overline{v}_x\cos(n_0, x) + \overline{v}_y\cos(n_0, y) + \overline{v}_z\cos(n_0, z)$$
$$+ f(x, y, z, -\boldsymbol{n}_0, t)\}.$$

We divide both sides by σ_0 and, keeping the direction of $\boldsymbol{n}_0$ constant, let $h \to 0$. Then the left-hand member clearly vanishes, and so, noting that

$$f(x, y, z, -\boldsymbol{n}, t) = -f(x, y, z, \boldsymbol{n}, t),$$

we find

$$f(x_0, y_0, z_0, \boldsymbol{n}, t) = [v_x\cos(n, x) + v_y\cos(n, y) + v_z\cos(n, z)]_{(x_0, y_0, z_0)}$$

$$= [v_n]_{(x_0, y_0, z_0)}$$

where $\boldsymbol{n}$ is the direction of the outward normal to a side S of the tetrahedron Ω, and concides respectively with $-\boldsymbol{i}, -\boldsymbol{j}, -\boldsymbol{k}, \boldsymbol{n}_0$ for the sides S_x, S_y, S_z, S_0.

We thus see that the function $f(x, y, \boldsymbol{n}, t)$ is the projection of a certain vector $\boldsymbol{v}$ on to the direction of $\boldsymbol{n}$.

Hence, for an arbitrary volume D, (1.18) becomes

$$\iiint_D \frac{\partial}{\partial t} (\varrho cT) \, dv \; = \; -\iint_S v_n \, dS + \iiint_D q \, dv.$$

This formula is somewhat similar to (1.15). The vector v, which is analogous to the velocity of a fluid flow, we shall call the *heat flow*.

The heat flow existing in a medium is bound up with the temperature distribution in the medium. Under natural conditions heat always flows from the parts with higher temperatures to those with lower temperatures. We shall select some small area dS in the medium and investigate how the temperature varies at points close to this area.

The increase of temperature is characterized by the quantity

$$\frac{\partial T}{\partial n} = n \, \text{grad} \, T.$$

Suppose the medium is isotropic, *i.e.*, it has the same properties in all directions. It is natural to assume that, if the temperature increases in a direction normal to a surface S, then the heat flow across S will be negative. In other words, the quantities

$$n \, \text{grad} \, T = \text{grad}_n \, T \quad \text{and} \quad v_n \, dS$$

must have opposite signs: and this must be true for any direction n. Hence the projections of the vectors $\text{grad} \, T$ and v in any direction must be opposite in sign, and this is possible only if these vectors have opposite directions. That is,

$$v = -k \, \text{grad} \, T$$

where k is some positive scalar quantity, which may depend on the properties of the medium, on the temperature, on the way the temperature changes, and so on. If the temperature does not change very sharply, then as a first approximation we may assume that k is a function of position in the medium only. This assumption agrees very well with experiment.

Substituting this expression for v in the last equation for the heat balance, we obtain

$$\iiint_D \frac{\partial}{\partial t} (\varrho cT) \, dv = \iint_S k \frac{\partial T}{\partial n} \, dS + \iiint_D q \, dv.$$

Applying Lemma 1 and taking into account that when the direction of the outward normal is replaced by that of the inward normal the sign of the derivative $\partial T/\partial n = \text{grad}_n \, T$ changes, we get

$$\frac{\partial}{\partial t} (\varrho cT) = \frac{\partial}{\partial x} \left(k \frac{\partial T}{\partial x} \right) + \frac{\partial}{\partial y} \left(k \frac{\partial T}{\partial y} \right) + \frac{\partial}{\partial z} \left(k \frac{\partial T}{\partial z} \right) + q. \quad (1.19)$$

If we assume ϱ, c and k to be constants, we get

$$\frac{\partial T}{\partial t} = a^2\,\nabla^2 T + q_1 \qquad (1.20)$$

where

$$a^2 = \frac{k}{\varrho c} = \text{const.}, \qquad q_1 = \frac{q}{\varrho c}.$$

Equation (1.19) or (1.20) is called the *equation of heat conduction*.

§ 6. Sound Waves

As a last example we consider the *equation for transmission of sound*.

Suppose a compressible fluid is moving with velocity v, having the components along the coordinate axes

$$v_x(t, x, y, z), \quad v_y(t, x, y, z), \quad v_z(t, x, y, z).$$

The trajectories of the fluid particles will be determined by the equations

$$\frac{\mathrm{d}x}{\mathrm{d}t} = v_x, \quad \frac{\mathrm{d}y}{\mathrm{d}t} = v_y, \quad \frac{\mathrm{d}z}{\mathrm{d}t} = v_z.$$

The accelerations are easily calculated. We have

$$\left.\begin{aligned}
\frac{\mathrm{d}^2 x}{\mathrm{d}t^2} &= \frac{\partial v_x}{\partial t} + \frac{\partial v_x}{\partial x}\frac{\mathrm{d}x}{\mathrm{d}t} + \frac{\partial v_x}{\partial y}\frac{\mathrm{d}y}{\mathrm{d}t} + \frac{\partial v_x}{\partial z}\frac{\mathrm{d}z}{\mathrm{d}t} \\
&= \frac{\partial v_x}{\partial t} + v_x\frac{\partial v_x}{\partial x} + v_y\frac{\partial v_x}{\partial y} + v_z\frac{\partial v_x}{\partial z} \\
\frac{\mathrm{d}^2 y}{\mathrm{d}t^2} &= \frac{\partial v_y}{\partial t} + v_x\frac{\partial v_y}{\partial x} + v_y\frac{\partial v_y}{\partial y} + v_z\frac{\partial v_y}{\partial z} \\
\frac{\mathrm{d}^2 z}{\mathrm{d}t^2} &= \frac{\partial v_z}{\partial t} + v_x\frac{\partial v_z}{\partial x} + v_z\frac{\partial v_z}{\partial y} + v_z\frac{\partial v_z}{\partial z}
\end{aligned}\right\} \qquad (1.21)$$

Suppose that at each point of the fluid there is a force acting which, per unit volume, we denote by F with components X, Y, Z along the axes. If $p(t, x, y, z)$ denotes the pressure at any point, then the force acting on a surface S enclosing a volume Ω will have a component along the x-axis given by

$$\iint_S p(t, x, y, z)\cos(\boldsymbol{n}, x)\,\mathrm{d}S.$$

We obtain the equation of motion by applying d'Alembert's principle; in the x-direction,

$$\iint_S p \cos(\boldsymbol{n}, x) \, dS + \iiint_\Omega X \, dv$$

$$- \iiint_\Omega \varrho \left(\frac{\partial v_x}{\partial t} + v_x \frac{\partial v_x}{\partial x} + v_y \frac{\partial v_x}{\partial y} + v_z \frac{\partial v_x}{\partial z} \right) dv = 0,$$

where dv is an element of volume.

Applying Lemma 1 the equation of motion becomes

$$\varrho \left(\frac{\partial v_x}{\partial t} + v_x \frac{\partial v_x}{\partial x} + v_y \frac{\partial v_x}{\partial y} + v_z \frac{\partial v_x}{\partial z} \right) + \frac{\partial p}{\partial x} - X = 0. \quad (1.22)$$

Similarly we find

$$\varrho \left(\frac{\partial v_y}{\partial t} + v_x \frac{\partial v_y}{\partial x} + v_y \frac{\partial v_y}{\partial y} + v_z \frac{\partial v_y}{\partial z} \right) + \frac{\partial p}{\partial y} - Y = 0$$

$$\varrho \left(\frac{\partial v_z}{\partial t} + v_x \frac{\partial v_z}{\partial x} + v_y \frac{\partial v_z}{\partial y} + v_z \frac{\partial v_z}{\partial z} \right) + \frac{\partial p}{\partial z} - Z = 0.$$

These three equations contain five unknown functions: v_x, v_y, v_z, p and ϱ. Two more equations are needed to make the system determinate. We have already derived one other equation relating these quantities, namely, the equation of continuity. So, in the general case, we still have to seek one more equation.

First, however, we consider an incompressible, homogeneous fluid, for which we can put

$$\varrho = \text{const.},$$

and then we immediately have sufficient equations.

We can now verify what was said earlier about the potential flow of an incompressible fluid: namely, that

$$v = \text{grad } V,$$

$$\nabla^2 V = 0,$$

do actually satisfy the complete system of equations, if the function ϱ is defined correspondingly, and if further

$$X = \frac{\partial U}{\partial x}, \quad Y = \frac{\partial U}{\partial y}, \quad Z = \frac{\partial U}{\partial z},$$

i.e., if the external forces have a potential.

It suffices to show that if we take

$$v_x = \frac{\partial V}{\partial x}, \quad v_y = \frac{\partial V}{\partial y}, \quad v_z = \frac{\partial V}{\partial z},$$

then the equations (1.22) allow the function p to be constructed. When the expressions for v_x, v_y, v_z are substituted, these equations yield explicit expressions for

$$\frac{\partial p}{\partial x}, \frac{\partial p}{\partial y}, \frac{\partial p}{\partial z}.$$

And it is known from the theory of partial differential equations of the first order that the equations will be compatible provided that the mixed second-order derivatives

$$\frac{\partial^2 p}{\partial x\, \partial y}, \frac{\partial^2 p}{\partial y\, \partial z}, \frac{\partial^2 p}{\partial z\, \partial x}$$

determined from the different equations have the same values. It is left to the reader to verify this.

Returning to the general case of a compressible fluid, it is known from physics that the density and pressure in any fluid are related by a so-called equation of state, into which the absolute temperature T also enters. For an ideal gas, for example, the equation of state is

$$\varrho = \frac{p}{RT}$$

where R is the gas constant.

Since this equation has introduced another unknown function, T, it may be necessary in some cases to bring in yet another equation for the inflow of heat.

However, in a number of cases we can suppose that there is a functional relation between density and pressure:

$$\varrho = f(p) \tag{1.23}$$

where f is a given function.

Such a circumstance holds good if, for example, we consider processes occurring so rapidly that there is no time for heat to be transmitted from one particle to another. Such processes are said to be *adiabatic*.

In order to obtain from the general equations (1.16), (1.22), (1.23) the required equations for the transmission of sound, we now make certain simplifying assumptions. We shall suppose that the motion of the fluid consists of small vibrations about an equilibrium position. In the equilibrium condition the pressure p_0 and the density ϱ_0 are constants. The deviations $p - p_0$ and $\varrho - \varrho_0$ and also the velocity will be supposed small and, in particular, we shall suppose that terms such as $v_x \dfrac{\partial v_x}{\partial x}$, ... in (1.22) may be neglected. We then get

$$\varrho \frac{\partial v_x}{\partial t} + \frac{\partial p}{\partial x} = X, \quad \varrho \frac{\partial v_y}{\partial t} + \frac{\partial p}{\partial y} = Y, \quad \varrho \frac{\partial v_z}{\partial t} + \frac{\partial p}{\partial z} = Z.$$

Working to the same degree of accuracy, we may write these equations in the form

$$\frac{\partial}{\partial t} (\varrho v_x) + \frac{\partial p}{\partial x} = X, \quad \frac{\partial}{\partial t} (\varrho v_y) + \frac{\partial p}{\partial y} = Y, \quad \frac{\partial}{\partial t} (\varrho v_z) + \frac{\partial p}{\partial z} = Z.$$

Differentiating these equations with respect to x, y, z respectively and adding, we get

$$\frac{\partial}{\partial t} [\operatorname{div} (\varrho v)] + \frac{\partial^2 p}{\partial x^2} + \frac{\partial^2 p}{\partial y^2} + \frac{\partial^2 p}{\partial z^2} = \frac{\partial X}{\partial x} + \frac{\partial Y}{\partial y} + \frac{\partial Z}{\partial z}. \tag{1.24}$$

Using the equation of continuity, we can write the left-hand member of (1.24) as

$$\frac{\partial^2 p}{\partial x^2} + \frac{\partial^2 p}{\partial y^2} + \frac{\partial^2 p}{\partial z^2} - \frac{\partial^2 \varrho}{\partial t^2} = \frac{\partial^2 p}{\partial x^2} + \frac{\partial^2 p}{\partial y^2} + \frac{\partial^2 p}{\partial z^2}$$

$$- f'(p) \frac{\partial^2 p}{\partial t^2} - f''(p) \left(\frac{\partial p}{\partial t} \right)^2.$$

Finally, regarding $f'(p)$ as constant for a small change in p, so that

$$f'(p) = f'(p_0)$$

and denoting the right-hand member of (1.24), *i.e.*, $\dfrac{\partial X}{\partial x} + \dfrac{\partial Y}{\partial y} + \dfrac{\partial Z}{\partial z}$ by $\varPhi$ we have

$$\frac{\partial^2 p}{\partial x^2} + \frac{\partial^2 p}{\partial y^2} + \frac{\partial^2 p}{\partial z^2} - \frac{1}{a^2} \frac{\partial^2 p}{\partial t^2} = \varPhi, \tag{1.25}$$

where a is a constant defined by $1/a = \sqrt{f'(p_0)}$.

Equation (1.25) may be written in symbolic form as

$$\nabla^2 p - \frac{1}{a^2} \cdot \frac{\partial^2 p}{dt^2} = \varPhi.$$

We might have obtained an equation for transmission of sound by taking as the unknown function the density ϱ instead of the pressure p: in which case we should have obtained a partial differential equation for ϱ of exactly the same form as (1.25).

The various equations which we have obtained are sufficiently characteristic. We could go on to adduce other examples, but our main purpose is not the derivation of the equations of mathematical physics but their investigation and solution. We shall therefore limit ourselves to these examples and pass on to a consideration of various problems connected with these equations.

THE FORMULATION OF PROBLEMS
OF MATHEMATICAL PHYSICS.
HADAMARD'S EXAMPLE

§ 1. Initial Conditions and Boundary Conditions

We know from the theory of ordinary differential equations that the solution of such an equation is not uniquely determined by the equation itself.

The solution of an equation of the form

$$F(x, y, y', ..., y^{(n)}) = 0 \qquad (2.1)$$

depends, in general, on n arbitrary constants:

$$y = \varphi(x, c_1, c_2, ..., c_n). \qquad (2.2)$$

Often we can take as these constants the initial values of the unknown function and its derivatives:

$$[y]_{x=0} = y_0, \quad [y']_{x=0} = y_0^{(1)}, \, ..., \, [y^{(n-1)}]_{x=0} = y_0^{(n-1)}. \qquad (2.3)$$

A solution of the form (2.2) is called a general solution if it is possible to satisfy the condition (2.3) with arbitrary values of $y_0, y_0^{(1)}, ..., y^{(n-1)}$, by choosing appropriate values for the constants $c_1, ..., c_n$. To do this it is usually necessary to solve a system of finite (not differential) equations. In particular, if the equation (2.1) is a linear homogeneous equation, then the general integral (2.2) will have the extremely simple form

$$y = c_1 y_1 + c_2 y_2 + \cdots + c_n y_n,$$

where the functions $y_1, y_2, ..., y_n$ are n linearly independent particular integrals.

The situation is similar for partial differential equations: such an equation has not a unique solution. Its solution will depend, in general, on a number of arbitrary functions. For example, the general solution of the equation

$$\frac{\partial u}{\partial y} = 0$$

in two independent variables x and y will be

$$u = f(x)$$

where $f(x)$ is a completely arbitrary function.

To make a solution determinate it is usually necessary to specify certain supplementary conditions; for example, to require that the unknown function, and often also some of its derivatives, or certain combinations of the function and its derivatives, shall take certain specified values on various manifolds.

It is conceivable, generally speaking, that we might pose the problem of finding a general form of solution for a partial differential equation, similar to the corresponding problem for an equation of the form (2.1). However, although such general solutions do indeed exist, a knowledge of them, with rare exceptions, in no way helps us to solve important particular problems; for instead of having a system of finite equations for finding c_1, c_2, ..., c_n as was the case for ordinary differential equations, we get for the solution of these particular problems so complicated a system of functional relations for the arbitrary functions that it is practically impossible to find these functions.

Any problem of mathematical physics presents itself as a problem of solving some equation, such as (1.9), (1.14), (1.17), (1.19), or (1.25) for example, with definite supplementary conditions which in most cases are dictated by the requirements in the physical formulation of the problem.

We now point out some of the possible ways of stating the problem for the equations which we have already obtained.

In the problem of a vibrating string, for example, it is natural to consider a length of string, $0 \leqslant x \leqslant l$, fixed at both ends. Hence we must seek a solution of equation (1.9) satisfying the conditions

$$[u]_{x=0} = 0 \quad \text{and} \quad [u]_{x=l} = 0. \tag{2.4}$$

If the ends are not fixed but are set in motion according to some definite law, then the conditions (2.4) are replaced by

$$[u]_{x=0} = f_1(t) \quad \text{and} \quad [u]_{x=l} = f_2(t). \tag{2.5}$$

It is possible to specify other end-conditions, but we shall not particularize them.

It is not sufficient to specify the behaviour of the string at its ends in order to solve the problem. We need to know in addition, say, the value of the function u and of its rate of change $\partial u/\partial t$ at the initial moment of time, i.e.,

$$[u]_{t=0} = \varphi_0(x) \quad \text{and} \quad \left[\frac{\partial u}{\partial t}\right]_{t=0} = \varphi_1(x). \tag{2.6}$$

The conditions (2.5) and (2.6) together completely determine a solution of equation (1.9). We shall show later that, subject to certain provisos about the smoothness of the functions $f_1, f_2, \varphi_0, \varphi_1$, such a solution always exists, and consequently none of these conditions is superfluous.

The problem of a vibrating membrane can be put in an exactly similar way. In order to determine the motion it is sufficient to specify that

$$[u]_{t=0} = \varphi_0(x, y), \quad \left[\frac{\partial u}{\partial t}\right]_{t=0} = \varphi_1(x, y) \tag{2.7}$$

and the values of the function u on the boundary S

$$[u]_S = f(S, t), \tag{2.8}$$

where the function $f(S, t)$ depends on the point of the boundary S and on the time t.

Instead of the values of u on the boundary, the linear combination

$$\alpha[u]_S + \beta\left[\frac{\partial u}{\partial n}\right]_S = f(S, t) \tag{2.9}$$

may sometimes be given.

As we shall show later, the conditions (2.7) and (2.9) make the problem completely determinate. A solution satisfying them always exists, and consequently neither of the conditions (2.7) or (2.9) is superfluous.

The problem of transmission of sound in a region of space bounded by a surface S becomes completely determinate if, in addition to equation (1.25), we take into account the initial conditions

$$[p]_{t=0} = \varphi_0(x, y, z), \quad \left[\frac{\partial p}{\partial t}\right]_{t=0} = \varphi_1(x, y, z) \tag{2.7'}$$

and the boundary condition

$$[p]_S = f(S, t) \tag{2.8'}$$

where φ_0, φ_1, and f are given functions.

For the solution of the problem of heat conduction in some body [see (1.19)] it suffices to know the body's initial temperature

$$[T]_{t=0} = \varphi(x, y, z) \tag{2.10}$$

and the conditions at the boundary of the body

$$\alpha[T]_S + \beta\left[\frac{\partial T}{\partial n}\right]_S = f(S) \tag{2.11}$$

where $f(S)$ denotes a specified function for points on the surface.

With all the equations that we have discussed — the equations of a vibrating string or membrane, the equation of heat conduction — it is by no

means necessary to restrict attention to a finite body. We can also consider
a line, plane, or space, of infinite extent: the conditions (2.4), (2.5), (2.8),
(2.9), and (2.11), which we call generally the *boundary conditions*, then just
drop out. A problem without such conditions is known as a *Cauchy problem*:
the conditions (2.6), (2.7), and (2.10) are called the *initial conditions* (or the
Cauchy data).

Consider next a body whose surface has been exposed for a long time
to various constant influences, so that inside the body the temperature has
attained a value constant at each point but varying from one point to another.
This means that the heat distribution within the body has reached a steady
state. From equation (1.20), putting $\partial T/\partial t = 0$, we get

$$\frac{\partial^2 T}{\partial x^2} + \frac{\partial^2 T}{\partial y^2} + \frac{\partial^2 T}{\partial z^2} = -\frac{q}{k}. \tag{2.12}$$

This equation can be solved subject to the condition (2.11). The two
simplest cases, when either $\alpha = 0$ or $\beta = 0$, are of especial interest. The
problem of finding a solution of (2.12) subject to the condition

$$[T]_S = f(S) \tag{2.13}$$

is known as *Dirichlet's problem*. The same problem can also be posed and
solved for an equation with two independent variables; the equation corresponding to (2.12) is then

$$\frac{\partial^2 T}{\partial x^2} + \frac{\partial^2 T}{\partial y^2} = -\frac{q}{k}. \tag{2.14}$$

If, further, $q \equiv 0$, then this equation also applies to the problem of the
equilibrium of a membrane. For, if we put $\partial u/\partial t \equiv 0$, and consequently
$\partial^2 u/\partial t^2 = 0$ in equation (1.14) for a vibrating membrane, we again obtain an
equation of the form (2.14).

It is interesting to note that, as we have seen, one and the same equation
may describe completely different physical processes or conditions. Thus
it appears that mathematical equations possess a high degree of generality
arising from the process of abstraction from the concrete physical properties
of this or that phenomenon.

The other problem, of finding a solution of the equation (2.12) subject
to the condition

$$\left[\frac{\partial T}{\partial n}\right]_S = f(S) \tag{2.15}$$

is called *Neumann's problem*. This too can be considered for the case of
two independent variables. The problem of the potential motion of an incompressible fluid leads to a Neumann problem [see (1.17)]. For if we want

to find the velocity of the fluid inside some region, it is natural to specify on the boundary of the region the magnitude of the normal component of velocity v_n, which characterizes the fluid flow across each point of the surface. If, for example, the surface S is an immovable, impenetrable wall, then we must specify that the fluid flow across this wall is to be zero. If the wall is moving, then the normal component of the velocity of the wall must coincide with the normal component of fluid velocity. But $v_n = \partial V/\partial n$, from which it is clear that the problem discussed is identical with a Neumann problem.

For equation (2.12), just as for the earlier equations, it is by no means obligatory to consider its solution only in a finite region. Very often it is important to solve it for an unbounded region. This happens, for example, when the dimensions of the region under consideration are very large compared with the scale of the phenomenon which is being investigated. It is natural, for example, when investigating the heat-transfer from a long, tubular conductor buried in the ground to regard the earth, not as a sphere, but as an infinite half-space, and to solve equation (2.12) in a half-space with a cylinder cut out of it.

When considering unbounded domains, the behaviour of the solution at points at a great distance is by no means a matter of indifference; indeed in many cases the problem becomes determinate only when definite assumptions are made about this behaviour. For example, if equation (2.12) is solved in the infinite space outside a sphere, then the solution must satisfy the supplementary condition that it should vanish at infinity. Otherwise the solution remains indeterminate.

There is still another supplementary consideration which plays an important role.

§ 2. The Dependence of the Solution on the Boundary Conditions. Hadamard's Example

The statement of problems of mathematical physics includes, as we have seen, certain functions which enter into the boundary and/or initial conditions, and the solution will depend on these functions. These functions are usually determined from experimental results and therefore cannot be found absolutely accurately.

A certain amount of error in the initial or boundary conditions is inescapable. This error will manifest itself in the solution too, and it will not always follow that the error in the solution will also be small.

We shall see very simple examples of problems where a small error in the data can entail a very large error in the result. When we investigate equations of mathematical physics, we should always particularly examine the question of the dependence of the solution on the boundary conditions.

Suppose some problem in mathematical physics reduces to finding a certain function $u(x, y, z, t)$ of four variables in some domain Ω of variation of these variables, which satisfies in this domain an equation

$$F\left(u, \frac{\partial u}{\partial x}, \frac{\partial u}{\partial y}, \frac{\partial u}{\partial z}, \frac{\partial u}{\partial t}, \frac{\partial^2 u}{\partial x^2}, \dots, \frac{\partial^m u}{\partial t^m}\right) = 0. \tag{2.16}$$

In the examples considered earlier, the domain Ω of the independent variables x, y, z and a certain interval of time $0 \leqq t \leqq T$ served as the domain.

Let the required solution also have to satisfy some supplementary conditions of the form

$$\left[\Psi_j\left(u, \frac{\partial u}{\partial x}, \frac{\partial u}{\partial y}, \frac{\partial u}{\partial z}, \frac{\partial u}{\partial t}, \dots, \frac{\partial^r u}{\partial t^r}\right)\right]_{S_i} = \varphi_j(S_i) \tag{2.17}$$

$$i = 1, 2, \dots, l; \quad j = 1, 2, \dots, p,$$

where S_i is some manifold of less than 4 dimensions, and $\varphi_j(S_i)$ is a given function defined on the manifold S_i (the S_i lie within or on the boundary of the domain Ω).

In our previous examples these conditions were set up on, *e.g.*, the manifolds $t = 0$ and on some surface S of space (x, y, z) for all values of t.

We are going to define the concept of continuous dependence of the solution u on the boundary data $\{\varphi_j(S_i)\}$. But before we can do this, we must in preparation introduce the ideas of: distance between any two systems of boundary functions $\{\varphi_j(S_i)\}_1$ and $\{\varphi_j(S_i)\}_2$, and also distance between any two functions u_1 and u_2.

Let us examine in more detail the familiar problem of measuring the distance between points of some set lying in a plane. Suppose, for example, that this set consists of points scattered over a certain region. With each pair of these points, A and B, can be associated a number $\varrho(A, B)$, the separation between them. In different circumstances it may be natural to take different numbers $\varrho(A, B)$ as this separation. Thus, for a helicopter pilot, the separation will be the length of the interval AB; but for a car-driver, it will be the shortest distance between the points A and B by road. However, the car-driver might instead regard the separation as being the shortest time necessary for the journey from A to B without counting the actual number of miles ("A is two hours' travel from B"). And on this basis, the separation on making a detour by a "By-pass" road might be shorter than that along an urban road coinciding with the rectilinear interval AB.

Again, consider for example a set of cross-roads in some city. Suppose that all the streets are parallel to one or other of the axes of a rectangular Cartesian coordinate system Oxy. We might take as the separation $\varrho(A, B)$ between two cross-roads $A(x_A, y_A)$ and $B(x_B, y_B)$ the length of the interval AB:

$$\varrho(A, B) = \sqrt{(x_A - x_B)^2 + (y_A - y_B)^2};$$

but for a pedestrian it would be natural to take the shortest distance he must walk to get from A to B:

$$\varrho(A, B) = |x_A - x_B| + |y_A - y_B|.$$

It might also be convenient, on occasion, to define the separation by the formula

$$\varrho(A, B) = \max\,(|x_A - x_B|, (y_A - y_B)).$$

We now take a certain class Φ of systems of functions $\{\varphi_j(S_i)\}$ which are defined on the manifolds S_i, and a certain class U of functions $u(x, y, z, t)$ defined in Ω. We assign to the difference between any two systems of functions $\{\varphi_j(S_i)\}$ and $\{\varphi_j{}^*(S_i)\}$ a certain positive number ϱ, by means of a suitable formula, and this number ϱ is defined to be the "separation" between these systems of functions:

$$\varrho_\Phi = \varrho_\Phi\{\varphi_j^*(S_i) - \varphi_j(S_i)\}.$$

In a similar way we introduce the separation between members of the function class U:

$$\varrho_U = \varrho_U(u^* - u).$$

Suppose that for a certain system of functions $\{\varphi_j(S_i)\}$ from Φ the problem (2.16), (2.17) has a unique solution u of U. We now change the right-hand member of (2.17) and put instead of the system of functions $\{\varphi_j(S_i)\}$ another system of functions $\{\varphi_j^*(S_i)\}$ from Φ:

$$\varphi_j^*(S_i) = \varphi_j(S_i) + \tilde{\varphi}_j(S_i). \tag{2.18}$$

We shall suppose that for any system of functions $\{\varphi_j^*\}$ sufficiently close to the system $\{\varphi_j(S_i)\}$ the problem has a unique solution, which we shall denote by u^*, and we write

$$u^* = u + \tilde{u}. \tag{2.19}$$

Suppose now that, given any $\varepsilon > 0$, we can find a δ such that the magnitude of the deviation of the solution, $\varrho_U(\tilde{u})$, is less than ε,

$$\varrho_U(\tilde{u}) < \varepsilon$$

provided only that the magnitude of the deviation of the boundary functions, $\varrho_\Phi(\{\tilde{\varphi}_j(S_i)\})$, is less than δ,

$$\varrho_\Phi(\{\tilde{\varphi}_j(S_i)\}) < \delta.$$

In this case we shall say that *in the region Ω, the solution u depends* continuously *on the boundary data.*

Such a definition of continuous dependence is clearly an immediate generalization of the familiar concept in analysis of the continuous dependence of a function on its argument.

In order to give this concept of the continuous dependence of a solution on the boundary functions a definite meaning, we must select the classes Φ and U, and specify formulae defining ϱ and ϱ_0.

We shall give two examples of possible definitions of continuous dependence, and we shall then show why it is natural to use them.

DEFINITION 1. The class Φ is defined to be the class of systems of functions $\{\varphi_j(S_i)\}$ which are continuous and have continuous partial derivatives up to a certain order k inclusive (the partial differentiation being with respect to any of the parameters on the manifold S_i).

The class U is defined to be the class of functions u which are continuous and have continuous partial derivatives up to a certain order p inclusive.

As the separation ϱ_Φ between two systems of functions we shall take the upper bound of the absolute values of the differences between these functions and between their partial derivatives up to order k inclusive on the manifolds S_i.

As the separation ϱ_U between two functions u and u^* we shall take the upper bound of the absolute values of the differences between these functions and between their partial derivatives up to order p inclusive in the domain Ω.

If, in the sense indicated, there is continuous dependence of the solution on the boundary data, we shall say that *the solution depends continuously to order (p, k) on the boundary data in the domain* Ω. If, conversely, for some $\varepsilon_0 > 0$ and for any $\delta > 0$ there exists a deviation of the system of boundary functions $\{\tilde{\varphi}_j(S_i)\}$ such that

$$\varrho_\Phi[\{\tilde{\varphi}_j(S_i)\}] < \delta,$$

but

$$\varrho_U(\tilde{u}) > \varepsilon_0,$$

then *the solution u in the domain Ω depends discontinuously to order (p, k) on the boundary data.*

DEFINITION 2. *Continuous dependence in the mean.*

The class Φ is defined to be the class of systems of functions $\{\varphi_j(S_i)\}$ such that the functions $\varphi_j(S_i)$ and their partial derivatives up to a certain order k inclusive have squares which are integrable on S_i.

The class U is defined to be the class of functions u which are such that the squares of these functions and of their partial derivatives up to a certain order p inclusive are integrable in the domain Ω.

As the square of the separation ϱ_Φ between two systems of functions $\{\varphi_j(S_i)\}$ and $\{\varphi_j^*(S_i)\}$ we shall take the sum of the integrals of the squares of the functions $\tilde{\varphi}_j(S_i) = \varphi_j^* - \varphi_j$ and of those of derivatives up to order p inclusive on the manifolds S_i.

As the square of the separation ϱ_U between the functions u^* and u we shall take the sum of the integrals in the domain Ω of the squares of the function $u^* - u$ and of those of its derivatives up to order p inclusive.

If a solution u of the problem (2.16), (2.17) depends continuously on the boundary conditions in the sense of this definition, we shall say that *the solution depends on the boundary conditions continuously in the mean to order* (p, k).

We may mention that continuous dependence to order (0, 0) admits of a simple physical interpretation. Thus in the problem (2.12), (2.13) where the temperature T of the body plays the part of u, and the function $f(S)$, the temperature on the boundary, plays the part of $\varphi_j(S_i)$, continuous dependence of order (0, 0) means that a small change of temperature at any point on the boundary would produce a small change of temperature at every point inside the body.

Continuous dependence in the mean can be interpreted just as easily. In the problem on transmission of sound (1.25), (2.7′), (2.8′), there is, as can easily be shown, continuous dependence in the mean of the solution on the functions φ_0 and φ_1 of orders (1, 1) and (0, 1) respectively. This property is the mathematical counterpart of the fact that a small change in the energy of vibration of the air, initially, produces a small change of energy subsequently.

Ordinarily in mathematical physics we use the idea of either continuous dependence of order (p, k) or continuous dependence in the mean of order (p, k). There are other natural definitions of continuous dependence apart from the two which we have introduced, but we shall not touch on them here.

We may remark that at one and the same time there may be continuous dependence in the sense of Definition 1 but not in the sense of Definition 2.

If for any system of functions $\{\varphi_j(S_i)\}$ from Φ it is possible to indicate a domain Ω', containing the manifolds S_i, such that the solution u of the problem (2.16), (2.17) depends in this domain on the boundary conditions continuously to order (p, k), or continuously in the mean to order (p, k), where p and k are any particular natural numbers, then we shall say the problem (2.16), (2.17) *is correctly formulated*. In the contrary case, *the problem is incorrectly formulated*. Sometimes, instead of an arbitrary system of functions, some particular fixed system of functions $\{\varphi_j(S_i)\}$ is considered, and we then speak of the correctness of the formulation of the problem for this system of functions. For non-linear problems, the idea of correct formulation may turn out to be substantially wider than that just expounded.

For ordinary differential equations the problem of integrating an equation

$$\frac{d^m u}{dx^m} = f\left(x, u, \frac{du}{dx}, \ldots, \frac{d^{m-1}u}{dx^{m-1}}\right)$$

with initial conditions

$$[u]_{x=0} = u_0, \quad \left[\frac{du}{dx}\right]_{x=0} = u_1, \cdots, \left[\frac{d^{m-1}u}{dx^{m-1}}\right]_{x=0} = u_{m-1}$$

is, as is well known, always correctly formulated, subject to the restrictions placed on the function f by the theorem of existence and uniqueness.

The same applies to a partial differential equation of the first order

$$\frac{\partial u}{\partial t} = f\left(t, x, y, z, \frac{\partial u}{\partial x}, \frac{\partial u}{\partial y}, \frac{\partial u}{\partial z}\right).$$

Cauchy's problem, *i.e.*, the problem of solving this equation with the initial data $[u]_{t=0} = u_0(x, y, z)$ is correctly formulated for this equation, since u depends on u_0 continuously to the order $(1,1)$.

But for a partial differential equation of order higher than the first this circumstance may no longer obtain; consequently, to pose Cauchy's problem for such an equation may not always be meaningful. To illustrate this, we choose an example due to Hadamard.

We shall find a solution of the equation

$$\frac{\partial^2 u}{\partial x^2} + \frac{\partial^2 u}{\partial y^2} = 0 \tag{2.20}$$

in the half-strip $y > 0$, $-\pi/2 \leqq x \leqq \pi/2$, satisfying the conditions

$$[u]_{x=-\pi/2} = 0 = [u]_{x=\pi/2} = [u]_{y=0}, \quad \left[\frac{\partial u}{\partial y}\right]_{y=0} = e^{-\sqrt{n}} \cos nx \tag{2.21}$$

where n is an odd integer.

It is not difficult to see that this solution will have the form

$$u = \frac{1}{n} e^{-\sqrt{n}} \cos nx \sinh ny. \tag{2.22}$$

It can be shown that the solution of the problem formulated is unique. It is easy to see that when n tends to infinity, the function $e^{-\sqrt{n}} \cos nx$ tends uniformly to zero, as do all its derivatives. But the solution (2.22) for any non-zero y has the form of a cosine-curve having an amplitude as great as we please.

It is clear that in this case continuous dependence in the sense of the first definition can never hold in any domain x, y which touches the axis $y = 0$.

We can also show that there is no continuous dependence in the sense of the second definition. It is easy to see that the integral

$$\int_0^y \int_{-\pi/2}^{\pi/2} \left\{\frac{1}{n} e^{-\sqrt{n}} \cos nx \sinh ny\right\}^2 dx \, dy$$

tends to infinity as $n \to \infty$ however small $y > 0$ may be, and this justifies our assertion.

Thus the problem just discussed on Laplace's equation is not formulated correctly.

If we look into the question of continuous dependence in an unbounded domain, we find that even for ordinary differential equations the question ceases to be a simple one. For these, continuous dependence on the initial data is known as "Lyapunov stability (or continuity)".

The related questions in the theory of partial differential equations could also fittingly be called the theory of stability. So far these questions have not been fully worked out; they are too difficult for an elementary course, and we shall not deal with them here.

The solution of problems which are incorrectly formulated has no practical value in the majority of cases. When analysing solutions of all the problems appearing in the course, we shall always stop to ascertain that the problem is correctly formulated.

In the next lecture we shall be concerned with a more detailed examination of the differences between the various equations which we have so far studied.

THE CLASSIFICATION OF LINEAR EQUATIONS
OF THE SECOND ORDER

§ 1. Linear Equations and Quadratic Forms. Canonical Form of an Equation

All the equations we have so far considered have been linear equations of the second order with real coefficients, *i.e.*, equations of the form

$$\sum_{i=1}^{n} \sum_{j=1}^{n} A_{ij} \frac{\partial^2 u}{\partial x_i \partial x_j} + \sum_{i=1}^{n} B_i \frac{\partial u}{\partial x_i} + Cu = F \tag{3.1}$$

where A_{ij}, B_i, C and F are given functions of $x_1, x_2, ..., x_n$.

To study the properties of these equations in more detail we shall investigate certain properties of their coefficients. We shall examine first of all how the coefficients of the equation (3.1) transform under an arbitrary change of the independent variables, or, what comes to the same thing, under any geometrical transformation of the space of the variables $x_1, x_2, ..., x_n$.

In place of $x_1, x_2, ..., x_n$ we introduce new variables $y_1, y_2, ..., y_n$. We assume that the functions $y_1(x_1, ..., x_n), ..., y_n(x_1, ..., x_n)$ have continuous second derivatives.

Then

$$\frac{\partial u}{\partial x_i} = \sum_{l=1}^{n} \frac{\partial u}{\partial y_l} \frac{\partial y_l}{\partial x_i} \tag{3.2}$$

$$\frac{\partial^2 u}{\partial x_i \partial x_j} = \frac{\partial}{\partial x_j}\left(\frac{\partial u}{\partial x_i}\right) = \sum_{k=1}^{n} \sum_{l=1}^{n} \frac{\partial^2 u}{\partial y_l \partial y_k} \frac{\partial y_l}{\partial x_i} \frac{\partial y_k}{\partial x_j} + \sum_{l=1}^{n} \frac{\partial u}{\partial y_l} \frac{\partial^2 y_l}{\partial x_i \partial x_j}. \tag{3.3}$$

Substituting the expressions (3.2) and (3.3) in the equation (3.1) we get

$$\sum_{k=1}^{n} \sum_{l=1}^{n} \frac{\partial^2 u}{\partial y_k \partial y_l} \left(\sum_{i=1}^{n} \sum_{j=1}^{n} A_{ij} \frac{\partial y_k}{\partial x_i} \frac{\partial y_l}{\partial x_j}\right)$$

$$+ \sum_{l=1}^{n} \frac{\partial u}{\partial y_l} \left(\sum_{i=1}^{n} \sum_{j=1}^{n} A_{ij} \frac{\partial^2 y_l}{\partial x_i \partial x_j} + \sum_{i=1}^{n} B_i \frac{\partial y_l}{\partial x_i}\right) + Cu = F. \tag{3.4}$$

Denoting the new coefficients of the second-order derivatives $\partial^2 u/\partial y_k \, \partial y_l$ in (3.4) by $\overline{A_{kl}}$, we have

$$\overline{A_{kl}} = \sum_{i=1}^{n} \sum_{j=1}^{n} A_{ij} \frac{\partial y_k}{\partial x_i} \frac{\partial y_l}{\partial x_j}. \tag{3.5}$$

If we fix on a definite point in space and write, at this point,

$$\frac{\partial y_k}{\partial x_i} = \alpha_{ki}, \tag{3.6}$$

then the transformation formulae (3.5) become

$$\overline{A_{kl}} = \sum_{i=1}^{n} \sum_{j=1}^{n} A_{ij} \, \alpha_{ki} \, \alpha_{lj}, \tag{3.7}$$

which are the same as the transformation formulae for the coefficients of the quadratic form

$$\sum_{i=1}^{n} \sum_{j=1}^{n} A_{ij} \, p_i \, p_j \tag{3.8}$$

under the change of variables

$$p_i = \sum_{k=1}^{n} \alpha_{ki} \, q_k \tag{3.9}$$

which takes (3.8) over into the form

$$\sum_{k=1}^{n} \sum_{l=1}^{n} \overline{A_{kl}} \, q_k \, q_l. \tag{3.10}$$

The determinant $|\alpha_{ki}|$ must, of course, be non-zero in order that the formulae (3.9) shall give a (1–1) change of variables.

If we want to simplify equation (4.1) by means of a change of variables, we can examine, instead of this problem, that of the simplification of the quadratic form (3.8) by means of the change of variables (3.9) with real coefficients α_{ki}.

A problem of this sort arises in analytical geometry in the simplification of equations of second-order surfaces. But our present problem is rather simpler, since we are not bound to orthogonal transformations, and we have no need to demand that in the space $p_1, p_2, \ldots, p_n$ the directions of the axes $q_1, q_2, \ldots, q_n$ shall be orthogonal.

It is proved in text-books on linear algebra that the coefficients of the transformation can always be chosen so that a quadratic form is reduced to a sum of squares (not necessarily with positive coefficients): (3.8) becomes

$$\sum_{i=1}^{n} \sum_{j=1}^{n} A_{ij} \, p_i \, p_j = \sum_{l=1}^{n} k_l \, q_l^2 \tag{3.11}$$

where the k_l are certain numbers which may be positive or negative or zero. We now prove the following theorem:

THEOREM 1. *(Rule of Inertia of Quadratic Forms)*

The number of positive coefficients and the number of negative coefficients among the $k_1, k_2, \ldots, k_n$ is independent of the particular choice of change of variables (3.9) which transforms the quadratic form (3.8) into the form (3.11), provided only that the determinant $|\alpha_{ij}|$ is non-zero.

It follows, incidentally, that the number of non-zero coefficients k_n must also be the same in all representations such as (3.11).

Proof.

If we transform our quadratic form in two ways into the sum of squares, we shall have

$$\sum_{l=1}^{n} k_l \, q_l^2 = \sum_{s=1}^{n} \varrho_s \, m_s^2 \tag{3.12}$$

where m_s and q_l are the space coordinates chosen in the two cases, and are linearly related

$$m_s = \sum_{l=1}^{n} \beta_{sl} \, q_l.$$

Suppose now that the number of positive k_l is, say, greater than the number of positive ϱ_s. We equate to zero all those q_l whose squares have non-positive coefficients and also all those m_s whose squares have positive coefficients. If we suppose that

$$k_1 \geq k_2 \geq \cdots \geq k_r > 0 \geq k_{r+1} \geq k_{r+2} \geq \cdots \geq k_n,$$

$$\varrho_1 \geq \varrho_2 \geq \cdots \geq \varrho_t > 0 \geq \varrho_{t+1} \geq \varrho_{t+2} \geq \cdots \geq \varrho_n,$$

then we shall put

$$q_{r+1} = q_{r+2} = \cdots = q_n = 0$$

$$m_1 = m_2 = \cdots = m_t = 0.$$

Now among the relations between m_s and q_l written above, those whose left-hand members have been put equal to zero may be regarded as a system of homogeneous linear equations for determining $q_1, q_2, \ldots, q_r$, and the remainder as equalities fixing the values of $m_{t+1}, m_{t+2}, \ldots, m_n$. Since the number of these equations is less than the number of unknowns, we can always find a solution of the system in which not all the $q_1, q_2, \ldots, q_r$ vanish. Substituting these values of $q_1, q_2, \ldots, q_r$ in (3.12), we see that the left-hand member will be positive, but the right-hand member will be non-positive; consequently (3.12) does not hold, and our supposition is contradicted. Hence the theorem.

This theorem enables a classification of partial differential equations of the second order to be made.

At each point of the space of the variables $(x_1, x_2, ..., x_n)$ we can carry out a change of the independent variables p_i so that at this point the form

$$\sum_{i=1}^{n} \sum_{j=1}^{n} A_{ij} p_i p_j$$

transforms into a sum of squares of the form

$$\sum_{i=1}^{n} k_i r_i^2.$$

Further, the simple substitution

$$q_i = \sqrt{|k_i|}\, r_i$$

transforms our form into the expression

$$\sum_{i=1}^{r} q_i^2 - \sum_{i=r+1}^{m} q_i^2 \tag{3.13}$$

where r is the number of positive coefficients k_i
and m is the total number of non-zero coefficients.

Suppose now that the substitution (3.9) which reduces the basic form (3.8) to the form (3.13) has been found, and that it is written as

$$p_i = \sum_{k=1}^{n} \alpha_{ik}^{*} q_k.$$

We introduce a linear change of the independent variables $x_1, x_2, ..., x_n$ in our space by means of the formula

$$y_k = \sum_{k=1}^{n} \alpha_{ki}^{*} x_i.$$

This means that in formula (3.6)

$$\alpha_{ki} = \alpha_{ki}^{*}$$

and so, at the point of space considered, all the $\overline{A_{ki}}$ with $i \neq k$ will vanish and the $\overline{A_{ii}}$ will be either ± 1 or 0. The terms of the equation which contain second-order derivatives will take the form

$$\sum_{i=1}^{r} \frac{\partial^2 u}{\partial y_i^2} - \sum_{i=r+1}^{m} \frac{\partial^2 u}{\partial y_i^2}.$$

If the equation (3.1) has constant coefficients, then under a linear change of variables it will transform into an equation with coefficients constant again. Consequently we shall, in this case, be able everywhere in the space to bring the equation into such a form that in the new coordinates the coefficients of the mixed second-order derivatives vanish and the coefficients of those of the derivatives $\partial^2 u/\partial x^2$ which do not vanish are equal to ± 1. We shall call this the *canonical form* of the equation.

As we shall see later, the character of a second-order equation is completely determined by the number r of positive coefficients and the number s of negative coefficients of the second-order derivatives after such a transformation. If we consider an equation with variable coefficients, we find that it is impossible by a single such transformation to bring it into canonical form for the whole space at once; all we can do is to put it into canonical form at each point of space separately.

Moreover, in doing this, we may obtain different values of the numbers r and s at different points. In such cases it is convenient to divide the space into parts over which r and s are constant, and to investigate the equation in each part in turn.

We now introduce the concept of types of partial differential equations of the second order. In domains where r and s keep constant values, we shall say that the *equation belongs to the type* (r, s). It is clear that the types (r, s) and (s, r) are essentially the same, since by a change in sign of all coefficients r and s change places. The equation for a vibrating string is of type $(1,1)$ with $n = 2$. The equation of a vibrating membrane belongs to type $(2,1)$ with $n = 3$. The equation of heat conduction belongs to type $(3,0)$ with $n = 4$: and Laplace's equation to type $(3,0)$ with $n = 3$.

We shall call type $(n, 0)$ the *elliptical type*; type (r, s) with $r + s = n$, $r > 0$, $s > 0$, the *hyperbolic type*; and of these, the type $(n - 1, 1)$ is the *normal-hyperbolic type*. Types (r, s) with $r + s < n$ are called *parabolic types*. Of these, the type $(n - 1, 0)$ is called the *normal-parabolic type*. Parabolic types where $s = 0$ are called *elliptico-parabolic types*, and those with $r > 0$ and $s > 0$ are called *hyperbolo-parabolic types*.

The equations for a vibrating string and membrane belong to the normal-hyperbolic type [equations (1.9) and (1.13)]; the equation of heat conduction (1.19) belongs to the normal-parabolic, and Laplace's equation to the elliptic type.

In this course we shall restrict ourselves to the normal-hyperbolic, normal-parabolic, and elliptic types. All the equations which we considered in Lecture 1 had canonical form and belonged to one of these types. We thus see that none of these equations can be reduced to another by a real transformation of coordinates and that each of them is typical of a certain species.

Note. The coordinate system in which an equation takes canonical form is not unique.

Laplace's equation

$$\nabla^2 u = \frac{\partial^2 u}{\partial x_1^2} + \frac{\partial^2 u}{\partial x_2^2} + \cdots + \frac{\partial^2 u}{\partial x_n^2}$$

remains invariant under any change of origin and any orthogonal transformation of coordinates.

For, the coefficients A_{kl} after an orthogonal rotation will have the form

$$\overline{A_{kl}} = \sum_{i=1}^{n} \alpha_{ik}\alpha_{il}$$

where $\alpha_{ik} = \dfrac{\partial y_k}{\partial x_i} = \cos(y_k, x_i)$,

and by the orthogonality conditions

$$\sum_{i=1}^{n} \alpha_{ik}\alpha_{il} = \begin{cases} 1 & \text{for } k = l \\ 0 & \text{for } k \neq l \end{cases},$$

so that

$$\overline{A_{kl}} = \begin{cases} 1 & \text{for } k = l \\ 0 & \text{for } k \neq l \end{cases},$$

as was to be shown.

The transformations of the independent variables which leave the wave equation invariant are known as *Lorentz transformations*; they play an important role in the theory of relativity.

§ 2. Canonical Form of Equations in Two Independent Variables

As we have seen, an equation with variable coefficients can always be brought into canonical form at an isolated point in space. If the number of independent variables is greater than two, then only in exceptional circumstances can the equation be brought into canonical form throughout the whole space in which the equation is given. However, an equation in two independent variables can always be brought into canonical form throughout the whole region of variation of the independent variables.

We shall examine this question in more detail. When there are two independent variables the sum of the terms containing second-order derivatives may be written in the form

$$Lu \equiv A\frac{\partial^2 u}{\partial x^2} + 2B\frac{\partial^2 u}{\partial x \partial y} + C\frac{\partial^2 u}{\partial y^2}.$$

We shall call this sum the *principal term of the second-order operator*.

On transition to new variables $\xi = \xi(x, y)$, $\eta = \eta(x, y)$, the operator Lu takes the form

$$Lu = \left[A\left(\frac{\partial\xi}{\partial x}\right)^2 + 2B\frac{\partial\xi}{\partial x}\frac{\partial\xi}{\partial y} + C\left(\frac{\partial\xi}{\partial y}\right)^2 \right]\frac{\partial^2 u}{\partial\xi^2}$$

$$+ 2\left[A\frac{\partial\xi}{\partial x}\frac{\partial\eta}{\partial x} + B\left(\frac{\partial\xi}{\partial x}\frac{\partial\eta}{\partial y} + \frac{\partial\eta}{\partial x}\frac{\partial\xi}{\partial y}\right) + C\frac{\partial\xi}{\partial y}\frac{\partial\eta}{\partial y} \right]\frac{\partial^2 u}{\partial\xi\partial\eta}$$

$$+ \left[A\left(\frac{\partial\eta}{\partial x}\right)^2 + 2B\frac{\partial\eta}{\partial x}\frac{\partial\eta}{\partial y} + C\left(\frac{\partial\eta}{\partial y}\right)^2 \right]\frac{\partial^2 u}{\partial\eta^2} + \frac{\partial u}{\partial\xi}L\xi + \frac{\partial u}{\partial\eta}L\eta$$

or

$$Lu = L_1 u + \frac{\partial u}{\partial\xi}L\xi + \frac{\partial u}{\partial\eta}L\eta.$$

$L_1 u$ denotes the principal term of the operator in the new variables. We assume as regards $\xi(x, y)$ and $\eta(x, y)$, in addition to their smoothness, that the curves $\xi(x, y) = $ const. and $\eta(x, y) = $ const. do not meet tangentially at their points of intersection.

In order that the equation may have canonical form after the transformation of coordinates, it is necessary that:

(1) The coefficient of the mixed partial derivative

$$B_1 = A\frac{\partial\xi}{\partial x}\frac{\partial\eta}{\partial x} + B\left(\frac{\partial\xi}{\partial x}\frac{\partial\eta}{\partial y} + \frac{\partial\eta}{\partial x}\frac{\partial\xi}{\partial y}\right) + C\frac{\partial\xi}{\partial y}\frac{\partial\eta}{\partial y}$$

must vanish and (2) If the equation is of elliptical or hyperbolic type, then must

$$A_1 = \pm C_1,$$

and if the equation is of parabolic type, then must either

$$A_1 = 0 \quad \text{or} \quad C_1 = 0.$$

It is known from analytical geometry that the canonical form of the quadratic form

$$Ap^2 + 2Bpq + Cq^2$$

whose coefficients transform like those of the operator Lu is determined by the sign of its discriminant

$$\Delta = B^2 - AC,$$

and if $\Delta < 0$, the form is elliptical and transforms into $r_1^2 + r_2^2$

if $\Delta > 0$, the form is hyperbolic and transforms into $r_1^2 - r_2^2$

if $\Delta = 0$, the form is parabolic and transforms into r_1^2.

In the last case we shall exclude the possibility of complete degeneracy of the form ($A = B = C = 0$). We shall examine the parabolic case first. If $\Delta = 0$, then $B^2 = AC$ and, since we exclude the completely degenerate case, this implies that at least one of the coefficients A and C is not equal to zero. Suppose $A \neq 0$. Put

$$k = \frac{B}{A} = \frac{C}{B}.$$

(If $B = 0$, then $C = 0$ also, and the equation is already in canonical form.) We have:

$$A_1 = A\left(\frac{\partial \xi}{\partial x}\right)^2 + 2B\frac{\partial \xi}{\partial x}\frac{\partial \xi}{\partial y} + C\left(\frac{\partial \xi}{\partial y}\right)^2 = A\left(\frac{\partial \xi}{\partial x} + k\frac{\partial \xi}{\partial y}\right)^2$$

$$C_1 = A\left(\frac{\partial \eta}{\partial x}\right)^2 + 2B\frac{\partial \eta}{\partial x}\frac{\partial \eta}{\partial y} + C\left(\frac{\partial \eta}{\partial y}\right)^2 = A\left(\frac{\partial \eta}{\partial x} + k\frac{\partial \eta}{\partial y}\right)^2$$

$$B_1 = A\frac{\partial \xi}{\partial x}\frac{\partial \eta}{\partial x} + B\left(\frac{\partial \xi}{\partial x}\frac{\partial \eta}{\partial y} + \frac{\partial \eta}{\partial x}\frac{\partial \xi}{\partial y}\right) + C\frac{\partial \xi}{\partial y}\frac{\partial \eta}{\partial y}$$

$$= A\left(\frac{\partial \xi}{\partial x} + k\frac{\partial \xi}{\partial y}\right)\left(\frac{\partial \eta}{\partial x} + k\frac{\partial \eta}{\partial y}\right).$$

In order that the coefficients B_1 and C_1 should vanish simultaneously it is sufficient to put

$$\frac{\partial \eta}{\partial x} + k\frac{\partial \eta}{\partial y} = 0. \tag{3.14}$$

(3.14) is an equation of the first order for the unknown function $\eta(x, y)$. We can take any solution of this equation as the new independent variable η. In the new variables ξ, η the principal term of the equation will take the form

$$A_1\frac{\partial^2 u}{\partial \xi^2}.$$

The coefficient A_1 cannot vanish, since neither A nor $\partial \xi/\partial x + k\,\partial \xi/\partial y$ vanishes. For $A \neq 0$ by hypothesis, and the expression $\partial \xi/\partial x + k\,\partial \xi/\partial y$ could vanish only at points where the curves $\xi = $ const. and $\eta = $ const. touch, and there are no such points by hypothesis. Dividing the equation by A_1 throughout, the principal term becomes $\partial^2 u/\partial \xi^2$. It is useful to note that equation (3.14) does not define a single curve but only a family of curves $\eta = $ const. and so a certain arbitrariness remains for the choice of the function $\xi(x, y)$.

In the case when our second-order partial differential equation is of elliptic or hyperbolic type, it is easy to show that the coefficient B_1 vanishes.

The complete reduction to canonical form of an equation of elliptic type is a complicated problem, and we shall not stop to deal with it. An equation of hyperbolic type, as we shall show in a moment, can always be easily transformed to canonical form.

We select first the case when a linear second-order equation in two independent variables does not belong to the parabolic type and the coefficient A does not vanish (the case $C \neq 0$, $A = 0$ can be treated in a similar way; the case $A = C = 0$ we shall deal with separately). Let then $A \neq 0$. We put

$$\xi = x, \quad \eta = \varphi(x, y).$$

Then $\partial \xi / \partial y = 0$, $\partial \xi / \partial x = 1$, and the condition that the coefficient B_1 should vanish becomes

$$A \frac{\partial \eta}{\partial x} + B \frac{\partial \eta}{\partial y} = 0. \tag{3.15}$$

Taking once more for η any solution of the equation (3.15), we obtain a system of coordinates in which the equation has the required form. This form will not be canonical, because the coefficients of the second-order derivatives will not, in general, be ± 1.

If $A = C = 0$, then the equation will have a principal term of the form

$$\frac{\partial^2 u}{\partial x \, \partial y}$$

and will be hyperbolic. It is easy to see in this case that by the substitution

$$\xi = x + y, \quad \eta = x - y$$

the equation takes the canonical form

$$\frac{\partial^2 u}{\partial \xi^2} - \frac{\partial^2 u}{\partial \eta^2} + \cdots.$$

It is clear from this that the reduction of a hyperbolic equation to the form

$$\frac{\partial^2 u}{\partial x \partial y} + \cdots$$

would solve our problem completely. Moreover, in many problems it is precisely this form which is most convenient, and we shall therefore devote a separate section to it.

§ 3. Second Canonical Form of Hyperbolic Equations in Two Independent Variables

The second canonical form for a second-order partial differential equation of hyperbolic type in two independent variables is one where the equation does not contain the partial derivatives $\partial^2 u/\partial x^2$ and $\partial^2 u/\partial y^2$.

It follows from the formulae of the previous section that, in order to bring an equation to this form we must find functions $\xi(x, y)$ and $\eta(x, y)$ satisfying the equations

$$A_1 \equiv A\left(\frac{\partial \xi}{\partial x}\right)^2 + 2B \frac{\partial \xi}{\partial x} \frac{\partial \xi}{\partial y} + C\left(\frac{\partial \xi}{\partial x}\right)^2 = 0$$

$$C_1 \equiv A\left(\frac{\partial \eta}{\partial x}\right)^2 + 2B \frac{\partial \eta}{\partial x} \frac{\partial \eta}{\partial y} + C\left(\frac{\partial \eta}{\partial y}\right)^2 = 0.$$

These two equations are necessarily one and the same, and we have to find ξ and η as two different solutions of the one equation, which we may write in the form

$$A\left(\frac{\dfrac{\partial \zeta}{\partial x}}{\dfrac{\partial \zeta}{\partial y}}\right)^2 + 2B\left(\frac{\dfrac{\partial \zeta}{\partial x}}{\dfrac{\partial \zeta}{\partial y}}\right) + C = 0 \qquad (3.16)$$

where ζ stands for either of the functions ξ or η.

Along a curve $\zeta = $ const. we have

$$\frac{dy}{dx} = -\frac{\dfrac{\partial \zeta}{\partial x}}{\dfrac{\partial \zeta}{\partial y}}$$

and therefore (3.16) can be rewritten in the form of a quadratic in y'

$$Ay'^2 - 2By' + C = 0.$$

The condition $B^2 - AC > 0$ ensures the existence of two distinct real roots

$$y'_1 = \left[B + \sqrt{B^2 - AC}\right]/A, \quad y'_2 = \left[B - \sqrt{B^2 - AC}\right]/A. \quad (3.17)$$

By taking the integral curves of the first and second of the equations (3.17) as the curves $\xi = $ const. and $\eta = $ const. respectively, we obtain the solution of our problem.

§ 4. Characteristics

The introduction of suitably chosen variables ξ and η eliminated the terms in $\partial^2 u / \partial \xi^2$ and $\partial^2 u / \partial \eta^2$ in an equation of hyperbolic type in two independent variables.

In the general case of n variables we could also select new coordinates $y_1, y_2, \ldots, y_n$ so that there was no term containing $\partial^2 u / \partial y_1^2$, for example, in the equation. We have seen that the coefficient of this derivative is given by

$$\overline{A_{11}} = \sum_{i=1}^{n} \sum_{j=1}^{n} A_{ij} \, \alpha_{1i} \, \alpha_{1j} = \sum_{i=1}^{n} \sum_{j=1}^{n} A_{ij} \frac{\partial y_1}{\partial x_i} \frac{\partial y_1}{\partial x_j}. \tag{3.18}$$

As regards the function $y_1(x_1, x_2, \ldots, x_n)$, we shall assume that on the surface $y_1 = 0$ and in a certain neighbourhood of it the inequality

$$\sum_{i=1}^{n} \left(\frac{\partial y_1}{\partial x_i} \right)^2 > 0$$

holds good; this ensures that there are no singular points on the surfaces $y_1(x_1, x_2, \ldots, x_n) = \text{const}$. This condition can always be fulfilled if the surface $y_1 = 0$ is sufficiently smooth and if the family of surfaces $y_1 = C$, for varying C, fills a certain part of space adjoining the surface $y_1 = 0$. Any smooth surface having an equation $\varphi(x_1, x_2, \ldots, x_n) = 0$ can be taken as the surface $y_1 = 0$.

We examine the condition that the coefficient $\overline{A_{11}}$ shall vanish on the surface $y_1 = 0$. Applying a familiar result in the differential calculus, the cosine of the angle (n, x_i) which the normal n to the surface $y_1 = 0$ makes with the axis x_i is given by

$$\cos (n, x_i) = \frac{\dfrac{\partial y_1}{\partial x_i}}{\sqrt{\displaystyle\sum_{k=1}^{n} \left(\frac{\partial y_1}{\partial x_k} \right)^2}}.$$

Hence the equation $\overline{A_{11}} = 0$ can be written in the form

$$\sum_{i=1}^{n} \sum_{j=1}^{n} A_{ij} \cos (n, x_i) \cos (n, x_j) = 0.$$

This equation shows that the vanishing of $\overline{A_{11}}$ on the surface $y_1 = 0$ is an intrinsic property of this surface and that it in no way depends on the choice of the variables $y_2, y_3, \ldots, y_n$.

DEFINITION. *A surface*

$$y_1(x_1, x_2, \ldots, x_n) = c_0$$

is called a characteristic of the equation (3.1) *if, on changing from the variables* $x_1, x_2, \ldots, x_n$ *to new variables* $y_1, y_2, \ldots, y_n$, *where* $y_2, \ldots, y_n$ *are arbitrary functions of* $x_1, x_2, \ldots, x_n$ *and all the* y_i *are continuous and have first-order derivatives and a non-zero Jacobian in the neighbourhood of the surface under consideration, it happens that the coefficient* $\overline{A_{11}}$ *of* $\partial^2 u/\partial y_1^2$ *vanishes on this surface.*

It is not difficult to see that an equation of elliptic type can have no real characteristic, since $\overline{A_{11}}$ here appears as a positive definite quadratic form in α_{1i} and therefore cannot vanish.

If the surface $x_1 = 0$ is a characteristic for our equation (3.1),

$$\sum_{i=1}^{n} \sum_{j=1}^{n} A_{ij} \frac{\partial^2 u}{\partial x_i \partial x_j} + \sum_{i=1}^{n} B_i \frac{\partial u}{\partial x_i} + Cu = F,$$

i.e., if $[A_{11}]_{x_1=0} = 0$,

then this equation becomes a differential equation relating $[u]_{x_1=0}$ and $[\partial u/\partial x_1]_{x_1=0}$. For, if $x_1 = 0$, we can rewrite it in the form

$$2 \sum_{i=2}^{n} \left[A_{1i} \right]_{x_1=0} \frac{\partial}{\partial x_i} \left(\left[\frac{\partial u}{\partial x_1} \right]_{x_1=0} \right) + \sum_{i=2}^{n} \sum_{j=2}^{n} \left[A_{ij} \right]_{x_1=0} \frac{\partial^2}{\partial y_i \partial x_j} [u]_{x_1=0}$$

$$+ [B_1]_{x_1=0} \left(\left[\frac{\partial u}{\partial x_1} \right]_{x_1=0} \right) + \sum_{i=2}^{n} [B_i]_{x_1=0} \frac{\partial}{\partial x_i} ([u]_{x_1=0})$$

$$+ [C]_{x_1=0} ([u]_{x_1=0}) = [F]_{x_1=0}. \tag{3.19}$$

The problem of finding, for the second-order equation in general form, a solution which satisfies on a certain surface S the two conditions

$$[u]_S = \varphi_0 \quad \text{and} \quad \left[\frac{\partial u}{\partial n} \right]_S = \varphi_1$$

is known as Cauchy's problem.

As we saw earlier, particular instances of this problem arise in the investigation of vibrations of a string or membrane when the position and velocity of the particles of the vibrating body are given at the initial instant of the motion.

It may be usefully remarked that, in general, it is not essential that the direction along which the derivative of u is specified should be normal to the surface S; $[\partial u/\partial n]_S$ itself need not be given. For, the specification on S of the function u itself enables all its derivatives in any tangent plane to be determined, while a knowledge of the normal derivative enables the value of the gradient of the function to be found at all points on the surface S. But the derivative of the function in an arbitrary direction is the projection of its gradient in this direction. So that our aim, a knowledge of the gradient, will be achieved if we are given at each point of the surface the value of the derivative of u along any non-tangential direction.

It follows from equation (3.19) that in the case when the surface $x_1 = 0$ is a characteristic, $[u]_{x_1=0}$ and $[\partial u/\partial x_1]_{x_1=0}$ are not independent functions of the variables $x_2, x_3, \ldots, x_n$, and Cauchy's problem for this surface $x_1 = 0$ becomes impossible; for $[u]_{x_1=0}$ and $[\partial u/\partial x_1]_{x_1=0}$ cannot be specified arbitrarily and at the same time satisfy (3.19).

This argument has shown that it is impossible to prescribe arbitrarily on characteristic surfaces both the value of a function and that of any component of its gradient not lying in a tangent plane.

THE EQUATION FOR A VIBRATING STRING AND ITS SOLUTION BY D'ALEMBERT'S METHOD

§ 1. D'Alembert's Formula. Infinite String

The equation for free vibrations of a string, *i.e.*, for its vibrations when there are no external transverse forces, is

$$\frac{\partial^2 u}{\partial x^2} - \frac{1}{a^2} \frac{\partial^2 u}{\partial t^2} = 0 \tag{4.1}$$

where $a = \sqrt{T/\varrho}$. To bring it to the second canonical form we put

$$\xi = x - at, \quad \eta = x + at. \tag{4.2}$$

Then equation (4.1) becomes

$$\frac{\partial^2 u}{\partial \xi \partial \eta} = 0. \tag{4.3}$$

The general solution of equation (4.3) is easily obtained. Moreover, in contrast to what, as we said in Lecture 2, usually happens with such solutions, this general solution can easily be applied to various concrete problems.

From (4.3) we have

$$\frac{\partial}{\partial \xi} \frac{\partial u}{\partial \eta} = 0$$

whence

$$\frac{\partial u}{\partial \eta} = \psi_2'(\eta), \tag{4.4}$$

where $\psi_2'(\eta)$ is an arbitrary function.

From (4.4) we get

$$u = \psi_2(\eta) + \psi_1(\xi),$$

where $\psi_1(\xi)$ is an arbitrary function. Returning to the variables x, t, we get

$$u = \psi_2(x - at) + \psi_2(x + at) \tag{4.5}$$

This solution depends on two arbitrary functions ψ_1 and ψ_2: it is called d'Alembert's solution.

To solve any particular problem on free vibrations of a string, we have only to determine ψ_1 and ψ_2 for that particular case.

We consider first of all Cauchy's problem for an infinite string, *i.e.*, the problem of finding a solution satisfying the conditions

$$[u]_{t=0} = \varphi_0(x), \quad \left[\frac{\partial u}{\partial t}\right]_{t=0} = \varphi_1(x). \tag{4.6}$$

Substituting formula (4.5) in (4.6) we have

$$\psi_1(x) + \psi_2(x) = \varphi_0(x)$$

$$-a\psi_1'(x) + a\psi_2'(x) = \varphi_1(x),$$

and from the second of these we get

$$-a\psi_1(x) + a\psi_2(x) = \int_0^x \varphi_1(y) \, dy + aC$$

where C is an arbitrary constant, and y, of course, is merely a dummy variable. Hence

$$\psi_1(x) = \frac{1}{2}\left[\varphi_0(x) - \frac{1}{a}\int_0^x \varphi_1(y) \, dy - C\right]$$

$$\psi_2(x) = \frac{1}{2}\left[\varphi_0(x) + \frac{1}{a}\int_0^x \varphi_1(y) \, dy + C\right].$$

Hence

$$u = \frac{1}{2}\left[\varphi_0(x - at) - \frac{1}{a}\int_0^{x-at} \varphi_1(y) \, dy - C + \varphi_0(x + at)\right.$$

$$\left. + \frac{1}{a}\int_0^{x+at} \varphi_1(y) \, dy + C\right]$$

or finally we get the formula

$$u = \frac{1}{2}\left[\varphi_0(x - at) + \varphi_0(x + at) + \frac{1}{a}\int_{x-at}^{x+at} \varphi_1(y) \, dy\right]. \tag{4.7}$$

It is clear that this solution satisfies equation (4.1) and also the initial conditions. The method of deriving (4.7) shows the uniqueness of solution of the problem. Further, there is no doubt that the problem was correctly formulated. For any $\varepsilon > 0$ we can find an η such that, if we replace $\varphi_0(x)$ and $\varphi_1(x)$ by $\varphi_0^*(x)$ and $\varphi_1^*(x)$ such that

$$\left|\varphi_0(x) - \varphi_0^*(x)\right| < \eta, \quad \left|\varphi_1(x) - \varphi_1^*(x)\right| < \eta,$$

then the absolute value of the difference between the new solution and the original one will be less than ε in any finite and previously given time-interval. Thus the solution of the problem depends continuously to order $(0, 0)$ on the initial data.

We may profitably examine the vibrations of an infinite string in somewhat greater detail. We select one or two simple cases.

CASE 1. The function $\varphi_1(x)$ is identically zero, and the function $\varphi_0(x)$ is zero except in a finite interval $-k \leq x \leq k$.

The solution (4.7) then becomes

$$u = \frac{1}{2} [\varphi_0(x - at) + \varphi_0(x + at)].$$

The term $\frac{1}{2}\varphi_0(x - at)$ represents a disturbance of constant shape travelling with velocity a in the positive x-direction. This is evident from the fact that, if we put the origin of a moving coordinate system ξ at the point $x = at$, i.e., if we put $\xi = x - at$, then we should see a constant disturbance in the moving system. Similarly, the term $\frac{1}{2}\varphi_0(x + at)$ represents a disturbance of the same shape travelling with the same speed in the opposite direction. These disturbances are called *waves*. The first is the direct wave, the second the reverse wave. The waves are initially superimposed, and then they separate and move in opposite directions further and further away from each other. At points lying within the region of the initial disturbance the string

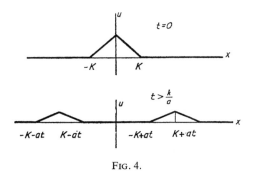

FIG. 4.

returns to its rest position after the passage of both waves, and at other points after the passage of one wave. Fig. 4 shows the disturbed string at two instants of time.

CASE 2. The function $\varphi_0(x)$ is identically zero, and the function $\varphi_1(x)$ is zero except in a finite interval $-k \leq x \leq k$. In such a case we may say that the string has an initial impulse but no initial disturbance.

Consider the function $\Phi_1(x)$, the indefinite integral of $\varphi_1(x)$. It is zero for values of x in the interval $-\infty < x \leq -k$. For $x \geq k$ it will be equal to some constant, in general non-zero. This constant is evidently equal to

$$\int_{-k}^{k} \varphi_1(x)\, dx.$$

Formula (4.7) gives

$$u = \frac{1}{2a}\left[\Phi_1(x + at) - \Phi_1(x - at)\right].$$

In this case again, two waves travel along the string, one forward, one reverse. They differ only in sign.

Where both the forward and reverse waves have already passed, the string will have reached a state of rest, but it will not, in general, have re-

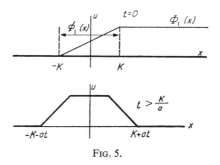

Fig. 5.

turned to its original position. For at sufficiently large values of the time, $x + at > k$ and then $\Phi_1(x + at)$ will be equal to the constant, but $x - at < -k$ and $\Phi_1(x - at)$ will be zero.

A so-called residual displacement will remain in the string. The shape of the displaced string at a particular time is shown in Fig. 5.

It will readily be seen that it is possible to produce a wave travelling in one direction only by giving the string a suitable initial disturbance and impulse. It is only necessary to ensure that the reverse waves evoked by the initial displacement and by the initial impulse differ only in sign.

§ 2. String with Two Fixed Ends

We next consider a string with fixed ends and seek a solution subject to the conditions (4.6) and also

$$[u]_{x=0} = 0 \quad \text{and} \quad [u]_{x=l} = 0.$$

Obviously the functions $\varphi_0(x)$ and $\varphi_1(x)$ in this case will be defined only in the interval $0 \leqq x \leqq l$.

Returning to formula (4.5), we see that $\psi_1(x)$ must be defined in the interval from $-\infty$ to l, and $\psi_2(x)$ in the interval from 0 to $+\infty$. Putting $x = 0$ and $x = l$ in the solution (4.5) we get

$$\left.\begin{aligned} \psi_1(-at) + \psi_2(at) &= 0 \\ \psi_1(l - at) + \psi_2(l + at) &= 0 \end{aligned}\right\} \tag{4.8}$$

or

$$\left.\begin{aligned} \psi_1(-x) + \psi_2(x) &= 0, \quad 0 \leqq x < \infty \\ \psi_1(-x) + \psi_2(x + 2l) &= 0, \quad -l \leqq x < \infty. \end{aligned}\right\} \tag{4.8'}$$

Any two functions $\psi_1(x)$ and $\psi_2(x)$ which satisfy (4.8') will give a solution of the problem.

The first of the conditions (4.8') enables $\psi_2(x)$ for positive x to be expressed in terms of $\psi_1(x)$ for negative values of the argument. Substituting the value of $\psi_2(x)$ from the first condition into the second, we get

$$\psi_1(-x) - \psi_1(-x - 2l) = 0. \tag{4.9}$$

This formula asserts that the function $\psi_1(x)$ must be periodic with period $2l$ throughout the region in which it is of interest to us.

Regarding (4.8') as a pair of equations in two unknown functions, we see that the first of them can be regarded as a definition of $\psi_2(x)$. It is evident that this pair of equations is exactly equivalent to the single equation (4.9), since if (4.9) is satisfied then so are both of (4.8). Let us formulate the result achieved.

We have shown that any solution of the equation (4.1) subject to the conditions (4.8) may be expressed in terms of an arbitrary periodic function $\psi_1(x)$ with period $2l$, defined in the interval $-\infty < x \leqq l$:

$$u = \psi_1(x - at) - \psi_1(-x - at). \tag{4.10}$$

If $\psi_1(x)$ continues to be periodic with period $2l$ along the whole straight line $-\infty < x < \infty$, the function (4.10) will not change for any x in $0 \leqq x \leqq l$ from what it was before. This is the only interval that interests us in the present problem. However, (4.10) would in that case also be the solution of some problem on the vibrations of an infinite string.

Putting $t = 0$ in (4.10), we see that

$$[u]_{t=0} = \psi_1(x) - \psi_1(-x).$$

$$\left[\frac{\partial u}{\partial t}\right]_{t=0} = -a[\psi_1'(x) - \psi_1'(-x)].$$

But $\psi_1(x) - \psi_1(-x)$ is, as may easily be seen, an odd function, which, by what has been proved, has period $2l$. Such a function can easily be constructed in full from its values in the interval $0 \leqq x \leqq l$, where we have

$$\psi_1(x) - \psi_1(-x) = \varphi_0(x), \quad 0 \leqq x \leqq l.$$

Similarly, starting from the equality

$$-a\psi_1'(x) + a\psi_1'(-x) = \varphi_1(x) \quad \text{for} \quad 0 \leqq x \leqq l,$$

it is easy to construct $-a[\psi_1'(x) - \psi_1'(-x)]$ everywhere as an odd, periodic function. This is equivalent to continuing the functions $\varphi_0(x)$ and $\varphi_1(x)$ along the whole straight line as odd, periodic functions.

When the functions $\varphi_0(x)$ and $\varphi_1(x)$ have been constructed along the whole straight line, $\psi_1(x)$ can be found without difficulty in the same way as we did in the case of the infinite string.

It follows from this that the solution to our problem is expressed by the same formula (4.7) if in it we take $\varphi_0(x)$ and $\varphi_1(x)$ to be odd, periodic functions. The formula (4.7) does indeed give the required solution, since the fact that the initial conditions are odd implies that $[u]_{x=0} = 0$; and an odd function of period $2l$ is obviously odd relative to the points $x = kl$ $(k = 0, 1, -1, 2, -2, ...)$. Hence $[u]_{x=l}$ is also equal to 0.

One important point must be mentioned. Strictly speaking, a function given by the formula (4.5) will satisfy the equation provided that $\psi_1(x)$ and $\psi_2(x)$, together with their second-order derivatives, are all continuous functions. The question arises whether the solution we have found satisfies this condition.

For the condition to be satisfied, we require that $\varphi_0(x)$ and $\varphi_1(x)$ after their continuation along the whole straight line as odd, periodic functions shall be continuous and have continuous second-order derivatives. As the reader can verify, this means that we must have

$$\varphi_0(l) = \varphi_0''(l) = \varphi_0(0) = \varphi_0''(0) = 0,$$

$$\varphi_1(l) = \varphi_1''(l) = \varphi_1(0) = \varphi_1''(0) = 0.$$

However, it is also possible to contemplate functions which do not satisfy all these conditions. It then becomes necessary to generalize in an appropriate way the class of admissible solutions of equation (4.1). We shall consider this question later.

§ 3. Solution of the Problem for a Non-Homogeneous Equation and for More General Boundary Conditions

We consider next a more general problem. Suppose we require to solve the equation

$$\frac{\partial^2 u}{\partial x^2} - \frac{1}{a^2} \frac{\partial^2 u}{\partial t^2} = p(x, t) \tag{4.11}$$

subject to the conditions

$$\alpha[u]_{x=0} + \beta\left[\frac{\partial u}{\partial x}\right]_{x=0} = f_1(t) \left.\vphantom{\begin{array}{c}a\\a\\a\\a\end{array}}\right\}$$

$$\gamma[u]_{x=l} + \delta\left[\frac{\partial u}{\partial x}\right]_{x=l} = f_2(t) \qquad (4.12)$$

and

$$[u]_{t=0} = \varphi_0(x), \quad \left[\frac{\partial u}{\partial t}\right]_{t=0} = \varphi_1(x). \qquad (4.13)$$

A solution is sought in the domain defined by the inequalities

$$0 < x < l \text{ and } 0 < t.$$

We may first remark that it is easy to construct a particular solution of equation (4.11), though not, it is true, one satisfying the conditions (4.12) and (4.13).

By introducing new variables

$$\xi = x - at, \quad \eta = x + at \qquad (4.14)$$

we can transform (4.11) into the form

$$4\frac{\partial^2 u}{\partial\xi\partial\eta} = Q(\xi, \eta). \qquad (4.15)$$

ξ and η will vary over the domain defined by the inequalities

$$\eta - \xi > 0 \qquad (4.16)$$

and

$$0 < \eta + \xi < 2l; \qquad (4.17)$$

this is the half-strip ending in GF and going through E and D, as shown in Fig. 6. We continue the function $Q(\xi, \eta)$ quite arbitrarily beyond the two straight lines given by

$$\xi + \eta = 0$$

$$\xi + \eta = 2l,$$

thus defining it over the whole half-plane given by (4.16). If we find a solution of (4.15) over the whole half-plane, then clearly it will also satisfy our equation in the strip defined by the inequalities (4.16), (4.17). We shall try to find a particular solution v_1 of equation (4.15). Putting

$$\frac{\partial v_1}{\partial\eta} = z(\xi, \eta) \qquad (4.18)$$

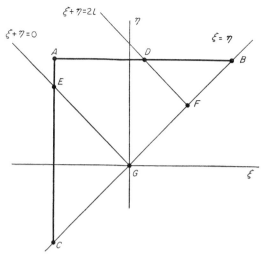

Fig. 6.

we get from (4.15)

$$\frac{\partial z}{\partial \xi} = \frac{1}{4} Q(\xi, \eta); \tag{4.19}$$

and a particular solution of (4.19) has the form

$$z(\xi, \eta) = \frac{1}{4} \int_{\eta}^{\xi} Q(\alpha, \eta) \, d\alpha. \tag{4.20}$$

Further, a particular solution of (4.18) will be

$$v_1 = \int_{\xi}^{\eta} z(\xi, \beta) \, d\beta \tag{4.21}$$

or, combining (4.20) and (4.21), a particular solution of (4.15) will have the form

$$v_1 = \frac{1}{4} \int_{\xi}^{\eta} \left[\int_{\beta}^{\xi} Q(\alpha, \beta) \, d\alpha \right] d\beta = -\frac{1}{4} \int_{\eta}^{\xi} \left[\int_{\beta}^{\xi} Q(\alpha, \beta) \, d\alpha \right] d\beta \tag{4.22}$$

The integral (4.22) is taken over the domain

$$\xi < \beta < \alpha < \eta,$$

which is the triangle ABC in Fig. 6, the point A being the point (ξ, η). In other words,

$$v_1 = -\frac{1}{4} \iint_{\triangle ABC} Q(\alpha, \beta) \, d\alpha \, d\beta$$

and this integral may be divided into the sum of the same integral taken over the areas $\varDelta CEG$, $\varDelta DBF$, and $ADFGE$:

$$v_1 = -\frac{1}{4} \iint_{\varDelta CEG} Q(\alpha, \beta)\, d\alpha\, d\beta - \frac{1}{4} \iint_{\varDelta DBF} Q(\alpha, \beta)\, d\alpha\, d\beta$$

$$- \frac{1}{4} \iint_{ADFGE} Q(\alpha, \beta)\, d\alpha\, d\beta.$$

The triangle CEG depends only on the coordinate ξ of the point A, and the triangle DBF only on the coordinate η of A. The point A is taken inside the half-strip. Consequently,

$$-\frac{1}{4} \iint_{\varDelta CEG} Q(\alpha, \beta)\, d\alpha\, d\beta = w_1(\xi), \qquad -\frac{1}{4} \iint_{\varDelta DBF} Q(\alpha, \beta)\, d\alpha\, d\beta = w_2(\eta).$$

If the function v_1 satisfies equation (4.15) then the function

$$v = v_1 - w_1(\xi) - w_2(\eta) = -\frac{1}{4} \iint_{ADFGE} Q(\alpha, \beta)\, d\alpha\, d\beta \qquad (4.23)$$

will also satisfy the same equation. If we transform the variables in the integral (4.23) back to x and t, then, since

$$\frac{D(\alpha, \beta)}{D(x, t)} = \begin{vmatrix} 1 & -a \\ 1 & a \end{vmatrix} = 2a,$$

we get

$$v = -\frac{a}{2} \iint_{ADFGE} p(x, t)\, dx\, dt \qquad (4.24)$$

where $ADFGE$ has the shape shown in Fig. 7.

We shall try to find a solution of our problem in the form of a particular solution of the non-homogeneous equation and a solution of the homo-

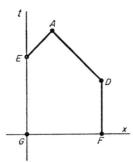

Fig. 7.

geneous equation which satisfies the boundary conditions. To do this we introduce instead of u a new function w where

$$w = u - v. \tag{4.25}$$

The function w will then satisfy a homogeneous equation for a vibrating string and conditions of the same type as (4.12) and (4.13).

In dealing with this problem we can immediately take $p(x, t) = 0$ and suppose the conditions (4.12), (4.13) transformed according to (4.25).

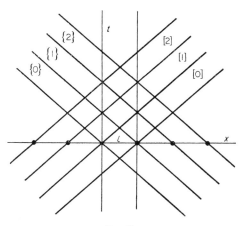

FIG. 8.

Our change of variable does not disturb the uniqueness and existence of the solution. If the problem was correctly formulated initially, then it will still be so after the change of variable. For, the solution w depends continuously on the initial conditions. The new conditions for w will depend continuously on v and consequently on $p(x, t)$. As a result, small deviations in $p(x, t), f_1, f_2$ will give small deviations in the solution u.

A solution of the equation

$$\frac{\partial^2 u}{\partial x^2} - \frac{1}{a^2} \frac{\partial^2 u}{\partial t^2} = 0 \tag{4.1}$$

may again be sought in the previous form

$$u = \psi_1(x - at) + \psi_2(x + at). \tag{4.5}$$

As in the first problem, the initial conditions (4.13) enable the functions $\psi_1(x)$ and $\psi_2(x)$ to be defined in the interval $0 \leq x \leq l$. This means that the function u is defined inside a triangle lying in the strip (see Fig. 8) and resting on the x-axis ($t > 0$).

The next question is: Can we satisfy the conditions (4.12) by a suitable choice of continuations of $\psi_1(x)$ and $\psi_2(t)$ over the whole domain of interest to us?

Substituting the expressions (4.5) in (4.12), we get

$$\left.\begin{array}{l}
\alpha\psi_1(-at) + \beta\psi_1'(-at) + \alpha\psi_2(at) + \beta\psi_2'(at) = f_1(t) \\[2mm]
\gamma\psi_1(l-at) + \delta\psi_1'(l-at) + \gamma\psi_2(l+at) + \delta\psi_2'(l+at) = f_2(t)
\end{array}\right\} \quad (4.26)$$

We can draw, in the (x, t) plane, two series of strips, parallel to the straight lines given by $x - at = 0$ and $x + at = 0$, in which the inequalities

$$-lk < x - at < -l(k-1), \quad [k]$$

$$ml < x + at < (m+1)l \quad \{m\}$$

hold good; and we shall denote each strip by its corresponding number placed inside either a square bracket [ ] or a curly bracket { } (see Fig. 8).

From the first of the conditions (4.26), the value of ψ_1 in the strip $[m]$ can be determined if the value of ψ_2 in the strip $\{m-1\}$ is known; and from the second of these conditions, the value of ψ_2 in the strip $\{k\}$ can be determined if the value of ψ_1 in the strip $[k-1]$ is known; in each case by solving an ordinary differential equation with constant coefficients. Thus, denote $-at$ in the first equation of (4.26) by ξ; then it becomes

$$\alpha\psi_1(\xi) + \beta\psi_1'(\xi) = -\alpha\psi_2(-\xi) - \beta\psi_2'(-\xi) + f_1\left(\frac{-\xi}{a}\right); \quad (4.27)$$

and if we replace $l - at$ by ξ in the second equation of (4.26), it becomes

$$\gamma\psi_1(\xi) + \delta\psi_1'(\xi) = -\gamma\psi_2(2l-\xi) - \delta\psi_2'(2l-\xi) + f_2\left(\frac{l-\xi}{a}\right). \quad (4.28)$$

Knowing ψ_2 in $\{0\}$ and solving (4.27) for ψ_1, we obtain the value of ψ_1 in [1]; and then, solving (4.28) for ψ_2, we get the value of ψ_2 in $\{2\}$; and so on. The arbitrary constants arising in the integrations are determined from the conditions for continuity of ψ_1 and ψ_2. In exactly the same way, starting from the value of ψ_1 in [0] and solving equation (4.28), we get the value of ψ_2 in $\{1\}$; and then, solving (4.27) for ψ_1, we find the value of ψ_1 in [2]; and so on. It is clear that in this way all the required conditions can be satisfied.

It is clear that in this process no doubt can arise about the existence and uniqueness of the solution nor about the correctness of formulation of the problem. The only thing which is not clear is whether the functions ψ_1 and ψ_2 which we have constructed will have continuous second-order derivatives. We shall leave the reader to establish the necessary and sufficient conditions

for this to be the case, and we merely remark that, since we have extended
the class of solutions of the equation (4.1), we can drop the requirement
that the functions ψ_1 and ψ_2 shall be continuous.

The equation for a vibrating string and equations of hyperbolic type
in two independent variables similar to it are often encountered in various
problems of mathematical physics.

RIEMANN'S METHOD

§ 1. The Boundary-Value Problem of the First Kind for Hyperbolic Equations

In constructing a solution for the equation of a vibrating string we made use of the fundamental property of the characteristic of this equation, allowing us to integrate the equation (4.3) immediately.

This property consists in the fact that, on the characteristic curves $\eta = $ const., the equation (4.3) becomes an ordinary differential equation in $\partial u/\partial \eta$ with ξ as the independent variable. Hence we were able to find the function u by quadrature. This same property of characteristics – the existence of equations with fewer variables, which express relations between the values of the unknown function and its derivatives – forms the basis of several important methods of integrating partial differential equations of hyperbolic type.

Consider the equation

$$\frac{\partial^2 u}{\partial x \, \partial y} + a(x, y) \frac{\partial u}{\partial x} + b(x, y) \frac{\partial u}{\partial y} + c(x, y) \, u = F(x, y); \quad (5.1)$$

as we have seen, any linear, hyperbolic equation in two independent variables can be put into this form.

We first prove this lemma:

Suppose the value of u is given on two intersecting straight lines parallel to the coordinate axes (i.e., on the characteristics):

$$[u]_{x=x_0} = \varphi_1(y); \quad y_0 \leqq y \leqq b$$
$$[u]_{y=y_0} = \varphi_2(x); \quad x_0 \leqq x \leqq a \quad (5.2)$$

and suppose that $\varphi_1(y_0) = \varphi_2(x_0)$. *Then in the rectangle defined by* $x_0 \leqq x \leqq a$, $y_0 \leqq y \leqq b$, *equation (5.1) has a unique solution satisfying the conditions (5.2).* (We assume that $\varphi_1(y)$ and $\varphi_2(y)$ have continuous, first-order derivatives.)

Put

$$\frac{\partial u}{\partial x} = v, \quad \frac{\partial u}{\partial y} = w. \quad (5.3)$$

Then (5.1) can be rewritten in the form

$$\frac{\partial v}{\partial y} = \frac{\partial w}{\partial x} = F(x, y) - a(x, y)\, v - b(x, y)\, w - c(x, y)\, u. \quad (5.4)$$

It follows at once that

$$
\left.
\begin{aligned}
v(x, y) &= v(x, y_0) + \int_{y_0}^{y} [F(x, y) - a(x,y)v - b(x, y)w - c(x, y)u]\, dy \\[2mm]
w(x,y) &= w(x_0, y) + \int_{x_0}^{x} [F(x, y) - a(x, y)v - b(x, y)w - c(x, y)u]\, dx \\[2mm]
u(x, y) &= u(x, y_0) + \int_{y_0}^{y} w(x, y)\, dy
\end{aligned}
\right\} (5.5)
$$

and by virtue of (5.2)

$$v(x, y_0) = \left[\frac{\partial u}{\partial x}\right]_{y = y_0} = \varphi_2'(x)$$

$$w(x_0, y) = \left[\frac{\partial u}{\partial y}\right]_{x = x_0} = \varphi_1'(y).$$

It follows from (5.4) and (5.5) that

$$
\left.
\begin{aligned}
v(x, y) &= \varphi_2'(x) + \int_{y_0}^{y} [F(x, y) - a(x, y)v - b(x, y)w - c(x, y)u]\, dy \\[2mm]
w(x,y) &= \varphi_1'(y) + \int_{x_0}^{x} [F(x, y) - a(x, y)v - b(x, y)w - c(x, y)u]\, dx \\[2mm]
u(x, y) &= \varphi_2(x) + \int_{y_0}^{y} w\, dy.
\end{aligned}
\right\} (5.6)
$$

Conversely, any solution of the system (5.6) will obviously satisfy the equations (5.4) and the second of the equations (5.3). Moreover,

$$\frac{\partial u}{\partial x} = \varphi_2'(x) + \int_{y_0}^{y} \frac{\partial w}{\partial x}\, dy$$

$$= \varphi_2'(x) + \int_{y_0}^{y} [F(x, y) - a(x, y)v - b(x, y)w - c(x, y)u]\, du$$

that is,

$$\frac{\partial u}{\partial x} = v.$$

Consequently, the first of the equations (5.3) is also satisfied. It also follows from (5.6) that

$$[u]_{y=y_0} = \varphi_2(x),$$

$$[u]_{x=x_0} = \varphi_2(x_0) + \int_{y_0}^{y} [w]_{x=x_0}\, dy = \varphi_2(x_0) + \int_{y_0}^{y} \varphi_1'(y)\, dy$$

$$= \varphi_2(x_0) + \varphi_1(y) - \varphi_1(y_0) = \varphi_1(y).$$

Thus, any solution of the system (5.6) is a solution of the problem proposed. It follows from all that has been stated that the system (5.6) is equivalent to the equation (5.1) with the conditions (5.2).

We shall find a solution of the system (5.6) by the method of successive approximations. Let

$$v_0 = \varphi_2'(x), \quad w_0 = \varphi_1'(y), \quad u_0 = \varphi_2(x)$$

and write

$$
\left.
\begin{aligned}
v_n &= \varphi_2'(x) + \int_{y_0}^{y} [F(x,y - a(x,y)\,v_{n-1} - b(x,y)\,w_{n-1} \\
&\qquad\qquad\qquad\qquad\qquad - c(x,y)\,u_{n-1}]\, dy \\
w_n &= \varphi_1'(y) + \int_{x_0}^{x} [F(x,y) - a(x,y)\,v_{n-1} - b(x,y)\,w_{n-1} \\
&\qquad\qquad\qquad\qquad\qquad - c(x,y)\,y_{n-1}]\, dx \\
u_n &= \varphi_2(x) + \int_{y_0}^{y} w_{n-1}\, dy \\
&\qquad (n = 1, 2, \ldots).
\end{aligned}
\right\} \quad (5.7)
$$

We show that the sequences u_n, v_n, and w_n converge. To do this, we assume that all the functions $\varphi_1(y)$, $\varphi_2(x)$, $\varphi_1'(y)$, $\varphi_2'(x)$, $F(x,y)$, $a(x,y)$, $b(x,y)$, $c(x,y)$ are bounded in the rectangle defined by (5.2). We have

$$
\left.
\begin{aligned}
v_{n+1} - v_n &= -\int_{y_0}^{y} [a(x,y)\,(v_n - v_{n-1}) + b(x,y)\,(w_n - w_{n-1}) \\
&\qquad\qquad\qquad\qquad + c(x,y)\,(u_n - u_{n-1})]\, dy \\
w_{n+1} - w_n &= -\int_{x_0}^{x} [a(x,y)\,(v_n - v_{n-1}) + b(x,y)\,(w_n - w_{n-1}) \\
&\qquad\qquad\qquad\qquad + c(x,y)\,(u_n - u_{n-1})]\, dx \\
u_{n+1} - u_n &= \int_{y_0}^{y} (w_n - w_{n-1})\, dy.
\end{aligned}
\right\} \quad (5.8)
$$

We shall show that $|u_n - u_{n-1}|$, $|v_n - v_{n-1}|$, $|w_n - w_{n-1}|$ satisfy the inequalities

$$\left.\begin{aligned}
\left|v_n - v_{n-1}\right| &\leq K^{n-1} A \frac{(x + y - x_0 - y_0)^{n-1}}{(n-1)!} \\[2mm]
\left|w_n - w_{n-1}\right| &\leq K^{n-1} A \frac{(x + y - x_0 - y_0)^{n-1}}{(n-1)!} \\[2mm]
\left|u_n - u_{n-1}\right| &\leq K^{n-1} A \frac{(x + y - x_0 - y_0)^{n-1}}{(n-1)!}
\end{aligned}\right\} \tag{5.9}$$

where

$$K > \left|a(x, y)\right| + \left|b(x, y)\right| + \left|c(x, y)\right|$$

and the A's are certain numbers, independent of n.

For $n = 1$ the inequalities (5.9) obviously are true if the constants A are chosen sufficiently large. We show that these inequalities remain true when n is replaced by $n + 1$. From (5.8) we have, for example,

$$\begin{aligned}
\left|v_{n+1} - v_n\right| &\leq \int_{y_0}^{y} \left(|a| + |b| + |c|\right) K^{n-1} A \frac{(x + y - x_0 - y_0)^{n-1}}{(n-1)!} \, dy \\[2mm]
&\leq A K^n \int_{y_0}^{y} \frac{(x + y - x_0 - y_0)^{n-1}}{(n-1)!} \, dy \\[2mm]
&= A K^n \left[\frac{(x + y - x_0 - y_0)^n}{n!} - \frac{(x - x_0)^n}{n!} \right] \\[2mm]
&\leq A K^n \frac{(x + y - x_0 - y_0)^n}{n!}.
\end{aligned}$$

The other differences in (5.9) may be estimated in the same way.

It follows from (5.9) that the series

$$u_0 + \sum_{n=1}^{\infty} (u_n - u_{n-1}), \quad v_0 + \sum_{n=1}^{\infty} (v_n - v_{n-1}), \quad w_0 + \sum_{n=1}^{\infty} (w_n - w_{n-1})$$

are absolutely and uniformly convergent, since their terms are in absolute magnitude less than the corresponding terms of the uniformly convergent series

$$A + A \sum_{n=1}^{\infty} K^{n-1} \frac{(x + y - x_0 - y_0)^{n-1}}{(n-1)!},$$

which, as is well known, is the function $A + A\, e^{K(x+y-x_0-y_0)}$.

Consequently, u_n, v_n, and w_n tend uniformly to definite limits in the rectangle given in (5.2). Passing to the limit in the formulae (5.7), we see that the limit functions u, v, and w satisfy (5.6), and our problem is solved. We may note that an exactly similar argument would apply in the case $x < x_0$, $y < y_0$.

It is not difficult to show that our solution is unique. To do this, it is sufficient to show that in the case when $F \equiv 0$, $\varphi_1(y) \equiv \varphi_2(x) \equiv 0$, the system (5.6) has no other bounded solutions than $u \equiv 0$, $v \equiv 0$, $w \equiv 0$. Suppose that there is some solution satisfying the conditions $|u| < A$, $|v| < A$, $|w| < A$. Then the functions u, v, w will satisfy the inequalities

$$\left.\begin{aligned}
|u| &\leqq K^{n-1} A \, \frac{(x + y - x_0 - y_0)^{n-1}}{(n-1)!} \\[2mm]
|v| &\leqq K^{n-1} A \, \frac{(x + y - x_0 - y_0)^{n-1}}{(n-1)!} \\[2mm]
|w| &\leqq K^{n-1} A \, \frac{(x + y - x_0 - y_0)^{n-1}}{(n-1)!}
\end{aligned}\right\} \tag{5.9'}$$

These inequalities are obtained in the same way as were those in (5.9), *i.e.*, by successive approximation.

The uniqueness of the solution immediately follows, since the only functions which can satisfy (5.9') for all n are $u = v = w = 0$.

§ 2. Adjoint Differential Operators

Consider the linear differential operator L of the second order such that

$$Lu \equiv \sum_{i=1}^{n} \sum_{j=1}^{n} A_{ij} \frac{\partial^2 u}{\partial x_i \, \partial x_j} + \sum_{i=1}^{n} B_i \frac{\partial u}{\partial x_i} + Cu \tag{5.10}$$

where A_{ij}, B_i, and C are given functions of x_1, x_2, ..., x_n and have second-order derivatives. Without loss of generality we may assume $A_{ij} = A_{ji}$; for, if this is not true initially, we can make it so by letting A_{ij} stand for half the sum of the original coefficients of $\partial^2 u/\partial x_i \, \partial x_j$ and $\partial^2 u/\partial x_j \, \partial x_i$.

We now introduce *the differential operator Mv which is adjoint to Lu:*

$$Mv \equiv \sum_{i=1}^{n} \sum_{j=1}^{n} \frac{\partial^2 (A_{ij}v)}{\partial x_i \, \partial x_j} - \sum_{i=1}^{n} \frac{\partial (B_i v)}{\partial x_i} + Cv. \tag{5.11}$$

It is easily shown that the relation between L and M is reciprocal: Lv is adjoint to Mu.

A differential operator which is identical with its adjoint is said to be *self-adjoint*.

Next we show that

$$v\,Lu - u\,Mv = \sum_{i=1}^{n} \frac{\partial}{\partial x_i} \left\{ \sum_{j=1}^{n} \left[vA_{ij} \frac{\partial u}{\partial x_j} - u \frac{\partial(A_{ij}v)}{\partial x_j} \right] + B_i uv \right\} \quad (5.12)$$

i.e., the expression $v\,Lu - u\,Mv$ is the sum of the partial derivatives with respect to the x_i of certain expressions P_i,

$$v\,Lu - u\,Mv = \sum_{i=1}^{n} \frac{\partial P_i}{\partial x_i}$$

where

$$P_i = \sum_{j=1}^{n} \left(vA_{ij} \frac{\partial u}{\partial x_j} - u \frac{\partial(A_{ij}v)}{\partial x_j} \right) + B_i uv. \quad (5.13)$$

Formula (5.12) follows at once on differentiating:

$$\sum_{i=1}^{n} \frac{\partial P_i}{\partial x_i} = \left[\sum_{i=1}^{n} \sum_{j=1}^{n} vA_{ij} \frac{\partial^2 u}{\partial x_j\,\partial x_i} + \sum_{i=1}^{n} vB_i \frac{\partial u}{\partial x_i} + Cuv \right]$$

$$- \left[\sum_{i=1}^{n} \sum_{j=1}^{n} u \frac{\partial^2(A_{ij}v)}{\partial x_j\,\partial x_i} - \sum_{i=1}^{n} u \frac{\partial(B_i v)}{\partial x_i} + Cuv \right]$$

$$+ \sum_{i=1}^{n} \sum_{j=1}^{n} \left(\frac{\partial u}{\partial x_j} \frac{\partial(A_{ij}v)}{\partial x_i} - \frac{\partial u}{\partial x_i} \frac{\partial(A_{ij}v)}{\partial x_j} \right).$$

The last sum vanishes, and we have

$$\sum_{i=1}^{n} \frac{\partial P_i}{\partial x_i} = v\,Lu - u\,Mv \quad (5.14)$$

as was to be proved.

Now consider some n-dimensional volume Ω, bounded by a piecewise smooth surface S. We assume that all the conditions of continuity of the functions and their derivatives which were discussed when we established formula (1.2) are fulfilled.

(If $n = 2$ or 1, the words "volume" and "surface" would be replaced by "domain" and "curve" or "interval" and "ends of the interval".)

On the basis of a formula similar to (1.2) we shall have

$$\int\!\!\int \cdots \int_{\Omega} (v\,Lu - u\,Mv)\,dx_1 \cdots dx_n = -\int \cdots \int_{S} \sum_{i=1}^{n} P_i \cos(\boldsymbol{n}, x_i)\,dS \quad (5.15)$$

where $\cos(\boldsymbol{n}, x_1)$, $\cos(\boldsymbol{n}, x_2)$, ... are the direction-cosines of the inward normal to S.

(5.15) is known as *Green's Formula*. We give two examples.

EXAMPLE 1. *Green's formula for the Laplace operator.*
Let

$$Lu = \nabla^2 u.$$

Then

$$Mv = \nabla^2 v$$

and

$$P_x = v\frac{\partial u}{\partial x} - u\frac{\partial v}{\partial x}, \quad P_y = v\frac{\partial u}{\partial y} - u\frac{\partial v}{\partial y}, \quad P_z = v\frac{\partial u}{\partial z} - u\frac{\partial v}{\partial z}.$$

Green's formula takes the form

$$\iiint_\Omega (v\nabla^2 u - u\nabla^2 v)\,dx\,dy\,dz$$

$$= -\iint_S \left[\left(v\frac{\partial u}{\partial x} - u\frac{\partial v}{\partial x} \right)\cos(\mathbf{n}, x) + \left(v\frac{\partial u}{\partial y} - u\frac{\partial v}{\partial y} \right)\cos(\mathbf{n}, y) \right.$$

$$\left. + \left(v\frac{\partial u}{\partial z} - u\frac{\partial v}{\partial z} \right)\cos(\mathbf{n}, z) \right] dS = \iint_S \left(u\frac{\partial v}{\partial n} - v\frac{\partial u}{\partial n} \right) dS \quad (5.16)$$

where, as usual, $\partial v/\partial n$ denotes the derivative of v along the normal and is equal to the projection of the vector grad $v = \{\partial v/\partial x, \partial v/\partial y, \partial v/\partial z\}$ in the direction of the inward normal. (5.16) is also known as *Green's formula for the Laplace operator.*

EXAMPLE 2. Consider the equation (5.1).
The adjoint operator Mv and the functions P_1 and P_2 for the operator Lu on the left-hand side of this equation will have the form

$$Mv \equiv \frac{\partial^2 v}{\partial x\,\partial y} - \frac{\partial(av)}{\partial x} - \frac{\partial(bv)}{\partial y} + cv$$

$$P_1 = \frac{1}{2}\left(v\frac{\partial u}{\partial y} - u\frac{\partial v}{\partial y} \right) + auv$$

$$P_2 = \frac{1}{2}\left(v\frac{\partial u}{\partial x} - u\frac{\partial v}{\partial x} \right) + buv.$$

Then Green's formula gives (with the inward normal)

$$\iint_\Omega (vLu - uMv)\,dx\,dy = -\int_S \left\{ \left[\frac{1}{2}\left(v\frac{\partial u}{\partial y} - u\frac{\partial v}{\partial y} \right) + auv \right]\cos(\mathbf{n}, x) \right.$$

$$\left. + \left[\frac{1}{2}\left(v\frac{\partial u}{\partial x} - u\frac{\partial v}{\partial x} \right) + buv \right]\cos(\mathbf{n}, y) \right\} dS \quad (5.17)$$

§ 3. Riemann's Method

An important method for finding various solutions of equation (5.1), based on Green's formula (5.17), was given by Riemann.

Consider the solution of Cauchy's problem for the equation (5.1). Suppose that we are given the values of u and $\partial u/\partial y$ on the curve $y = \mu(x)$ and assume that

$$\mu'(x) < 0 \qquad (5.18)$$

$$[u]_{y=\mu(x)} = \varphi_0(x) \qquad (5.19)$$

$$\left[\frac{\partial u}{\partial y}\right]_{y=\mu(x)} = \varphi_1(x) \qquad (5.20)$$

(the derivative in (5.20) is partial, and is *not* taken along the curve $y = \mu(x)$).

Differentiating (5.19) we have

$$\left[\frac{\partial u}{\partial x}\right]_{y=\mu(x)} + \left[\frac{\partial u}{\partial y}\right]_{y=\mu(x)} \mu'(x) = \varphi_0'(x)$$

so that

$$\left[\frac{\partial u}{\partial x}\right]_{y=\mu(x)} = \varphi_0'(x) - \varphi_1(x)\,\mu'(x) \qquad (5.21)$$

We now transform (5.17) into a rather more convenient form. Assuming that the boundary of the domain is traced out anticlockwise so as to keep the enclosed area always on the left, we have

$$dx = \cos(\boldsymbol{n}, y)\, dS, \quad dy = -\cos(\boldsymbol{n}, x)\, dS,$$

if we take dS to be always positive.

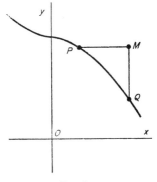

FIG. 9.

Using these relations, we get from (5.17)

$$\iint_\Omega (v\,Lu - u\,Mv)\,dx\,dy = \int_S \left[\frac{1}{2}\left(u\frac{\partial v}{\partial x} - v\frac{\partial u}{\partial x}\right) - buv\right]dx$$
$$- \int_S \left[\frac{1}{2}\left(u\frac{\partial v}{\partial y} - v\frac{\partial u}{\partial y}\right) - auv\right]dy \quad (5.22)$$

Draw through the point $M(x_0, y_0)$ two straight lines parallel to the coordinate axes to intersect the curve $y = \mu(x)$ in the points P and Q (see Fig. 9). Then, applying (5.22) to the triangle MPQ, we have

$$\iint_\Omega (v\,Lu - u\,Mv)\,dx\,dy = \int_P^Q \left\{\left[\frac{1}{2}\left(u\frac{\partial v}{\partial x} - v\frac{\partial u}{\partial x}\right) - buv\right]dx\right.$$
$$\left. - \left[\frac{1}{2}\left(u\frac{\partial v}{\partial y} - v\frac{\partial u}{\partial y}\right) - auv\right]dy\right\} + \int_Q^M \left[\frac{1}{2}\left(v\frac{\partial u}{\partial y} - u\frac{\partial v}{\partial y}\right) + auv\right]dy$$
$$+ \int_P^M \left[\frac{1}{2}\left(v\frac{\partial u}{\partial x} - u\frac{\partial v}{\partial x}\right) + buv\right]dx. \quad (5.23)$$

Transforming the last two integrals,

$$\int_Q^M \left[\frac{1}{2}\left(v\frac{\partial u}{\partial y} - u\frac{\partial v}{\partial y}\right) + auv\right]dy = \frac{1}{2}\,[uv]_Q^M + \int_Q^M \left(-u\frac{\partial v}{\partial y} + auv\right)dy$$
$$= \frac{1}{2}\,[uv]_Q^M + \int_Q^M u\left(-\frac{\partial v}{\partial y} + av\right)dy \quad (5.24)$$

$$\int_P^M \left[\frac{1}{2}\left(v\frac{\partial u}{\partial x} - u\frac{\partial v}{\partial x}\right) + buv\right]dx = \frac{1}{2}\,[uv]_P^M + \int_P^M u\left(-\frac{\partial v}{\partial x} + bv\right)dx.$$
$$(5.25)$$

These formulae enable our problem to be readily solved.

Let $v(x, y, x_0, y_0)$ be any function satisfying the conditions

$$Mv = 0$$
$$[v]_{x=x_0} = \exp\int_{y_0}^y a(x_0, y)\,dy, \quad [v]_{y=y_0} = \exp\int_{x_0}^x b(x, y_0)\,dx.$$

Then

$$v(x_0, y_0, x_0, y_0) = 1,$$
$$\left[\frac{\partial v}{\partial y}\right]_{x=x_0} = a(x_0, y)\exp\int_{y_0}^y a(x_0, y)\,dy = a(x_0, y)\,[v]_{x=x_0}$$
$$\left[\frac{\partial v}{\partial x}\right]_{y=y_0} = b(x, y_0)\exp\int_{x_0}^x b(x, y_0)\,dx = b(x, y_0)\,[v]_{y=y_0}.$$

We have already established in § 2 that such a function exists by our solution of the boundary problem of the first kind. A function v satisfying these conditions is called a *Riemann function*.

Since we have $y = y_0$ on the straight line PM, and $x = x_0$ on QM, the last terms in (5.24) and (5.25) vanish, and we get

$$\left.\begin{aligned}
\int_Q^M \left[\frac{1}{2} \left(v \frac{\partial u}{\partial y} - n \frac{\partial v}{\partial y} \right) + auv \right] dy &= \frac{1}{2} [uv]_Q^M \\[2mm]
\int_P^M \left[\frac{1}{2} \left(v \frac{\partial u}{\partial x} - u \frac{\partial v}{\partial x} \right) + buv \right] dx &= \frac{1}{2} [uv]_P^M.
\end{aligned}\right\} \qquad (5.26)$$

Substituting the expressions (5.26) in (5.23), we now have

$$\iint_\Omega vF(x, y)\, dx\, dy = [u]_M - \frac{1}{2} [uv]_P - \frac{1}{2} [uv]_Q + \Phi \qquad (5.27)$$

where Φ denotes the first integral on the right-hand side of (5.23) and is expressible entirely in terms of v and known functions, since on the curve PQ $u, \partial u/\partial x, \partial u/\partial y$ are all known, by (5.19), (5.20) and (5.21).

Hence (5.27) gives the so-called *Riemann formula*

$$u(x_0, y_0) = \frac{1}{2} [uv]_P + \frac{1}{2} [uv]_Q - \Phi + \iint_\Omega vF(x, y)\, dx\, dy. \qquad (5.28)$$

$M(x_0, y_0)$ is an arbitrary point, and thus (5.28) gives in explicit form the solution to our problem.

It follows from the very method of deriving Riemann's formula that the solution is unique, since we have obtained for the unknown function u an explicit and single-valued, definite expression, without making any assumptions whatever about it except that it exists. To complete the investigation, it remains only to show that the solution of the Cauchy problem which we have examined does actually exist.

We note first that it is sufficient to establish the existence of a solution of the equation (5.1) under the conditions that on the curve $y = \mu(x)$ the function u together with its first-order derivatives all vanish. For, we could introduce instead of u a new unknown function

$$w = u - \varphi_0(x) - [y - \mu(x)]\, \varphi_1(x).$$

The function w would satisfy the equation

$$\frac{\partial^2 w}{\partial x\, \partial y} + a(x, y) \frac{\partial w}{\partial x} + b(x, y) \frac{\partial w}{\partial y} + c(x, y)\, w = F_1(x, y) \qquad (5.29)$$

where

$$F_1(x,y) = F(x,y) - \varphi_1'(x) - a(x,y)\{\varphi_0'(x) + [y - \mu(x)]\,\varphi_1'(x) - \mu'(x)\varphi_1(x)\}$$
$$- b(x,y)\,\varphi_1(x) - c(x,y)\{\varphi_0(x) + [y - \mu(x)]\,\varphi_1(x)\},$$

with the homogeneous conditions

$$[w]_{y=\mu(x)} = \left[\frac{\partial w}{\partial y}\right]_{y=\mu(x)} = 0.$$

The equation (5.29) has exactly the same form as (5.1), but differs from it in its free member. If we prove the existence of a solution for this new problem formulated for the function w, then this will imply that a solution for the original problem exists. In fact,

$$u = w + \varphi_0(x) + [y - \mu(x)]\,\varphi_1(x)$$

would be the solution of the equation. Now in the case when the functions φ_0 and φ_1 vanish, Riemann's formula gives

$$u(x_0, y_0) = \iint_\Omega vF(x,y)\,\mathrm{d}x\,\mathrm{d}y. \tag{5.30}$$

It is sufficient for us to prove that the function $u(x_0, y_0)$ defined by (5.30) satisfies the equation (5.1) and that together with its first-order derivatives it vanishes when $y_0 = \mu(x_0)$. We verify the last statement first. When $y_0 = \mu(x_0)$ the domain Ω disappears and consequently $[u]_{y_0=\mu(x_0)} = 0$.

Further, on the basis of the rule stated earlier for the differentiation of integrals in a variable region, we have

$$\left.\begin{aligned}
\frac{\partial u}{\partial x_0} &= \int_Q^M vF(x_0, y_0)\,\mathrm{d}y + \iint_\Omega \frac{\partial v}{\partial x_0} F(x,y)\,\mathrm{d}x\,\mathrm{d}y \\
\frac{\partial u}{\partial y_0} &= \int_P^M vF(x, y_0)\,\mathrm{d}x + \iint_\Omega \frac{\partial v}{\partial y_0} F(x,y)\,\mathrm{d}x\,\mathrm{d}y
\end{aligned}\right\} \tag{5.31}$$

and in both formulae the integrals on the right-hand sides vanish when $y_0 = \mu(x_0)$.

It remains to show that $u(x_0, y_0)$ satisfies equation (5.1). We shall verify this at the end of the next section.

§ 4. Riemann's Function for the Adjoint Equation

In § 1 we found a solution of a boundary problem of the first kind by a method of successive approximations. Riemann's method enables the solution to be found in a more convenient form.

Suppose that, as in § 1, we wish to find a solution u of equation (5.1) subject to the conditions (5.2). In the plane (x, y) we draw through the point $S(x_0, y_0)$ the straight lines SP, SQ, given by $x = x_0, y = y_0$; and through the point $M(x_1, y_1)$ we draw MP, MQ parallel to the axes. We apply (5.22) to the rectangle $MPSQ$, taking as u the as yet unknown solution and as v, Riemann's function $v(x, y, x_1, y_1)$. We then have

$$\iint_\Omega (vLu - uMv)\, dx\, dy = \iint_\Omega vF\, dx\, dy$$
$$= \int_P^M \left[\frac{1}{2} \left(v \frac{\partial u}{\partial x} - u \frac{\partial v}{\partial x} \right) + buv \right] dx$$
$$+ \int_Q^M \left[\frac{1}{2} \left(v \frac{\partial u}{\partial y} - u \frac{\partial v}{\partial y} \right) + auv \right] dy$$
$$- \int_S^Q \left[\frac{1}{2} \left(v \frac{\partial u}{\partial x} - u \frac{\partial v}{\partial x} \right) + buv \right] dx$$
$$- \int_S^P \left[\frac{1}{2} \left(v \frac{\partial u}{\partial y} - u \frac{\partial v}{\partial y} \right) + auv \right] dy.$$

Transforming, as before, the integrals along the intervals PM and QM, and noting that on the intervals SQ and SP the integrands consist of known functions, we get

$$\iint_\Omega vF(x, y)\, dx\, dy = [u]_M - \frac{1}{2} [uv]_Q$$
$$- \int_S^Q \left[\frac{1}{2} \left(v\varphi_2' - \frac{\partial v}{\partial x} \varphi_2 \right) + bv\varphi_2 \right] dx - \frac{1}{2} [uv]_P$$
$$- \int_S^P \left[\frac{1}{2} \left(v\varphi_1' - \frac{\partial v}{\partial x} \varphi_1 \right) + av\varphi_1 \right] dy. \qquad (5.32)$$

It is convenient to eliminate the derivatives of the function v by integrating by parts both integrals on the right-hand side of (5.32). Then, rearranging the terms, we have

$$[u]_M = \frac{1}{2} [uv]_P + \frac{1}{2} [uv]_Q - \frac{1}{2} [uv]_S^Q - \frac{1}{2} [uv]_S^P + \int_S^Q v(\varphi_2' + b\varphi_2)\, dx$$
$$+ \int_S^P v(\varphi_1' + a\varphi_1)\, dy + \iint_\Omega vF(x, y)\, dx\, dy.$$

Thus

$$[u]_M = [uv]_S + \int_S^Q v(\varphi_2' + b\varphi_2)\,dx + \int_S^P v(\varphi_1' + a\varphi_1)\,dy + \iint_\Omega vF(x,y)\,dx\,dy \tag{5.33}$$

Formula (5.33) leads to an important result. Suppose the function u is the Riemann function for the adjoint equation $Mv = 0$, *i.e.*, u is a function which satisfies the equation $Lu = 0$ and the conditions

$$[u]_{y=y_0} = \exp\left\{-\int_{x_0}^x b(x,y_0)\,dx\right\},$$

$$[u]_{x=x_0} = \exp\left\{-\int_{y_0}^y a(x_0,y)\,dy\right\}.$$

Then all the terms on the right-hand side of (5.33) vanish, and we get, using the equality $[u]_S = 1$,

$$[u]_M = [v]_S \quad \text{or} \quad u(x_1,y_1,x_0,y_0) = v(x_0,y_0,x_1,y_1).$$

This relation is known as the theorem on the *symmetry of Riemann's function*. It may be expressed in words thus:

If in a Riemann function the variables x_0, y_0 are regarded as the current coordinates, and x_1, y_1 as the coordinates of the vertex, then the function becomes the Riemann function of the adjoint equation.

COROLLARY. *The Riemann function* $v(x,y,x_0,y_0)$ *defined above satisfies the equation*

$$\frac{\partial^2 v}{\partial x_0\,\partial y_0} + a(x_0,y_0)\frac{\partial v}{\partial x_0} + b(x_0,y_0)\frac{\partial v}{\partial y_0} + c(x_0,y_0)v = 0. \tag{5.34}$$

Using this result, we can easily verify that the function $u(x_0,y_0)$ defined by (5.30) does actually satisfy the equation (5.1) and hence that it is a solution of the Cauchy problem. On differentiating the first of the formulae (5.31) with respect to y_0, we get

$$\frac{\partial^2 u}{\partial x_0\,\partial y_0} = [v]_{(x_0,y_0)}F(x_0,y_0) + \int_Q^M \frac{\partial v}{\partial y_0}F(x_0,y)\,dy + \int_P^M \frac{\partial v}{\partial x_0}F(x,y_0)\,dx$$
$$+ \iint_\Omega \frac{\partial^2 v}{\partial x_0\,\partial y_0}F(x,y)\,dx\,dy.$$

Hence

$$\frac{\partial^2 u}{\partial x_0\,\partial y_0} + a(x_0,y_0)\frac{\partial u}{\partial x_0} + b(x_0,y_0)\frac{\partial u}{\partial y_0} + c(x_0,y_0)u(x_0,y_0) =$$

$$= F(x_0, y_0) + \int_Q^M \left[\frac{\partial v}{\partial y_0} + a(x_0, y_0)v \right] F(x_0, y) \, \mathrm{d}y$$

$$+ \int_P^M \left[\frac{\mathrm{d}v}{\mathrm{d}x_0} + b(x_0, y_0)v \right] F(x, y_0) \, \mathrm{d}x$$

$$+ \iint_\Omega \left[\frac{\partial^2 v}{\partial x_0 \, \partial y_0} + a(x_0, y_0) \frac{\partial v}{\partial x_0} + b(x_0, y_0) \frac{\partial v}{\partial y_0} \right.$$

$$\left. + c(x_0, y_0) \, v \right] F(x, y) \, \mathrm{d}x \, \mathrm{d}y.$$

From the last formula we get immediately

$$Lu = F,$$

which is what we had to prove.

§ 5. Some Qualitative Consequences of Riemann's Formula

From Riemann's formula may be derived certain results of general interest, which we shall examine a little more closely.

Let us investigate the behaviour of a solution to Cauchy's problem for equation (5.1) in relation to a change in the initial conditions. It is easy to see that the value of this solution at a certain point (x_0, y_0) will not depend at all on the Cauchy data outside the curvilinear triangle MPQ, formed by the two characteristics through this point and the curve carrying the initial data. If we were to change the initial data outside this triangle, the solution would change only outside this triangle. Thus any characteristic will separate the region where the solution remains unchanged from the region where it does change. We reach this conclusion: *to a given solution which is fixed inside the triangle MPQ we may attach along the characteristic, in general*, various *solutions which may be regarded as its continuation.*

Thus the characteristics are curves along which the domain of existence of a solution may be cut if we wish to replace one solution by another in certain parts of this domain and still obtain a valid solution over the whole domain. This important property of characteristics is closely linked with the fact that Cauchy's problem is, in general, insoluble for arbitrary initial data specified on the characteristics. For any other curve we could, knowing the solution on one side of the curve, find the values of the solution and its derivatives on this curve and solve the Cauchy problem on the other side of the curve. Thus across any non-characteristic curve there is a single-valued continuation of the solution.

MULTIPLE INTEGRALS:
LEBESGUE INTEGRATION

IN COURSES on elementary mathematical analysis it is mainly continuous functions of one or more independent variables that are dealt with. In problems in mathematical physics, however, continuous functions are not enough, and discontinuous and unbounded functions play an essential part. One of the most important concepts is that of an integral, and the application of this concept to discontinuous and unbounded functions is needed in mathematical physics. We shall therefore dwell particularly on the theory of integration of discontinuous and unbounded functions. We shall construct the theory of the Lebesgue integral for such functions, following in the main the ideas of the Russian mathematical school, and we shall prove for this integral all the fundamental theorems that are usually given in courses on mathematical analysis for integrals of continuous functions.

It should not be supposed that our purpose is merely to achieve the greatest generality possible in the theory of integration of discontinuous functions. The generalization of the concept of an integral gives to the theory of integration an intrinsic completeness, and, thanks to this, enables a whole series of important theorems to be obtained which have no place in the theory of ordinary integrals of continuous functions.

We shall later make use of the following three important results:

1. The criterion for the admissibility of passing to the limit under the integral sign.

2. The Lebesgue–Fubini theorem on the possibility of changing the order of integration in a multiple integral.

3. The criterion for the convergence in the mean of a sequence of functions.

We shall consider the integration of discontinuous and unbounded functions, making use mainly of what we may call the idea of exclusion of singularities.

If a function f of many variables $x_1, x_2, \ldots, x_n$ which is given in some domain Ω is discontinuous in this domain, but becomes continuous in the domain Ω' which remains after a certain partial domain σ, which may be as small as we please, has been excluded, then the natural way of finding the integral of this function is to consider first its integral in Ω' and then to pass to the limit as Ω' tends to Ω. This method is always used in the elementary theory of improper integrals.

We shall explain the theory of Lebesgue integrals using a generalized form of this procedure. The concept of a function continuous on a closed set will be of fundamental importance to us. The properties of closed sets, *i.e.*, of sets which contain their limit points, will be dealt with later. We merely mention at the moment that functions which are continuous on a closed set of points have a number of important properties; in particular, the integral of such a function can be constructed.

As well as closed sets we shall also consider open sets, *i.e.*, sets which do not contain any of their boundary points. We shall first define the concept of an integral for functions which are continuous on an open set. Using this concept we shall further develop the theory of integration of continuous functions on closed sets and then investigate the behaviour of an integral when the region of integration varies.

In the general case the Lebesgue integral of a discontinuous function is constructed as follows. From the domain Ω, in which the function f is given, a set σ is excluded so as to leave a closed set Ω' on which the function is continuous. The integral of f on the closed set Ω' is then calculated. Passing to the limit as Ω' tends to Ω, we obtain the integral of f on Ω.

In the theory of integration built up in this way we shall naturally try to preserve a number of fundamental properties of the ordinary integral, such as, for example, the possibility of termwise integration of the sum of a number of functions. This at the very outset imposes a limitation on the set of functions suitable for our consideration: we are led to the concept of summable functions. And it is with these summable functions that the further investigation is concerned; for them all the most important properties of ordinary integrals remain valid. We shall in this lecture touch on some of the questions in the theory of summable functions.

§ 1. Closed and Open Sets of Points

Before we enter into the exposition of the theory of the Lebesgue integral, we must go into certain properties of point sets in n-dimensional space.

We consider an n-dimensional space with coordinates $x_1, x_2, \ldots, x_n$. *By an open set of points in this space we mean a set of points M such that every point of the set is an internal point, i.e., there can be drawn, with its centre at this point, a sphere which belongs wholly to the set M.*

An open set may be connected, *i.e.*, it consists of a single piece; but it may also be unconnected, being formed of a finite or infinite number of separate pieces. An open connected set is usually called a *domain*.

EXAMPLE 1. The set of all internal points of a certain rectangular parallelepiped defined by the inequalities

$$0 < x_i < a_i \quad (i = 1, 2, \ldots, n)$$

is an open set. (The sign $<$ must not be replaced by the sign $\leqq$. The parallelepiped together with its boundaries does not form an open set.)

EXAMPLE 2. The set of all internal points of a certain sphere defined by the inequality

$$\sum_{i=1}^{n} x_i^2 < R^2$$

is an open set.

If we exclude the origin from this set and consider points defined by

$$0 < \sum_{i=1}^{n} x_i^2 < R^2,$$

we again obtain an open set.

By the *sum* or *union* of several sets we mean the set of all points which belong to at least one of the sets. We shall denote the sum of the sets E_1 and E_2 by $E_1 + E_2$.

The sum of a finite or infinite number of open sets is itself an open set.

For, each point of such a sum is an internal point of at least one of the open sets and consequently it will be an internal point for the sum.

Besides open sets, *closed sets* will play an important part, *i.e.*, *sets which contain all their boundary points.*

We give some examples of closed sets.

EXAMPLE 3. The set of all points of the sphere

$$\sum_{i=1}^{n} x_i^2 \leqq R^2$$

is a closed set. (The sign $\leqq$ must not be replaced by $<$.)

EXAMPLE 4. The set of all points of a certain parallelepiped $0 \leqq x_i \leqq a_i$ $(i = 1, 2, \ldots, n)$ is a closed set.

EXAMPLE 5. The set consisting of a single point $x_i = a_i$ $(i = 1, 2, \ldots, n)$ is a closed set. For, this point has no boundary points at all.

The sum of a finite number of closed sets is a closed set.
(For an infinite number of sets this assertion is no longer true.) The proof presents no great difficulty, and we leave it to the reader.

Let us agree to call a set which contains no point at all an *empty (null) set*. This convention serves to simplify the statement of a number of later theorems. The empty set will be regarded as being simultaneously both open and closed.

Let $E_1, E_2, \ldots, E_k, \ldots$ *be certain sets; the number of them may be either finite or infinite. By their* intersection *we mean the set E of all points each of which belongs to all these sets.* We shall denote the intersection E of the sets $E_1, E_2, \ldots, E_k$ by

$$E = E_1 E_2 \cdots E_k \quad \text{or} \quad E = \prod_k E_k.$$

The intersection F of a finite or infinite number of closed sets $F_1, F_2, \ldots, F_k, \ldots$ is a closed set. (In particular, it may be the empty set.) For, a boundary point of the set F is also a boundary point of each of the sets $F_1, F_2, \ldots, F_k \ldots$ and consequently it belongs to all of them and therefore must be contained in their intersection F.

If all points of a certain set E_1 belong to the set E_2, then we shall say that E_1 is contained in E_2 or is included in E_2. We shall express this property symbolically by

$$E_2 \supseteq E_1 \quad \text{or} \quad E_1 \subseteq E_2.$$

A bounded infinite set in n-dimensional space always has at least one limit point.

The truth of this assertion is established in the same way as for a single independent variable. Let the coordinates of the point $P^{(i)}$ $(i = 1, 2, \ldots)$ belonging to the set E be $(x_1^{(i)}, x_2^{(i)}, \ldots, x_n^{(i)})$. Consider the set of numbers $\{x_1^{(i)}\}$ which are the first coordinates of the points $P^{(i)}$. This set, being infinite and bounded, has at least one limit point (in particular, if any one of the numbers, say $x_1^{(0)}$, is repeated infinitely often in our sequence, then it may be taken as such a limit point).

We select from the sequence $P^{(i)}$ a sequence $P_1^{(j)}$ such that for it the sequence $x_1^{(j)}$ has a limit. Repeating the argument, we select from the sequence $P_1^{(j)}$ a sequence $P_2^{(k)}$ for which the sequences of each of the first two coordinates have limits; then we construct a sequence $P_3^{(l)}$, and so on. After n steps we reach a sequence $P_n^{(s)}$, in which all the coordinates converge, and which consequently has a limit point, as was to be shown.

THEOREM 1. *If a sequence of bounded closed sets is a* contracting *sequence, i.e., if*

$$F_1 \supseteq F_2 \supseteq \cdots \supseteq F_k \supseteq \cdots$$

and *if none of the sets F_k is empty (null), then their intersection also is not null.*

For, consider a sequence of points $P_1, P_2, \ldots, P_k, \ldots$ with $P_k \subseteq F_k$. The set of points of this sequence, being bounded, must have at least one limit point. This point, being a limit point for each F_k, belongs to all the F_k and consequently enters into their intersection. This intersection therefore cannot be null. Hence the theorem.

It may usefully be remarked that:
The intersection of a finite *number of open sets* $\Omega = \Omega_1 \Omega_2 \ldots \Omega_k$ *is an open set.*

For, any point of Ω is an interior point of each of the sets $\Omega_1, \Omega_2, \ldots, \Omega_k$, and in each of them it is the centre of some internal sphere. The smallest of these spheres will be an internal sphere in Ω.

Let E_1 and E_2 be two sets. Remove from E_1 all points belonging to their intersection $E_1 E_2$. Then we shall call the remaining set E the difference of E_1 and E_2, and write $E = E_1 - E_2$. By means of this operation of forming the difference of sets, we can construct further examples of open and closed sets.

If Ω is an open set, and F a closed set, then

$$\Omega_0 = \Omega - F$$

is an open set.

For, if there were a point belonging to Ω_0 which was not surrounded by a sphere internal to Ω_0, then in any neighbourhood of this point there would be points excluded from Ω and therefore belonging to F. Being thus a limit point for the set F, the point must itself belong to F, and consequently cannot belong to $\Omega_0 = \Omega - F$. This contradiction proves our assertion.

Further, $F_0 = F - \Omega$ *is a closed set.*

For, a point P which is a limit point for the set F_0 and a point of F would be excluded from F on the subtraction of Ω only if it belonged to Ω. But points of Ω cannot be limit points for F_0 because they, together with certain neighbourhoods round them, belong to Ω; and since these neighbourhoods belong to Ω they cannot contain points from F_0. This means that no point P which is a limit point for F_0 is excluded from F on subtracting Ω. Consequently F_0 contains all its limit points and is therefore a closed set.

The closure $\bar{E}$ of any set E, i.e., the set obtained by the adjunction to E of all its limit points, is a closed set.

For, let P_0 be a limit point for some sequence of points $P_1, P_2, \ldots, P_n$ of $\bar{E}$. We shall show that P_0 is a limit point not only for $\bar{E}$, but also for E, and consequently enters into $\bar{E}$. It will follow from this that the set $\bar{E}$ contains all its limit points and is therefore closed. Inside every sphere σ_k described about P_k with radius $1/2^k$ there is at least one point Q_k belonging to E. The sequence Q_k obviously again converges to P_0, and consequently P_0 is a limit point of the set E, as was to be proved.

The boundary C of an open set Ω, i.e., the set of limit points of this open set which are not internal points, is a closed set.

For, by definition of the set C we have $C = \bar{\Omega} - \Omega$, where $\bar{\Omega}$ is the closure of the set Ω. Hence by the previous propositions the set C is closed.

By using the operations which have just been discussed on sets of closed and open sets of simple structure we can obtain closed sets of quite complex structure, as the following two examples show.

EXAMPLE 6. The set F_* of all points of the closed interval $[0, 1]$, whose co-ordinates can be expressed in the decimal system without using the digit 5, is a closed set. It is to be allowed here to write a number as a recurring decimal; *e.g.*, $0 \cdot 5 = 0 \cdot 4999\ldots$ Hence, if the decimal fraction expressing the coordinate of any point is finite and contains only a single figure 5 in the last decimal place, then that point belongs to the set F_*.

The easiest way to show that this set is closed is to note that it can be expressed as the difference

$$[0, 1] - \Omega_*$$

where $[0, 1]$ is a closed set, and Ω_* is an open set consisting of all points belonging to the open intervals Ω_i which are given by inequalities of the form

$$0 \cdot 5 < x < 0 \cdot 6: \quad 0 \cdot 15 < x < 0 \cdot 16: \quad 0 \cdot 25 < x < 0 \cdot 26:$$

$$\Omega_* = \sum \Omega_i.$$

In fact, any point whose coordinate cannot be written without using the digit 5 belongs to one of the intervals Ω_i. All the Ω_i are open sets and therefore Ω_* is also open, and this implies that F_* is closed.

A set of points E such that in any open set Ω_1 there is an open subset Ω_2 which does not contain a single point of E is said to be nowhere dense.

The set F_* considered in Example 6 is an example of a closed, nowhere dense set.

EXAMPLE 7. Consider in some open set a continuous function f. The sets $E(f > a)$ and $E(f < a)$, where a is any number, are open sets.

We prove this, *e.g.*, for $E(f > a)$. For, if at some point P the value of the continuous function f is greater than a, then the inequality $f > a$ will also hold in some sufficiently small sphere round P. Hence the set $E(f > a)$ consists only of interior points and so is an open set.

EXAMPLE 8. Now let f be a function which is continuous in the closure $\overline{\Omega}$ of an open set Ω: $\overline{\Omega} = \Omega + C$, where C is the boundary of the set Ω. The sets $E(f \geqq a)$ and $E(f \leqq a)$ for any a are closed sets.

We give the proof for $E(f \geqq a)$.

If for some sequence of points $P_1, P_2, \ldots, P_k, \ldots$, tending to P_0, the inequality $f(P_k) \geqq a$ holds, then $\lim_{k \to \infty} f(P_k) \geqq a$. But since f is continuous, $\lim_{k \to \infty} f(P_k) = f(P_0)$. Hence the point P_0 belongs to the set $E(f \geqq a)$, and this implies that the set $E(f \geqq a)$ is closed.

In courses on mathematical analysis, functions are considered which are given, as a rule, in open or closed domains. We shall have to consider later on functions which are given on sets of a more general character.

A function which is given on some set E is said to be *continuous at the point P_0* (belonging to E) if, given any positive number ε, a neighbourhood (however small) can be found round the point P_0 such that for any point P

(of the set E) belonging to this neighbourhood the following inequality holds good:

$$\left| f(P_0) - f(P) \right| < \varepsilon.$$

In particular, it follows from the definition that at isolated points of E *any* function is continuous.

Functions which are continuous at all points of a certain set are said to be *continuous on this set*.

Let

$$\Omega = \Omega_1 + \Omega_2 + \cdots + \Omega_k + \cdots$$

be the sum of a finite or infinite number of open sets. If a function f, given on Ω, is continuous on each set Ω_k, then it is continuous on Ω also.

For, any point of Ω together with a certain neighbourhood round it belongs to one or other of the Ω_k and in it is a point of continuity of f.

We shall consider in more detail functions which are given and continuous on closed sets, and first of all we shall prove two important theorems.

THEOREM 2. *(Weierstrass) A function f which is continuous on a bounded, closed set F is bounded on this set and attains on it its upper and lower bounds.*

We prove first that the function f is bounded on the set F. For, if it is not, we can obtain a sequence of points belonging to F at which the values of f increase without bound. This sequence will have at least one limit point belonging to the set F because the latter is bounded and closed. Denote this limit point by P_0. Then any neighbourhood of P_0 must contain points at which the values of the function f become as large in absolute value as we please. But this contradicts the continuity of f at the point P_0, since by definition of continuity, for points P sufficiently close to P_0 and belonging to F it must be true that

$$\left| f(P_0) - f(P) \right| < \varepsilon$$

for any positive ε. This proves that f is bounded on F.

The set of values taken by f on F, being bounded, has an upper bound M and a lower bound m. We now show that the function f attains the values M and m at some points of the set F. By the definition of an upper bound, for any positive ε there must be a point P_ε of F such that $M - f(P_\varepsilon) < \varepsilon$. Putting $\varepsilon = 1/n$, we obtain a sequence of points P_n such that $f(P_n) \to M$ as $n \to \infty$. Since the set F is closed and bounded, this sequence must have at least one limit point P_0 in F. Since f is continuous, the equality $f(P_0) = M$ must hold at the point P_0, as was to be shown.

In exactly the same way it may be shown that f attains its lower bound at some point of F.

THEOREM 3. *(Weierstrass) A function f which is continuous on a bounded, closed set F is uniformly continuous on it; i.e., given any positive number ε, a* *number $\delta = \delta(\varepsilon)$ can be found such that, for any two points P and Q of the set F,*

$$|f(P) - f(Q)| < \varepsilon$$

provided only that the distance between P and Q is less than δ.

If, as is usual, we call the difference between the greatest and least values of the function on a certain closed set the *oscillation* of the function on this set, then this theorem of Weierstrass can be restated thus:

For any $\varepsilon > 0$ there is a number $\delta(\varepsilon)$ such that the oscillation of the function f does not exceed ε on the intersection of F with any sphere of radius $\delta(\varepsilon)$.

We give an indirect proof. Suppose, if possible, that f is not uniformly continuous on F. Then a positive number ε_0 can be chosen such that, for any $\delta > 0$, two points P_δ and Q_δ can be found such that the distance between them is less than δ but the absolute value of the difference between the values of the function at these points is greater than ε_0. Putting $\delta = 1/n$, and choosing from the sequence P_n a sequence P_{n_k} converging to some point P_0 (the corresponding sequence of points Q_{n_k} will obviously also converge to this point), we reach the conclusion that in some neighbourhood of P_0 the oscillation of f is not less than ε_0, contrary to the hypothesis that f is continuous on F. Hence the theorem.

We mention one application of these theorems. Let F be a closed set and P_0 some fixed point. The distance $r(P, P_0)$ from the variable point P of the closed set F to the point P_0 is a continuous function of the point P for a given point P_0. By the first theorem of Weierstrass, this function attains its lower bound. Put $\delta(P_0) = \min_P r(P, P_0)$. The number $\delta(P_0)$ is called the *distance of the point P_0 to the closed set F.* If P_0 does not belong to F, then this distance is different from zero.

§ 2. Integrals of Continuous Functions on Open Sets

As already mentioned, in constructing the general concept of an integral we shall broaden bit by bit the class of sets on which an integral is defined. We begin with integration in open sets.

Suppose a function $f(x_1, x_2, \ldots, x_n)$ is given, and is continuous and non-negative, on a certain open set. (It may, of course, be unbounded and not uniformly continuous, as, for example, the function

$$\frac{1}{r} = \frac{1}{\sqrt{x_1^2 + x_2^2 + \cdots + x_n^2}}$$

in the region $0 < r < 1$ obtained by removing the origin from the open sphere $0 \leq r < 1$.)

We divide all space into cubic cells $F_{k_1, k_2, \ldots, k_n}$ of side h defined by the inequalities

$$k_i h \leqq x_i \leqq (k_i + 1)h \quad (i = 1, 2, \ldots, n)$$

where the k_i are integers. We shall call such a division a *net*.

We now form from a *finite* number of cubes of the net some set Φ_h lying wholly within Ω. We shall call such a set an *internal net set*. We shall then decrease h, by dividing each cube of the net into a whole number of smaller cubes, and again form an internal net set. We shall say that such a system of internal net sets is *exhaustive for the domain* Ω if two conditions are satisfied:

(a) the sequence Φ_h expands, *i.e.*,

$$\Phi_{h_1} \subseteqq \Phi_{h_2} \subseteqq \cdots \subseteqq \Phi_{h_k} \subseteqq \cdots,$$

(b) every point of the domain Ω falls strictly inside all the sets Φ_{h_k} starting from a certain k.

We remark that for any open set there is at least one exhaustive system of internal net sets. For, in forming the internal net set we can include in it all the cubes of the net, which, together with their boundaries, lie within Ω. The system of internal net sets obtained in this way will be exhaustive. In fact, any point P_0 of the domain Ω can be enclosed by a small sphere σ belonging to Ω. Consequently, for a sufficiently small net, when the greatest diagonal of the cube with side h becomes less than half the radius of the sphere σ, one of the cubes together with the point P_0 and all the cubes bordering on this one will enter into Φ_h, as was to be shown.

Put

$$J_k = \int_{\Phi_{h_k}} \cdots \int f(x_1, x_2, \ldots, x_n) \, dx_1 \, dx_2 \ldots dx_n,$$

where the Φ_{h_k} form an exhaustive system of net sets. Since the function f is non-negative, the magnitude of J_k does not decrease with increasing k. If the sequence J_k remains bounded as $k \to \infty$, then obviously J_k will tend to a certain limit. In this event we shall say that the function f is *integrable in the domain* Ω, and we shall call the limit of the sequence J_k *the integral of the function* $f(x_1, x_2, \ldots, x_n)$ *in the domain* Ω:

$$\int_{\Omega} \cdots \int f(x_1, \ldots, x_n) \, dx_1 \ldots dx_n = \lim_{k \to \infty} \int_{\Phi_{h_k}} \cdots \int f(x_1, \ldots, x_n) \, dx_1 \ldots dx_n.$$

As we shall show, this limit does not depend on the method of choosing the exhaustive system nor on how the coordinate axes are situated in the space.

We shall write the integral $\int_{\Omega} \cdots \int f(x_1, x_2, \ldots, x_n) \, dx_1 \, dx_2 \ldots dx_n$ in the shorter forms $\int_{\Omega} f \, dv$ or $\int_{\Omega} f \, dx_1 \, dx_2 \ldots dx_n$.

If the sequence of expanding sets is not exhaustive, then

$$\lim \int_{\Phi_{h_k}} f \, dv \leqq \int_{\Omega} f \, dv.$$

Remark 1. A function f which is non-negative, bounded, and continuous, in a bounded, open set Ω, is integrable in this set. For, the integrals over an exhaustive system of net sets are bounded in their set, and hence they converge to a definite limit.

LEMMA 1. *Consider an expanding system of open sets*

$$\Omega_1 \subseteqq \Omega_2 \subseteqq \cdots \subseteqq \Omega_k \subseteqq \cdots,$$

and let

$$\Omega = \Omega_1 + \Omega_2 + \cdots + \Omega_k + \cdots.$$

Then any closed, bounded set F lying entirely within Ω will fall inside all Ω_k, starting from a certain k.

For, suppose this assertion is false. Then for any k we could find a point P_k belonging to F but not to Ω_k. The set of points P_k has at least one limit point P_0 which obviously will belong to F. It is easy to see that P_0 does not belong to any one of the Ω_k, for in any neighbourhood round it there are points P_k with k as large as we please. This, however, is impossible, since the set F belongs to the sum Ω, and consequently each of its points must belong to one of the terms forming the sum Ω. This contradiction proves the lemma.

We now prove that *an integral on an open set does not depend either on the direction of the coordinate axes in space or on the method of setting up the exhaustive system of net sets.*

Consider an open set Ω_k consisting of all interior points of the sets Φ_{h_k}. The sum of these open sets is an open set Ω, and this implies that any bounded, closed set F lying inside Ω will be contained in all the net sets Φ_{h_k} starting from some k. Let Φ_{h_k} and Ψ_{h_k} be two exhaustive systems of net sets, not necessarily given in the same system of coordinates. By Lemma 1, for a suitable choice of $k_2 = k_2(k_1)$ and $k_3 = k_3(k_2)$ and for any k_1, we shall have

$$\Phi_{h_{k_1}} \subseteqq \Psi_{h_{k_2}} \subseteqq \Phi_{h_{k_3}}.$$

Consequently

$$\int_{\Phi_{h_{k_1}}} f \, dv \leqq \int_{\Psi_{h_{k_2}}} f \, dv \leqq \int_{\Phi_{h_{k_3}}} f \, dv$$

from which it is clear that the limits for the systems Φ_h and Ψ_h are equal. Hence the independence of the integral from the choice of the exhaustive system of net sets and from the direction of the axes is proved.

Any non-negative function which is integrable on an open set Ω is integrable on any open subset Ω_1. For, integrals on an exhaustive system of net sets for Ω_1 will be majorized by the integrals on the corresponding net sets constructed for Ω. Hence they form a bounded, and therefore convergent, sequence.

Remark 2. If $\Omega_2 \subset \Omega_1$ and the function f is non-negative, then

$$\int_{\Omega_2} f \, dv \leq \int_{\Omega_1} f \, dv.$$

For, the exhaustive sets for Ω_2 can be taken so that they enter into the exhaustive sets for Ω_1.

Remark 3. If f_1 and f_2 are any two non-negative functions which are continuous and integrable in Ω, then their sum $f_1 + f_2$ is integrable in Ω, and

$$\int_{\Omega} (f_1 + f_2) \, dv = \int_{\Omega} f_1 dv + \int_{\Omega} f_2 dv.$$

The proof of this assertion presents no difficulty. It follows at once on passing to the limit from the obvious equality

$$\int_{\Phi_h} (f_1 + f_2) \, dv = \int_{\Phi_h} f_1 dv + \int_{\Phi_h} f_2 dv.$$

Remark 4. If f is non-negative, continuous and integrable in Ω, and if a is a non-negative constant, then af has the same properties as f, and

$$\int_{\Omega} af \, dv = a \int_{\Omega} f \, dv.$$

Again the proof presents no difficulty.

THEOREM 4. *Let Ω_1 and Ω_2 be two open sets, which may intersect each other, and let f be a continuous, non-negative function in the set $\Omega_1 + \Omega_2$ and be integrable on this set. Then*

$$\int_{\Omega_1} f \, dv + \int_{\Omega_2} f \, dv = \int_{\Omega_1 + \Omega_2} f \, dv + \int_{\Omega_1 \Omega_2} f \, dv. \tag{6.1}$$

Construct in Ω_1 and Ω_2 exhaustive net sets $\Phi_h^{(1)}$ and $\Phi_h^{(2)}$. Clearly,

$$\int_{\Phi_h^{(1)}} f \, dv + \int_{\Phi_h^{(2)}} f \, dv = \int_{\Phi_h^{(1)} + \Phi_h^{(2)}} f \, dv + \int_{\Phi_h^{(1)} \Phi_h^{(2)}} f \, dv. \tag{6.2}$$

It is readily seen that the sets $\Phi_h^{(1)} + \Phi_h^{(2)}$ form an exhaustive system for the set $\Omega_1 + \Omega_2$, for any point of this set sooner or later falls into some set $\Phi_h^{(1)}$ or $\Phi_h^{(2)}$; and $\Phi_h^{(1)} \Phi_h^{(2)}$ forms an exhaustive system for $\Omega_1 \Omega_2$, since sooner or later any point of this set falls in both the net sets $\Phi_h^{(1)}$ and $\Phi_h^{(2)}$. Hence, passing to the limit in (6.2) we get (6.1).

COROLLARY. *The integral of a non-negative function taken over the sum of a finite number of open sets does not exceed the sum of the integrals taken over each of these sets:*

$$\int_{\Omega} f \, dv \leqq \int_{\Omega_1} f \, dv + \int_{\Omega_2} f \, dv + \cdots + \int_{\Omega_k} f \, dv.$$

Hence it follows that if all the integrals on the open sets Ω_k exist, then so does the integral on Ω.

THEOREM 5. *Given an expanding sequence of open sets*

$$\Omega_1 \subseteqq \Omega_2 \subseteqq \cdots \subseteqq \Omega_k \subseteqq \cdots,$$

let Ω_0 denote their sum $\Omega_1 + \Omega_2 + \cdots + \Omega_k + \cdots$
Let f be any non-negative, continuous, and integrable function in Ω_0. Then

$$\int_{\Omega_0} f \, dv = \lim_{k \to \infty} \int_{\Omega_k} f \, dv.$$

Proof. Clearly,

$$\int_{\Omega_0} f \, dv \geqq \int_{\Omega_k} f \, dv$$

hence

$$\int_{\Omega_0} f \, dv \geqq \lim_{k \to \infty} \int_{\Omega_k} f \, dv. \qquad (*)$$

Construct a system of exhaustive net sets Φ_h for the open set Ω_0. By Lemma 1 any such net set will be wholly included in a certain Ω_n. We shall therefore have

$$\int_{\Phi_h} f \, dv \leqq \int_{\Omega_n} f \, dv \leqq \lim_{k \to \infty} \int_{\Omega_k} f \, dv$$

and passing to the limit as $h \to 0$, we get

$$\int_{\Omega_0} f \, dv \leqq \lim_{k \to \infty} \int_{\Omega_k} f \, dv.$$

Comparing this inequality with (*), we see that

$$\int_{\Omega_0} f \, dv = \lim_{k \to \infty} \int_{\Omega_k} f \, dv$$

as was to be proved.

COROLLARY. *Given a sequence (not necessarily expanding) of open sets $\Omega_1, \Omega_2, \ldots, \Omega_k, \ldots$ and f a non-negative continuous function; if the sums*

$$\sum_{k=1}^{N} \int_{\Omega_k} f \, dv \quad (N = 1, 2, \ldots)$$

are bounded in the aggregate, then the integral exists over the sum Ω_0 of the sets Ω_k, and

$$\int_{\Omega_0} f \, dv = \lim_{k \to \infty} \int_{\Omega_1 + \Omega_2 + \cdots + \Omega_k} f \, dv$$

$$\leq \lim_{k \to \infty} \left[\int_{\Omega_1} f \, dv + \int_{\Omega_2} f \, dv + \cdots + \int_{\Omega_k} f \, dv \right].$$

Remark 5. Let Ω be an open set and Φ_{h_0} be a net set contained in it, and let f be a non-negative function continuous and integrable in Ω. Then

$$\int_{\Omega - \Phi_{h_0}} f \, dv + \int_{\Phi_{h_0}} f \, dv = \int_{\Omega} f \, dv.$$

To prove this, we include Φ_{h_0} strictly within a net set Φ_{h_ε} so that

$$\int_{\Phi_{h_\varepsilon}} f \, dv < \int_{\Phi_{h_0}} f \, dv + \varepsilon.$$

The sets $\Phi_{h_\varepsilon} + \Psi_h$, where Ψ_h is an exhaustive system of interior net sets for $\Omega - \Phi_{h_0}$, will be exhaustive for Ω and consequently

$$\int_{\Omega} f \, dv = \lim_{h \to 0} \int_{\Phi_{h_\varepsilon} + \Psi_h} f \, dv \leq \lim_{h \to 0} \left[\int_{\Phi_{h_\varepsilon}} f \, dv + \int_{\Psi_h} f \, dv \right]$$

$$\leq \int_{\Phi_{h_0}} f \, dv + \int_{\Omega - \Phi_{h_0}} f \, dv + \varepsilon,$$

from which our assertion follows.

THEOREM 6. *Suppose that a contracting sequence of open sets is given:*

$$\Omega_1 \supseteq \Omega_2 \supseteq \cdots \supseteq \Omega_k \supseteq \cdots$$

and suppose that their intersection $\Omega_0 = \Omega_1 \Omega_2 \ldots \Omega_k \ldots$ is also an open set (in particular, it may be empty).

Let f be a continuous, non-negative function, integrable on Ω_1. Then

$$\int_0 f \, dv = \lim_{k \to \infty} \int_{\Omega_k} f \, dv$$

(if Ω_0 is the empty set, then this limit is zero).

Proof. In each set Ω_k we introduce a corresponding interior net set $\Phi_h^{(k)}$ such that

$$\int_{\Omega_k} f \, dv - \int_{\Phi_h^{(k)}} f \, dv < \frac{\varepsilon}{2^k}.$$

We set up the open sets $\Omega_k' = \Omega_k - \Phi_h^{(k+1)}$. Then it is not difficult to see that the open set Ω_1 coincides with the sum $\Omega_0 + \Omega_1' + \cdots + \Omega_k' + \ldots$. By the corollary to Theorem 5

$$\int_{\Omega_1} f \, dv \leqq \int_{\Omega_0} f \, dv + \int_{\Omega_1'} f \, dv + \cdots + \int_{\Omega_k'} f \, dv + \cdots.$$

But by virtue of the choice of $\Phi_h^{(k+1)}$ and Remark 5,

$$\int_{\Omega_k'} f \, dv = \int_{\Omega_k} f \, dv - \int_{\Phi_h^{(k+1)}} f \, dv \leqq \int_{\Omega_k} f \, dv - \int_{\Omega_{k+1}} f \, dv + \frac{\varepsilon}{2^{k+1}}$$

and this implies

$$\int_{\Omega_1} f \, dv \leqq \int_{\Omega_0} f \, dv + \left(\int_{\Omega_1} f \, dv - \int_{\Omega_2} f \, dv + \frac{\varepsilon}{2^2} \right)$$

$$+ \left(\int_{\Omega_2} f \, dv - \int_{\Omega_3} f \, dv + \frac{\varepsilon}{2^3} \right) + \cdots$$

$$= \int_{\Omega_0} f \, dv + \lim_{k \to \infty} \left(\int_{\Omega_1} f \, dv - \int_{\Omega_k} f \, dv \right) + \frac{\varepsilon}{2},$$

whence

$$\lim_{k \to \infty} \int_{\Omega_k} f \, dv \leqq \int_{\Omega_0} f \, dv + \frac{\varepsilon}{2}.$$

And the theorem now follows from the obvious fact that

$$\lim_{k \to \infty} \int_{\Omega_k} f \, dv \geqq \int_{\Omega_0} f \, dv.$$

In this section only non-negative functions have been considered. In § 4 we shall show that the integration of functions which change sign reduces to the integration of non-negative functions. This will be proved for an integral understood in a more general sense, namely for the Lebesgue integral, particular cases of which are the integrals of continuous functions on open sets (§ 2) and on closed sets (§ 3).

§ 3. Integrals of Continuous Functions on Bounded Closed Sets

We now pass on to the definition of the concept of an integral of a continuous function on a bounded closed set. It is natural to define it in this way. Let the function f be defined on some bounded closed set F and be continuous on it. We enclose F in some open set. Suppose we have succeeded in continuing f in a continuous manner on to the open set Ω and that f is found to be integrable on this set. We further form the open set $\Omega - F$. We then define $\int_F f \, dv$ by writing

$$\int_F f \, dv = \int_\Omega f \, dv - \int_{\Omega - F} f \, dv.$$

But in this definition much still remains to be explained. It is not clear, firstly, whether the function f can always be continued in a continuous manner and whether an integrable function will result from doing this; and secondly, it is not known whether such a definition of an integral will lead to an unambiguous result.

In order to establish the validity of our definition we now prove certain auxiliary propositions.

Let f be any non-negative function, continuous and integrable in an open set Ω_1. Suppose further that Ω is contained in Ω_1, and that F is a closed set included in Ω. Then we shall have

$$\Omega + (\Omega_1 - F) = \Omega_1, \quad \Omega(\Omega_1 - F) = \Omega - F.$$

Hence, by Theorem 4,

$$\int_\Omega f \, dv + \int_{\Omega_1 - F} f \, dv = \int_{\Omega_1} f \, dv + \int_{\Omega - F} f \, dv$$

i.e.,

$$\int_\Omega f \, dv - \int_{\Omega - F} f \, dv = \int_{\Omega_1} f \, dv - \int_{\Omega_1 - F} f \, dv.$$

Let f be non-negative, continuous and integrable on the open sets Ω_1 and Ω_2, and let $F \subseteq \Omega_1 \Omega_2$. We put $\Omega = \Omega_1 \Omega_2$. Then from what has been shown,

$$\int_\Omega f \, dv - \int_{\Omega - F} f \, dv = \int_{\Omega_1} f \, dv - \int_{\Omega_1 - F} f \, dv = \int_{\Omega_2} f \, dv - \int_{\Omega_2 - F} f \, dv.$$

We have obtained the important result that: *the difference*

$$\int_\Omega f \, dv - \int_{\Omega - F} f \, dv$$

does not depend on the choice of the open set Ω.

LEMMA 2. *If a continuous function f vanishes at points of a certain bounded closed set F lying within an open set Ω, then*

$$\int_\Omega f \, dv = \int_{\Omega - F} f \, dv.$$

Suppose first that the set Ω is bounded.

From Remark 2 (p. 82),

$$\int_{\Omega-F} f \, dv \leqq \int_{\Omega} f \, dv. \tag{**}$$

We consider an open set Ω_ε of those points at which $f < \varepsilon$, where ε is a small positive number. Since Ω and consequently all Ω_ε are also bounded, we see that

$$\lim_{\varepsilon \to 0} \int_{\Omega_\varepsilon} f \, dv = 0,$$

By the corollary to Theorem 4,

$$\int_{\Omega} f \, dv \leqq \int_{\Omega-F} f \, dv + \int_{\Omega_\varepsilon} f \, dv$$

which implies

$$\int_{\Omega-F} f \, dv \geqq \int_{\Omega} f \, dv.$$

Comparing this with (**), we obtain the asserted result.

In the general case, we may take Ω as the sum of expanding, bounded, open sets Ω_k. Passing to the limit as $k \to \infty$ in the equality

$$\int_{\Omega_k} f \, dv = \int_{\Omega_k-F} f \, dv,$$

we get the result asserted in the lemma.

If two functions f_1 and f_2 are equal at all points of the closed set, F, then

$$\int_{\Omega} f_1 \, dv - \int_{\Omega-F} f_1 \, dv = \int_{\Omega} f_2 \, dv - \int_{\Omega-F} f_2 \, dv.$$

For, from Lemma 2 we get immediately the equivalent equality

$$\int_{\Omega} (f_1 - f_2) \, dv = \int_{\Omega-F} (f_1 - f_2) \, dv.$$

The result just proved implies that in the definition given at the beginning of this section of the integral of a function f, continuous on a closed set F, taken over this set, the value of the integral does not depend on the way in which the function is continued outside F. We still have to demonstrate the possibility of such a continuation.

LEMMA 3. *Consider an open set Ω and its boundary C. Let F be a closed set lying in the set $\Omega + C$. Let some continuous, non-negative function f be given on F. Then a continuous, non-negative function φ can be constructed in*

$\Omega + C$ which will be equal to f at the points of F: and the upper and lower bounds of φ will be equal respectively to the upper and lower bounds of f.

As was shown on p. 79, any point P of the domain Ω which does not belong to F is situated at a finite positive distance $\delta(P)$ from the set F. We construct a sphere of radius R with its centre at a given point P and consider the maximum value of the function f in this sphere. Let

$$\max_{r \leq R} f = M_P(R).$$

If there are no points of F inside this sphere, we shall take $M_P(R)$ to be zero.

$M_P(R)$ increases monotonically and consequently is an integrable function (in the usual sense of the word) of the variable R†.

As the function φ we may take, for example,

$$\varphi(P) = \frac{1}{\delta(P)} \int_{\delta(P)}^{2\delta(P)} M_P(R) \, dR.$$

At points of the set F we put $\varphi(P) = f(P)$.

We prove that the function $\varphi(P)$ so defined is continuous. For, at a point P_1 at a distance $\delta > 0$ from some point P_0 of the set F, the value of this function will be between

$$\min_{r \leq 3\delta} f(P) \quad \text{and} \quad \max_{r \leq 3\delta} f(P),$$

where we have taken a sphere of radius 3δ with centre at the point P_0 of the set F and consequently $f(P_1)$ tends to $f(P_0)$ as $\delta \to 0$ since f is continuous.

Next, for two points P_1 and P_2 which do not belong to F and which are at a distance h from one another, we have

$$M_{P_1}(R) \leq M_{P_2}(R + h) \quad \text{and} \quad \delta(P_1) \leq \delta(P_2) + h,$$
$$M_{P_2}(R) \leq M_{P_1}(R + h) \quad \text{and} \quad \delta(P_2) \leq \delta(P_1) + h.$$

We shall take

$$h < \min \left\{ \frac{\delta(P_1)}{3}, \frac{\delta(P_2)}{3} \right\}.$$

Then

$$\varphi(P_1) - \varphi(P_2) = \frac{1}{\delta(P_1)} \int_{\delta(P_1)}^{2\delta(P_1)} M_{P_1}(R) \, dR - \frac{1}{\delta(P_2)} \int_{\delta(P_2)}^{2\delta(P_2)} M_{P_2}(R) \, dR$$

$$\leq \frac{1}{\delta(P_1)} \int_{\delta(P_1)}^{2\delta(P_1)} M_{P_1}(R) \, dR$$

$$- \frac{1}{\delta(P_1) + h} \int_{\delta(P_1)+h}^{2\delta(P_1)-2h} M_{P_1}(R - h) \, dR$$

† See, for example. V.I.Smirnov, *Course of Higher Mathematics*, Vol. 1, p. 299*ff*, Pergamon Press, 1964.

$$= \frac{1}{\delta(P_1)} \int_{\delta(P_1)}^{2\delta(P_1)-3h} M_{P_1}(R) \, dR$$

$$+ \frac{1}{\delta(P_1)} \int_{2\delta(P_1)-3h}^{2\delta(P_1)} M_{P_1}(R) \, dR$$

$$- \frac{1}{\delta(P_1)+h} \int_{\delta(P_1)}^{2\delta(P_1)-3h} M_{P_1}(R) \, dR$$

$$= \frac{h}{\delta(P_1)[\delta(P_1)+h]} \int_{\delta(P_1)}^{2\delta(P_1)-3h} M_{P_1}(R) \, dR$$

$$+ \frac{1}{\delta(P_1)} \int_{2\delta(P_1)-3h}^{2\delta(P_1)} M_{P_1}(R) \, dR$$

$$\leqq \frac{h}{[\delta(P)]^2} \int_{\delta(P_1)}^{2\delta(P_1)} M_{P_1}(R) \, dR$$

$$+ \frac{1}{\delta(P_1)} \int_{\delta(P_1)-3h}^{2\delta(P_1)} M_{P_1}(R) \, dR \leqq \frac{4hM}{\delta(P_1)},$$

where

$$M = \max f \geqq M_P(R).$$

For a sufficiently small h this quantity can be as small as we please, and this implies that, for any given $\varepsilon > 0$, we can make $\varphi(P_1) - \varphi(P_2) < \varepsilon$ by taking h sufficiently small. But the points P_1 and P_2 are on an equal footing, and so the inequality $\varphi(P_2) - \varphi(P_1) < \varepsilon$ also holds. Hence it follows that $\varphi(P)$ is continuous, and the lemma is proved.

We may remark that the case when the function f changes sign may readily be reduced to the foregoing one, by considering in place of f the function $f + c$, where c is a sufficiently large positive constant.

The concept of an integral on a closed set has now been firmly established, and we have at the same time shown that *any non-negative function f which is continuous on a bounded closed set F is integrable on this set.* For, by Theorem 2 this function is bounded on F, and so an immediate application of Lemma 3 shows that f is integrable.

For closed sets such as a parallelepiped or a net set, our definition of the integral obviously agrees exactly with the usual definition of a Riemann integral of a continuous function.

THEOREM 7. *If F_1 and F_2 are bounded closed sets and f is a non-negative function, continuous on both sets, then*

$$\int_{F_1} f \, dv + \int_{F_2} f \, dv = \int_{F_1+F_2} f \, dv + \int_{F_1 F_2} f \, dv.$$

This theorem is readily reduced to Theorem 4 if we enclose $F_1 + F_2$ in a domain Ω, and put $\Omega - F_1 = \Omega_1$ and $\Omega - F_2 = \Omega_2$. Then, noting that

$$\Omega_1 + \Omega_2 = \Omega - F_1 F_2, \quad \Omega_1 \Omega_2 = -(F_1 + F_2),$$

we get

$$\left.
\begin{aligned}
&\int_{F_1} f\,dv = \int_{\Omega} f\,dv - \int_{\Omega_1} f\,dv; \quad \int_{F_2} f\,dv = \int_{\Omega} f\,dv - \int_{\Omega_2} f\,dv; \\
&\int_{F_1 + F_2} f\,dv = \int_{\Omega} f\,dv - \int_{\Omega_1 \Omega_2} f\,dv; \\
&\int_{F_1 F_2} f\,dv = \int_{\Omega} f\,dv - \int_{\Omega_1 + \Omega_2} f\,dv.
\end{aligned}
\right\} \tag{6.3}$$

And the theorem follows from (6.1) and (6.3).

THEOREM 8. *Let $F_1 \supseteq F_2 \supseteq \cdots \supseteq F_k \supseteq \ldots$ be a contracting sequence of bounded closed sets, and let F_0 denote their intersection:*

$$F_0 = F_1 F_2 \cdots F_k \cdots.$$

Let f be some non-negative function continuous on F_1. Then

$$\int_{F_0} f\,dv = \lim_{k \to \infty} \int_{F_k} f\,dv.$$

This theorem follows from Theorem 5 in the preceding section, if we continue f on an open set Ω containing F_1 and put $\Omega_k = \Omega - F_k$. Then putting also $\Omega_0 = \Omega_1 + \Omega_2 + \cdots + \Omega_k + \cdots$ we have $\Omega - \Omega_0 = F_0$,

$$\int_{F_k} f\,dv = \int_{\Omega} f\,dv - \int_{\Omega_k} f\,dv \quad (k = 1, 2, \ldots).$$

Applying Theorem 5 and passing to the limit, we get the stated result.

THEOREM 9. *If the intersection of a contracting sequence of open sets*

$$\Omega_1 \supseteq \Omega_2 \supseteq \cdots \supseteq \Omega_k \supseteq \cdots$$

is a closed set F:

$$F = \Omega_1 \Omega_2 \ldots \Omega_k \ldots,$$

and if f is a non-negative, continuous function, integrable on Ω_1, then

$$\int_F f\,dv = \lim_{k \to \infty} \int_{\Omega_k} f\,dv.$$

For, let

$$\Omega_k' = \Omega_k - F,$$

then

$$\int_F f\,dv = \int_{\Omega_k} f\,dv - \int_{\Omega_k'} f\,dv.$$

Passing to the limit on both sides of this equality, and noting that

$$\lim_{k\to\infty} \int_{\Omega_k'} f\,dv = 0$$

by Theorem 6, since the intersection Ω_k' is null,
we get

$$\int_F f\,dv = \lim_{k\to\infty} \int_{\Omega_k} f\,dv,$$

as was to be shown.

We may remark that any closed set F can always be taken to be the intersection of a "nest" of open sets Ω_k. For we may take as Ω_k the set of points whose distance to the closed set F is less than ε_k, where $\varepsilon_k \to 0$ as $k \to \infty$. It is clear that the intersection of such sets Ω_k will contain only points whose distance to F is zero, and so it will coincide with the closed set F.

THEOREM 10. *Let $F_1 \subseteqq F_2 \subseteqq \cdots \subseteqq F_k \subseteqq \ldots$ be an expanding sequence of closed sets such that their sum E is bounded and is either a closed set or an open set. Let f be a non-negative function, continuous on all the F_k and on E, and integrable on E. Then*

$$\int_E f\,dv = \lim_{k\to\infty} \int_{F_k} f\,dv.$$

(The supplementary condition that f be integrable on E is not needed if E is a closed set.)

If E is an open set, say $E = \Omega$, then this theorem reduces to Theorem 9, if we consider the open sets $\Omega_k = \Omega - F_k$; and it reduces to Theorem 6, if the set E is closed and $= F$ say.

We now examine the particular case when $f = 1$. We shall call the integrals

$$\int_\Omega dv \quad \text{and} \quad \int_F dv$$

respectively the *Lebesgue measure* of the open set Ω and of the closed set F, and we shall denote them by $m\Omega$ and mF respectively.

All the theorems already proved, when applied to the case $f = 1$, at once give similar theorems about the measure of open or closed sets. Measure appears on the scene as a natural generalization of "volume". We shall come back to the properties of measure later.

Of particular importance for its applications is the following theorem.

THEOREM 11. *(Mean-value theorem)*

Let E be an open set or a bounded closed set on which a function f is continuous. When E is an open set, we further suppose that E has a finite measure and that f is integrable on E. Suppose that, everywhere on E,

$$M_1 \leqq f \leqq M_2$$

where M_1, M_2 are constants. Then

$$M_1 mE \leqq \int_E f \, dv \leqq M_2 mE.$$

The proof follows immediately from the fact that

$$\int_E (f - M_1) \, dv \geqq 0, \quad \int_E (M_2 - f) \, dv \geqq 0.$$

§ 4. Summable Functions

We now pass on to a consideration of the general theory of the Lebesgue integral.

Let f be an arbitrary non-negative function given in a bounded open set Ω. We consider all the closed sets $F \subseteq \Omega$ on which f is continuous, and determine the upper bound of the integrals of f taken over the sets F:

$$\sup_F \int_F f \, dv.$$

We shall call this upper bound, if it exists and is finite, the *inner integral on the set Ω of the function f*, and we shall denote it by

$$(\text{in}) \int_\Omega f \, dv.$$

The concept of the inner integral is of particular importance in cases where the function f is *measurable*.

A function f given on some bounded open set Ω is said to be measurable on the set Ω if there are closed sets F_δ, with measure as close as we please to the measure of Ω, on which the function f is continuous, such that

$$m\Omega - mF_\delta \leqq \delta,$$

where δ is any positive number.

If a non-negative measurable function f has an inner integral in the open set Ω, then it is said to be *integrable, or summable, in the Lebesgue sense on the domain* Ω, and its inner integral is called simply its *integral* or its *Lebesgue integral*. In this case we shall use the ordinary integral sign.

Remark. If a function f is measurable on a bounded open set Ω, *then it is measurable also on any open set* Ω' *which is wholly contained in* Ω.

To prove this it is sufficient to take as the set F'_ε the intersection of the set F_ε with an inner net set Φ'_ε belonging to an exhaustive system for Ω'.

COROLLARY. *If a function f is summable on a bounded open set* Ω, *then it is summable also on any open subset* Ω' *of* Ω.

Remark. It is easily proved that *the sum and the product of two measurable functions are measurable.*

As we shall show later, the Lebesgue integral has a number of extremely important properties, which remarkably simplify its use. The following example shows that, for a non-measurable function, the inner integral, even if it exists, does not have some of the most important properties of an ordinary integral.

Consider a function f which is positive in an open set Ω and has an inner integral on Ω, but which is not measurable (it can be shown that there are such functions).

Since f is not measurable, closed sets on which f is continuous will have a measure substantially less than $m\Omega$, *i.e.*, $\sup_F mF = a < m\Omega$.

It is obvious that the closed sets in which f and $f + 1$ are continuous will be identical. For any one of them we shall have

$$\int_F (f + 1)\, dv - \int_F f\, dv = mF$$

(see Remark 3, p. 82).

Hence

$$\int_F (f + 1)\, dv = \int_F f\, dv + mF \leqq \sup_{F_1} \int_{F_1} f\, dv + a$$

and

$$(\text{in}) \int_\Omega (f + 1)\, dv \leqq (\text{in}) \int_\Omega f\, dv + a.$$

Consequently,

$$(\text{in}) \int_\Omega (f + 1)\, dv \neq (\text{in}) \int_\Omega f\, dv + (\text{in}) \int_\Omega 1\, dv,$$

since

$$(\text{in}) \int_\Omega 1\, dv = m\Omega > a.$$

The last inequality shows that, in general, the additive property does *not* hold for inner integrals of non-measurable functions.

LEMMA 4. *Let f be a measurable, non-negative function in the bounded open set Ω, and let F_ε be some system of closed sets on which f is continuous such that*

$$m\Omega - mF_\varepsilon < \varepsilon.$$

If the limit

$$\lim_{\varepsilon \to 0} \int_{F_\varepsilon} f \, dv$$

exists and is finite, then the function f is summable, and this limit is equal to $\int_\Omega f \, dv$.

If even one of these assertions were false, we should have

$$\lim_{\varepsilon \to 0} \int_{F_\varepsilon} f \, dv < \sup_{F} \int_{F} f \, dv,$$

since, by definition, either

$$\sup_{F} \int_{F} f \, dv = \infty \quad \text{or} \quad \sup_{F} \int_{F} f \, dv = \int_\Omega f \, dv.$$

We should therefore be able to find a closed set F_0 such that, for any $\varepsilon > 0$,

$$\int_{F_0} f \, dv - \int_{F_\varepsilon} f \, dv \geqq \eta > 0,$$

with f continuous on F_0, and η independent of ε.

Put $F'_\varepsilon = F_0 F_\varepsilon$. Then using the previous inequality we have

$$\int_{F_0} f \, dv - \int_{F'_\varepsilon} f \, dv \geqq \eta > 0.$$

The function f is bounded on F_0. Let $f \leqq M$. Then

$$\int_{F_0} f \, dv - \int_{F'_\varepsilon} f \, dv = \left\{ \int_{F_0} (f - M) \, dv - \int_{F'_\varepsilon} (f - M) \, dv \right\} + M(mF_0 - mF'_\varepsilon).$$

But the expression inside the curly brackets is clearly non-positive, and so

$$mF_0 - mF'_\varepsilon \geqq \frac{\eta}{M}.$$

But by Theorem 7,

$$mF_0 - mF'_\varepsilon = m(F_0 + F_\varepsilon) - mF_\varepsilon < \varepsilon,$$

and we get $\eta/M < \varepsilon$. Hence our hypothesis is wrong, and the lemma is proved.

COROLLARY. *Suppose that in the bounded open set Ω an expanding system of closed sets is given:*

$$F_1 \subseteqq F_2 \subseteqq \cdots \subseteqq F_k \subseteqq \cdots,$$

and $mF_k > m\Omega - \delta_k$, where $\delta_k \to 0$ as $k \to \infty$.

Let f be a non-negative measurable function which is continuous on all the F_k. If the integrals $\int_{F_k} f \, dv$ are all bounded by the same number A, then the function f is summable in Ω and its integral is given by

$$\int_\Omega f \, dv = \lim_{k \to \infty} \int_{F_k} f \, dv.$$

This corollary follows at once from the lemma, if we note that the non-decreasing sequence of values of $\int_{F_k} f \, dv$ has a limit.

THEOREM 12. *Let f be a summable and non-negative function in the bounded open set Ω. Then, given any $\varepsilon > 0$, there is a $\delta(\varepsilon)$ such that, for an open set Ω_δ with $\Omega_\delta \subseteqq \Omega$ and $m\Omega_\delta < \delta$, the inequality $\int_{\Omega_\delta} f \, dv \leqq \varepsilon$ holds.*

For, suppose the theorem false. Then a number $\varepsilon_0 > 0$ and open sets Ω_δ can be found with measure as small as we like, $m\Omega_\delta < \delta$, for which

$$\int_{\Omega_\delta} f \, dv \geqq \varepsilon_0. \tag{6.4}$$

Using the open sets Ω_δ we can construct a contracting sequence of open sets Ω_k for which $m\Omega_k < \delta_k$, $\delta_k \to 0$ as $k \to \infty$ and the inequality (6.4) holds. To do this it is sufficient, for example, to choose from the Ω_δ some sequence of sets Ω'_s such that $m\Omega'_s < 1/2^{s+1}$ and to put $\Omega_k = \sum_{s=k}^{\infty} \Omega'_s$. By the corollary to Theorem 5, $m\Omega_k < 1/2^k$, so that we can take $\delta_k = 1/2^k$. Let F_k be an expanding system of exhaustive sets for f in Ω with $m(\Omega - F_K) < \delta_k$†. Then the closed set $F_k^* = F_k - \Omega_k$ has the property that

$$mF_k^* \geqq m\Omega - 2\delta_k.$$

This follows from the fact that $\Omega - F_k^* \subseteqq (\Omega - F_k) + \Omega_k$ and so

$$m(\Omega - F_k^*) \leqq m(\Omega - F_k) + m\Omega_k \leqq 2\delta_k.$$

From the corollary to Lemma 4,

$$\lim_{k \to \infty} \int_{F_k^*} f \, dv = \int_\Omega f \, dv.$$

---

† We shall call the system of closed sets F_δ lying in Ω *exhaustive for f in Ω* if the function f is continuous on each F_δ and if for all $\delta > 0$ an F_δ can be found such that $m(\Omega - F_\delta) < \delta$. The nomenclature "exhaustive (integral)" is justified by Theorem 12.

Hence, for sufficiently large k,

$$\int_{F_k^*} f \, dv > \int_{\Omega} f \, dv - \frac{\varepsilon_0}{3}.$$

On the other hand, we can find inside the set Ω_k a closed set F_k^{**} on which f is continuous and for which $\int_{F_k^{**}} f \, dv > \frac{2\varepsilon_0}{3}$.

Now the F_k^* have no points in common with the F_k^{**}, so that the function f is continuous on the closed set $F_k^* + F_k^{**}$ and therefore

$$\int_{F_k^* + F_k^{**}} f \, dv > \int_{\Omega} f \, dv + \frac{\varepsilon_0}{3}.$$

But this is impossible, since $\int_F f \, dv = \sup_F \int_F f \, dv$. The contradiction proves the theorem.

The property which we have just proved and which is of fundamental importance is known as the *absolute continuity of the Lebesgue integral*. It characterizes completely the dependence of the Lebesgue integral on the set on which the integration takes place. We shall make use of this property on more than one occasion in the sequel.

We shall now concern ourselves with a more detailed explanation of the relation between the Lebesgue integral and the function which is being integrated.

LEMMA 5. *If a function f_2 is measurable and non-negative, and if a function f_1 is summable, and if*

$$f_2 \leqq f_1,$$

then the function f_2 is also summable, and

$$\int_{\Omega} f_2 \, dv \leqq \int_{\Omega} f_1 \, dv.$$

For consider an expanding system of closed sets $F_\varepsilon^{(1)}$ on which the function f is continuous and which have measure satisfying $mF_\varepsilon^{(1)} \geqq m\Omega - \varepsilon$; and consider also a similar system $F_\varepsilon^{(2)}$ for the function f_2.

Then, from Theorem 7, the measure of the sets $F_\varepsilon^{(3)} = F_\varepsilon^{(1)} F_\varepsilon^{(2)}$ will tend towards $m\Omega$:

$$mF_\varepsilon^{(3)} \geqq m\Omega - 2\varepsilon.$$

We have

$$\int_{F_\varepsilon^{(3)}} f_2 \, dv \leqq \int_{F_\varepsilon^{(3)}} f_1 \, dv \leqq \int_{\Omega} f_1 \, dv.$$

The sequence $\int_{F_\varepsilon^{(3)}} f_2 \, dv$ does not decrease, is bounded, and consequently has a limit. So by Lemma 4 the function f_2 is summable. And, passing to the limit, we see that the integrals of f_1 and f_2 do satisfy the stated inequality.

THEOREM 13. *If f_1 and f_2 are non-negative and summable in Ω, then their sum $f_1 + f_2$ is also summable and*

$$\int_\Omega (f_1 + f_2)\, dv = \int_\Omega f_1\, dv + \int_\Omega f_2\, dv. \qquad (*)$$

We construct sets $F_\varepsilon^{(1)}, F_\varepsilon^{(2)}, F_\varepsilon^{(3)}$ in the same way as in the previous lemma. The functions f_1 and f_2 will obviously be continuous on $F_\varepsilon^{(3)}$ and, moreover,

$$\int_{F_\varepsilon^{(3)}} (f_1 + f_2)\, dv = \int_{F_\varepsilon^{(3)}} f_1\, dv + \int_{F_\varepsilon^{(3)}} f_2\, dv. \qquad (6.5)$$

The right-hand side of (6.5) has the limit $\int_\Omega f_1\, dv + \int_\Omega f_2\, dv$. Hence the limit of the left-hand side exists, and so, applying Lemma 4, we see that $(f_1 + f_2)$ is integrable and the formula $(*)$ is valid.

So far we have considered only the integrals of, firstly, non-negative functions which are continuous on open sets and on closed sets, and secondly, non-negative functions which are measurable on bounded open sets. We now pass on to the integration of functions of variable sign.

Let f be a function which can take both positive and negative sign. We put

$$f^+ = \tfrac{1}{2}\,[|f| + f], \quad f^- = \tfrac{1}{2}\,[|f| - f].$$

We call f^+ and f^- respectively the *positive and negative parts of the function f*. The function f^+ coincides with f at those points where f is positive or zero, and it vanishes at those points where f is negative. On the other hand, f^- vanishes at those points where f is positive and is equal to $-f$ at those points where f is negative or zero. Clearly, $f = f^+ - f^-$.

If a function f is continuous on a closed set F, then both the functions f^+ and f^- will also be continuous on this set.

For, if f is positive at a point P belonging to F, then f will be positive at all points of the set F which are sufficiently close to P. Moreover, the function f^+ will coincide with f at these points and so will also be continuous. The continuity of f^- at those points where f is negative is shown in exactly the same way. At points where $f = 0$ both functions f^+ and f^- will also be continuous, since, for example, the values of f^+ as the points P_k tend to P on F will either coincide with the values of f at these same points or will vanish, but in either case will tend to a limiting value equal to zero. Hence follows

Remark. If the function f is measurable on an open set Ω then both the functions f^+ and f^- will also be measurable on Ω.

We define the integral of f on an open or closed set E by the formula

$$\int_E f \, dv = \int_E f^+ \, dv - \int_E f^- \, dv.$$

A function f is said to be summable if f^+ and f^- are summable.

From this definition follows:

THEOREM 14. *If a function f is summable in a bounded open set Ω, then, for any system of closed sets F_δ on which f is continuous and which are such that $mF_\delta > m\Omega - \delta$, $\delta \to 0$,*

$$\int_\Omega f \, dv = \lim_{\delta \to 0} \int_{F_\delta} f \, dv.$$

To prove this, it is sufficient to notice that

$$\int_{F_\delta} f \, dv = \int_{F_\delta} f^+ \, dv - \int_{F_\delta} f^- \, dv.$$

and pass to the limit.

We now have another lemma.

LEMMA 6. *If a function f, measurable in an open set Ω, is represented in two different ways as the difference between two non-negative functions*

$$f = f_1 - f_2 \quad \text{and} \quad f = f_3 - f_4,$$

then

$$\int_\Omega f_1 \, dv - \int_\Omega f_2 \, dv = \int_\Omega f_3 \, dv - \int_\Omega f_4 \, dv.$$

For, the result to be proved is equivalent to

$$\int_\Omega f_1 \, dv + \int_\Omega f_4 \, dv = \int_\Omega f_2 \, dv + \int_\Omega f_3 \, dv,$$

and this immediately follows from Theorem 13 if we note that $f_1 + f_4 = f_2 + f_3$. Hence the lemma.

It follows from this lemma, incidentally, that, however we represent the function f as the difference of two non-negative functions, we shall always have

$$\int_\Omega f_1 \, dv - \int_\Omega f_2 \, dv = \int_\Omega f^+ \, dv - \int_\Omega f^- \, dv.$$

If a function is summable, then the integral of its absolute value exists, and conversely.

For, if the integral of $|f| = f^+ + f^-$ exists, then by Lemma 5 each of f^+ and f^- is summable. Conversely, if f^+ and f^- are summable, then from Theorem 13, $|f|$ is summable.

Remark. If c is an arbitrary constant, and if f is a function summable on the open set Ω, then

$$\int_\Omega cf\,dv = c\int_\Omega f\,dv.$$

This is almost obvious; the proof follows at once by using the similar property for functions of constant sign.

An important general property of the Lebesgue integral follows from the previous remark.

If $f_1, f_2, \ldots, f_k$ are summable functions, and $a_1, a_2, \ldots, a_k$ are arbitrary constants, then

$$\int_\Omega (a_1 f_1 + a_2 f_2 + \cdots + a_k f_k)\,dv = a_1\int_\Omega f\,dv + a_2\int_\Omega f\,dv + \cdots + a_k\int_\Omega f\,dv.$$

This is proved by induction.

We notice one more important property of the Lebesgue integral.

THEOREM 15. *Let f be a summable function, and φ a measurable and bounded function so that $|\varphi| \leq M$. Then the product $f\varphi$ will be summable, and*

$$\left|\int_\Omega f\varphi\,dv\right| \leq M\int_\Omega |f|\,dv.$$

For, $|f\varphi| \leq M|f|$. The function $M|f|$ is summable; therefore so is $|f\varphi|$ and

$$\left|\int_\Omega f\varphi\,dv\right| \leq \int_\Omega |f\varphi|\,dv \leq M\int_\Omega |f|\,dv.$$

§ 5. The Indefinite Integral of a Function of One Variable. Examples

If $f(x)$ is a summable function in the interval $0 < x < 1$, then it will obviously be summable also in the interval $0 < x < y$, where $y \leq 1$. The integral

$$F(y) = \int_0^y f(x)\,dx$$

is called *an indefinite integral.*

By the very definition of measure, an isolated point has zero measure. Hence the integral over the open interval $0 < x < y$ is equal to the integral taken over any one of the following half-open intervals or the closed interval: $0 \leq x < y, 0 < x \leq y, 0 \leq x \leq y$. This justifies the use of the notation

$$\int_0^y f(x)\,dx$$

and also allows us to write

$$\int_a^b f(x)\,dx + \int_b^c f(x)\,dx = \int_a^c f(x)\,dx$$

for any summable function $f(x)$.

We next prove:

THEOREM 16. *The derivative of an indefinite integral is equal to the function under the integral sign at all points where this function is continuous.*

For,

$$\frac{F(y + h) - F(y)}{h} = \frac{1}{h}\int_y^{y+h} f(x)\,dx.$$

Let m_h and M_h be the lower and upper bounds of the function $f(x)$ in the interval $y \leq x \leq y + h$ if $h > 0$, or in the interval $y + h \leq x \leq y$ if $h < 0$. Then

$$m_h \leq \frac{F(x + h) - F(y)}{h} \leq M_h$$

and consequently

$$\lim_{h \to 0} \frac{F(y + h) - F(y)}{h} = f(y),$$

as was to be shown.

A measurable function may not be summable.

Let us look at a few examples.

EXAMPLE 8. The function

$$f = \frac{1}{R^{n-a}} \text{ where } R = \sqrt{x_1^2 + x_2^2 + \cdots + x_n^2}$$

is measurable in the sphere given by $R < 1$. For, if we exclude from this sphere the interior of a small concentric inner sphere $R \leq \varepsilon$, f will be continuous in the remaining part.

This function will be summable provided $a > 0$. To prove this, we notice that

$$\int_{\frac{1}{2^k} \leq R \leq 1} \cdots \int \frac{dx_1 \ldots dx_n}{R^{n-a}} = \sum_{n=0}^{k-1} \psi_m,$$

where

$$\psi_m = \int_{\frac{1}{2^{m+1}} \leq R \leq \frac{1}{2^m}} \cdots \int \frac{dx_1 \ldots dx_n}{R^{n-a}}.$$

We change the variables in ψ_m by $x_i = \xi_i/2^m$ $(i = 1, \ldots, n)$. Then

$$\psi_m = \frac{1}{2^{ma}} \int\limits_{\frac{1}{2} \leq \varrho \leq 1} \cdots \int \frac{d\xi_1 \ldots d\xi_n}{\left[\sqrt{\sum_{i=1}^{n} \xi_i^2}\right]^{n-a}} = \frac{1}{2^{ma}} \psi_0,$$

where

$$\varrho = \sqrt{\sum_{i=1}^{n} \xi_i^2}.$$

Hence we have

$$\sum_{m=0}^{k-1} \psi_m \leq \sum_{m=0}^{\infty} \frac{1}{2^{ma}} \psi_0 = \psi_0 \frac{2^a}{2^a - 1} \text{ if } a > 0,$$

as was to be shown.

EXAMPLE 9. The function

$$f = \frac{1}{r^{n-a}} \quad \text{where} \quad r = \sqrt{(x_1 - y_1)^2 + \cdots + (x_n - y_n)^2},$$

is measurable in the $2n$-dimensional domain $0 < x_i < 1$, $0 < y_i < 1$, $(i = 1, \ldots, n)$. It will be summable if $a > 1$.

For, if the set $r \leq \varepsilon$ is excluded from this cube, then f will be continuous on the remaining part. The volume of the excluded domain, as may easily be seen, is a small quantity of the order ε^n. That f is summable if $a > 1$ may be established as in Example 8.

EXAMPLE 10. A function f in the cube v_0, $-1 < x_i < 1$ $(i = 1, \ldots, n)$, which is equal to unity at all points whose coordinates are rational numbers and to zero at all other points, is summable.

All rational points (*i.e.*, points with rational coordinates) can be enumerated, *i.e.*, put into a 1–1 correspondence with the natural numbers. For, suppose the coordinates of any positive rational point are expressed by the fractions

$$\frac{p_1}{q_1}, \frac{p_2}{q_2}, \ldots, \frac{p_n}{q_n}.$$

We write these $2n$ integers one after the other: $p_1, p_2, \ldots, p_n, q_1, q_2, \ldots, q_n$. Each rational point will correspond to many such combinations of integers, since the coordinates of one and the same rational point can be represented by many different fractions, if these have not been reduced to their lowest jerms. On the other hand, to each combination of $2n$ integers corresponds tust one well-defined rational point. All such combinations can be enumerated one after the other using the sequence of even natural numbers. To do this, we write first the combination of integers whose sum is equal to zero, then those whose sum is equal to 1, then to 2, and so on (there will obviously be a finite number of each of these). In this way all the positive rational points will be enumerated, though to each point will correspond infinitely many different natural numbers. If we wish to make the correspondence 1-1, then when numbering the points in sequence we must reject points to which a number has already been assigned. As regards the negative rational points, we can include these in our enumeration by using the sequence of odd natural numbers.

Each of the rational points $P_1, P_2, \ldots, P_k, \ldots$ can be included within a certain sphere Ω_k with centre at the given point and such that

$$m\Omega_k \leqq \frac{1}{2^{k+t}}.$$

The sum $\Omega_0^{(t)} = \Omega_1 + \cdots + \Omega_k + \cdots$ will be an open set with measure $\leqq 1/2^t$ and therefore as small as we please.

We exclude from the cube v_0 those points whose distance from the boundary is less than $1/t$; and we denote the remaining closed set by v_1. The closed set $F_t = v_1 - \Omega_0^{(t)}$ will have a measure as close as we please to that of v_0. On it f will be continuous and equal to zero. Consequently,

$$\int_{-1 < x_i < 1} \cdots \int f \, dv = 0.$$

We often have to consider functions which are given in an unbounded domain Ω.

We consider a system of spheres $R < N$, and let Ω_N be the part of the open set Ω lying within the sphere $R < N$. We shall say that say a non-negative function f is *measurable in* Ω *if it is measurable in any* Ω_N, and that it is *summable in* Ω *if the integrals* $\int_{\Omega_N} f \, dv$ *are bounded in the aggregate*.

Then by definition

$$\int_\Omega f \, dv = \lim_{N \to \infty} \int_{\Omega_N} f \, dv.$$

Integrals of functions of variable sign were defined on p. 98. For such integrals all the usual properties are valid: we enumerate

$$1. \quad \int_\Omega (f_1 + f_2) \, dv = \int_\Omega f_1 \, dv + \int_\Omega f_2 \, dv,$$

$$2. \quad \int_\Omega af \, dv = a \int_\Omega f \, dv,$$

$$3. \quad \int_\Omega \varphi f \, dv \leqq \max |\varphi| \cdot \int_\Omega |f| \, dv,$$

and moreover the existence of the right-hand members of these relations implies the existence of the left-hand members. The proof of these properties is obtained in an obvious way from a passage to the limit.

EXAMPLE 11. Let

$$f = \frac{1}{R^{n+a}}, \quad R = \sqrt{\sum_{i=1}^{n} x_i^2},$$

in the domain $R > 1$. The function f is clearly measurable in Ω. It will be summable in Ω if $a > 0$.

For,

$$\int_{R>1} \cdots \int \frac{dx_1 \ldots dx_n}{R^{n+a}} = \sum_{m=0}^{\infty} \psi_m$$

where

$$\psi_m = \int_{2^m < R < 2^{m+1}} \cdots \int \frac{dx_1 \ldots dx_n}{R^{n+a}}.$$

Put

$$x_i = 2^m \xi_i \quad (i = 1, \ldots, n) \quad \text{and} \quad \varrho = \sqrt{\sum_{i=1}^{n} \xi_i^2},$$

then

$$\psi_m = \frac{1}{2^{ma}} \cdot \int_{1<\varrho<2} \cdots \int \frac{d\xi_1 \ldots d\xi_n}{\varrho^{n+a}} = \frac{1}{2^{ma}} \psi_0,$$

or

$$\sum_{m=0}^{\infty} \psi_m = \frac{2^a}{2^a - 1} \psi_0,$$

as was to be shown.

It follows, incidentally, from Examples 8 and 11 that the integrals

$$\int_{R<\delta} \cdots \int \frac{dx_1 \ldots dx_n}{R^{n-a}} \quad \text{and} \quad \int_{R>\frac{1}{\delta}} \cdots \int \frac{dx_1 \ldots dx_n}{R^{n+a}}$$

tend to zero as $\delta \to 0$.

§ 6. Measurable Sets. Egorov's Theorem

With a view to the further study of the properties of summable functions, we now introduce the concept of a measurable set. Let Ω be a bounded open set, and let E be any point set included in it. We construct the *characteristic function* $\xi_E(P)$ for the set E, *i.e.*, a function which is equal to unity at the points of E and to zero at points not belonging to E.

If the function $\xi_E(P)$ is measurable (and consequently also summable, because of the boundedness of $\xi_E(P)$ and Ω), then the set E is said to be *measurable*, and the integral

$$\int_{\Omega} \xi_E(P) \, dv$$

is called the *measure of E* and is denoted by mE.

It is not difficult to verify that, in the cases when E is an open set or a closed set, the new definition of measure agrees with the one given earlier (see p. 91), since mE is constructed in exactly the same way as previously in these cases.

If the set E is measurable, then so is $\Omega - E$. For, if the function $\xi_E(P)$ is measurable, the function $1 - \xi_E(P) \equiv \xi_{\Omega - E}(P)$ will also be measurable, and in this case

$$m(\Omega - E) = \int_{\Omega} [1 - \xi_E(P)] \, dv = m\Omega - mE.$$

For any two measurable sets E_1 and E_2, the following relation holds:

$$\xi_{E_1}(P) + \xi_{E_2}(P) = \xi_{E_1 + E_2}(P) + \xi_{E_1 E_2}(P) \tag{6.6}$$

Further, $\xi_{E_1 E_2} = \xi_{E_1} \xi_{E_2}$. Evidently, $\xi_{E_1 E_2}(P)$ is measurable and consequently $\xi_{E_1 + E_2}$ is measurable. Hence we see that the sum and intersection of two measurable sets are always measurable.

THEOREM 17. *The necessary and sufficient condition for a set E lying in an open set Ω to be measurable is that there should be a sequence of closed sets F_k included in E and a sequence of open sets Ω_k containing E, such that*

$$m(\Omega_k - F_k) \to 0 \quad as \ k \to \infty.$$

Proof. Suppose that the function $\xi_E(P)$ is measurable. Then closed sets F_k can be found with measure as close as we please to $m\Omega$ and on which $\xi_E(P)$ is continuous. Each such set F_k may be decomposed into two subsets having no common points:

$$F_k = F_k^{(1)} + F_k^{(2)},$$

where $F_k^{(1)} = F_k E$, $F_k^{(2)} = F_k - E$. On $F_k^{(1)}$ the function $\xi_E(P)$ is equal to 1 and on $F_k^{(2)}$ it is zero.

Each of the sets, as may easily be seen, will be closed. For, a limit point for a sequence of points P_n from $F_k^{(1)}$ is a point of the set F_k and consequently belongs either to the set $F_k^{(1)}$ or to $F_k^{(2)}$. But it cannot belong to $F_k^{(2)}$ because the function $\xi_E(P)$ would then be discontinuous at this point on F_k, and this contradicts the choice of F_k. In exactly the same way it can be shown that $F_k^{(2)}$ is closed.

According to the premises we have

$$mF_k = mF_k^{(1)} + mF_k^{(2)} > m\Omega - \delta_k, \quad \text{where } \delta_k \to 0.$$

Hence

$$mF_k^{(1)} + m\Omega - m(\Omega - F_k^{(2)}) > m\Omega - \delta_k,$$

or

$$m(\Omega - F_k^{(2)}) - mF_k^{(1)} < \delta_k.$$

The sets $F_k^{(1)}$ and $(\Omega - F_k^{(2)})$ are, then, the sets whose existence was asserted in the theorem.

Conversely, suppose there are sets F_k and Ω_k having the properties indicated in the theorem. We can always replace the sets Ω_k by sets Ω_k^{**} such that the sets $\Omega - \Omega_k^{**}$ will be closed. To do this we join to the set Ω_k a set $\Omega_{1/k}^*$ consisting of those points of the set whose distance from the boundary C of the set Ω is less than $1/k$. Obviously, $m\Omega_{1/k}^* \to 0$ as $k \to \infty$, and therefore if we put $\Omega_k^{**} = \Omega_k + \Omega_{1/k}^*$, the difference $m\Omega_k^{**} - mF_k$ will as before be as small as we please for sufficiently large k. The set $\Omega - \Omega_k^{**}$ will be closed since it is the difference between the closed set $\Omega - \Omega_{1/k}^{**}$ and the open set $\Omega_k + \Omega_{1/k}^*$.

We can easily satisfy ourselves that the function $\xi_E(P)$ is continuous on F_k and on $\Omega - \Omega_k^{**}$. Consequently it is continuous on the set $F_k + (\Omega - \Omega_k^{**})$, since a function which is continuous on two closed sets is continuous on the sum of these sets. We evaluate the measure of this set, Φ_k say. Clearly,

$$\Phi_k \equiv F_k + (\Omega - \Omega_k^{**}) = \Omega - (\Omega_k^{**} - F_k).$$

Further,

$$m\Omega = m[\Phi_k + \Omega_k^{**} - F_k] \leq m\Phi_k + m(\Omega_k^{**} - F_k).$$

Consequently, $m\Phi_k \geq m\Omega - \delta_k$, where $\delta_k \to 0$ as $k \to \infty$.

Thus, $\xi_E(P)$ is continuous on the closed sets Φ_k with measure as close as we please to the measure of Ω, and consequently it is measurable, as we had to show.

It is not difficult to show that if a set E is measurable, then

$$mE = \sup_{F \subseteq E} mF = \inf_{E \subseteq \Omega} m\Omega$$

where the F are closed sets contained in E, and the Ω are open sets containing E.

Let E be a measurable set, and f be a function summable in Ω. Then we define the integral of f on the set E by the equation

$$\int_E f \, dv = \int_\Omega \xi_E(P) f(P) \, dv.$$

It is easy to see that this definition does not depend on the choice of the open set Ω which contains the set E. If the set E were included in any other open set, the value of the integral would remain unchanged.

This definition is equivalent to the former ones when $E = F$ is a closed set on which f is continuous, and when $E = \Omega$ is an open set on which f is continuous.

For, let $F_\delta^{(1)}$ and $F_\delta^{(2)}$ be two closed sets having no points in common, on which $\xi_E(P)$ is continuous and takes the values 1 and 0 respectively, with

$$mF_\delta^{(1)} + mF_\delta^{(2)} > m\Omega - \delta, \quad \delta \to 0.$$

The existence of such sets was shown in the proof of Theorem 17.

If $E = F$ is a closed set and f is continuous on F, then we may take as the set $F_\delta^{(1)}$ the set F itself. Then, by Theorem 14,

$$\int_\Omega \xi_E(P)\, f(P)\, dv = \lim_{\delta \to 0} \int_{F_\delta^{(1)} + F_\delta^{(2)}} \xi_E(P)\, f(P)\, dv = \int_F f\, dv.$$

If $E = \Omega_0$ is an open interval and Φ_δ is a system of exhaustive sets for f in Ω_0, then the set $F_\delta^{(1)}$ may be replaced by Φ_δ, for

$$[\Omega - (\Phi_\delta + F_\delta^{(2)})] \subseteqq [\Omega - (F_\delta^{(1)} + F_\delta^{(2)})] + (\Omega_0 - \Phi_\delta)$$

and this implies

$$m[\Omega - (\Phi_\delta + F_\delta^{(2)})] \leqq m[\Omega - (F_\delta^{(1)} + F_\delta^{(2)})] + m(\Omega_0 - \Phi_\delta) \leqq 2\delta.$$

Thus, $\Phi_\delta + F_\delta^{(2)}$ will be an exhaustive system for the function $\xi_E(P)f(P)$. Hence

$$\int_\Omega \xi_E(P)\, f(P)\, dv = \lim_{\delta \to 0} \int_{\Phi_\delta + F_\delta^{(2)}} \xi_E(P)\, f(P)\, dv = \lim_{\delta \to 0} \int_{\Phi_\delta} \xi_E(P)\, f(P)\, dv$$

$$= \lim_{\delta \to 0} \int_{\Phi_\delta} f(P)\, dv = \int_{\Omega_0} f\, dv,$$

as was to be shown.

From formula (6.6) we have

$$\int_{E_1} f\, dv + \int_{E_2} f\, dv = \int_{E_1 + E_2} f\, dv + \int_{E_1 E_2} f\, dv.$$

In particular, if $f = 1$, we get

$$mE_1 + mE_2 = m(E_1 + E_2) + m(E_1 E_2).$$

If for a certain set E there are open sets Ω_k containing E and such that $\inf m\Omega_k = 0$, then (by the remark following Theorem 17) $mE = 0$. Conversely, *any set of zero measure can be included in an open set with measure as small as we please.*

THEOREM 18. *If a function f is different from zero only on a set E of zero measure, then the integral of this function on an open set Ω containing E is zero.*

For, let F_k be an exhaustive system of closed sets for f, and Ω_k be a system of open sets, with measure tending to zero, which contain E. The measure of the closed sets $\Phi_k = F_k - \Omega_k$ will tend towards the measure of Ω, since $\Omega - \Phi_k = (\Omega - F_k) + \Omega_k$; hence $m(\Omega - \Phi_k) \leqq m(\Omega - F_k) + m\Omega_k \to 0$. Thus the system of closed sets Φ_k will be exhaustive. The assertion of the theorem follows from the fact that for this system all the integrals

$$\int_{\Phi_k} f \, dv$$

are zero.

COROLLARY. If two summable functions f_1 and f_2 differ from each other only on a set of points of measure zero, then

$$\int_{\Omega} f_1 \, dv = \int_{\Omega} f_2 \, dv.$$

For,

$$\left| \int_{\Omega} f_1 \, dv - \int_{\Omega} f_2 \, dv \right| = \left| \int_{\Omega} (f_1 - f_2) \, dv \right| \leqq \int_{\Omega} |f_1 - f_2| \, dv = 0.$$

We shall say that a certain assertion holds good *almost everywhere* if it holds good for all points except perhaps those of some set E of measure zero. As we have just shown, if two functions coincide almost everywhere, then their integrals are equal, and the integral of the modulus of their difference is zero. We shall say that such functions are *equivalent*.

We now consider another question of importance for the sequel. Let E be some measurable, bounded set, and let $F^{(1)}$ be an expanding system of closed sets belonging to the set E:

$$F_1^{(1)} \subseteqq F_2^{(1)} \subseteqq F_3^{(1)} \subseteqq \cdots \subseteqq F_k^{(1)} \subseteqq \cdots$$

We shall say that a system $F^{(2)}$ consisting of the sets

$$F_1^{(2)} \subseteqq F_2^{(2)} \subseteqq F_3^{(2)} \subseteqq \cdots \subseteqq F_k^{(2)} \subseteqq \cdots$$

is interior in relation to $F^{(1)}$ if for any set $F_k^{(2)}$ we can find a set $F_{n_k}^{(1)}$ such that $F_k^{(2)} \subseteqq F_{n_k}^{(1)}$.

Let us agree to express this relation by $F^{(2)} \ll F^{(1)}$ or $F^{(1)} \gg F^{(2)}$. (The relation $F^{(1)} \gg F^{(2)}$ does not exclude $F^{(1)} \ll F^{(2)}$).

$F^{(1)} \gg F^{(2)}$ and $F^{(2)} \gg F^{(3)}$ implies $F^{(1)} \gg F^{(3)}$.

LEMMA 7. *Suppose that a sequence of expanding systems of closed sets on a set E is given:*

$$F^{(1)} \gg F^{(2)} \gg F^{(3)} \gg \cdots \gg F^{(S)} \gg \cdots$$

such that

$$\lim_{k \to \infty} (mE - mF_k^{(S)}) = 0, \quad S = 1, 2, \cdots.$$

Then a system $F^{(\omega)}$ exists which is interior in relation to all the $F^{(S)}$ and which is such that

$$\lim_{k \to \infty} (mE - mF_k^{(\omega)}) = 0.$$

Proof. In each system $F^{(S)}$ we choose a set $F_n^{(S)} = F^{(S)*}$ such that

$$mF_n^{(S)} \geqq mE - \frac{1}{2^S}.$$

Let

$$F_k^{(\omega)} = F^{(k)*} F^{(k+1)*} \cdots F^{(m)*} \cdots.$$

It is clear that

$$F_1^{(\omega)} \subseteqq F_2^{(\omega)} \subseteqq \cdots \subseteqq F_k^{(\omega)} \subseteqq \cdots.$$

The sets $F_k^{(\omega)}$ are closed sets.

It is not difficult to see that the system $F^{(\omega)}$ is interior in relation to any one of the systems $F^{(S)}$. This follows from the fact that $F_k^{(\omega)} \subseteqq F^{(S)*}$ for any $S \geqq k$.

It remains to show that $\lim\limits_{k \to \infty} mF_k^{(\omega)} = mE$. We have

$$mE - mF_k^{(\omega)} \leqq m(E - F^{(k)*}) + m(E - F^{(k+1)*}) + \cdots$$

$$+ m(E - F^{(k+m)*}) + \cdots \leqq \frac{1}{2^k} + \frac{1}{2^{k+1}} + \cdots = \frac{1}{2^{k-1}},$$

from which our assertion follows and the lemma is proved.

A consequence of this important lemma is:

THEOREM 19. *Suppose a sequence of functions $f_1, f_2, \ldots, f_k, \ldots$ which are continuous on a closed set F is given. If the sequence converges everywhere on F, it will converge uniformly on every set of a certain system of closed sets F_δ which are such that $\lim\limits_{\delta \to 0} mF_\delta = mF$.*

Proof. We consider closed sets $F_k(\varepsilon)$ having the property that on them

$$\left| f_{m_1} - f_{m_2} \right| \leqq \varepsilon$$

for any $m_1 > k$ and $m_2 > k$.

Whatever ε may be, any point of F will belong to at least one $F_k(\varepsilon)$. Hence

$$F = F_1(\varepsilon) + \cdots + F_k(\varepsilon) + \cdots.$$

Moreover,

$$F_1(\varepsilon) \subseteqq F_2(\varepsilon) \subseteqq \cdots \subseteqq F_k(\varepsilon) \subseteqq \cdots .$$

Applying Theorem 10, we shall have $\lim_{k \to \infty} mF_k(\varepsilon) = mF$.

Consider the sequence of systems $F^{(S)}$ consisting of $F_1(1/2^S), F_2(1/2^S), \cdots$. Obviously $F^{(1)} \gg F^{(2)} \gg \cdots$.

Applying Lemma 7, we see that a system $F^{(\omega)}$ exists which is interior in relation to all the $F^{(S)}$. On any set $F_k^{(\omega)}$ the sequence $f_1, f_2, \ldots$ converges uniformly. For, $F_k^{(\omega)}$ lies inside a certain $F_{n_S}^{(S)}$ for any (S) and so we get

$$\left| f_{m_1} - f_{m_2} \right| < \frac{1}{2^S} \quad \text{for} \quad m_1 > n_S, \ m_2 > n_S.$$

Hence the theorem.

From this theorem follows immediately the essential fact in the theory of measurable functions:

Egorov's Theorem. If a sequence of functions $f_1, f_2, \ldots, f_k, \ldots$, which are measurable on a bounded open set Ω, converges almost everywhere, then the limit function is measurable.

Proof. We select closed sets F_k on which f_k is continuous so that we have

$$m\Omega - mF_k \leqq \frac{\delta}{2^{k+1}}.$$

On the set $F_0 = F_1 F_2 \ldots F_k \ldots$ all the functions $f_1, f_2, \ldots$ are continuous. We evaluate the measure of F_0. We have

$$mF_0 = \lim_{k \to \infty} mF_1 F_2 \ldots F_k$$

$$mF_1 - mF_1 F_2 = m(F_1 + F_2) - mF_2 \leqq m\Omega - mF_2 \leqq \frac{\delta}{2^3}$$

$$mF_1 F_2 - mF_1 F_2 F_3 = m(F_1 F_2 + F_3) - mF_3 \leqq m\Omega - mF_3 \leqq \frac{\delta}{2^4}.$$

and so on. Hence

$$mF_1 F_2 \ldots F_k \geqq mF_1 - \frac{\delta}{4} = m\Omega - (m\Omega - mF_1) - \frac{\delta}{4} \geqq m\Omega - \frac{\delta}{2}$$

and consequently

$$mF_0 \geqq m\Omega - \frac{\delta}{2}.$$

Since the sequence $f_1, f_2, \ldots, f_k, \ldots$ converges almost everywhere, it follows that there are closed sets F_δ such that $mF_\delta \geqq m\Omega - \delta/2$ and on which the sequence converges. Then $F_\delta^* = F_0 F_\delta$ has the property

$$mF_\delta^* \geqq m\Omega - \delta.$$

By Theorem 19 the sequence $f_1, f_2, ..., f_k, ...$ converges uniformly on the set

$$F_\delta' \subseteq F_\delta^*, \quad \text{so that} \quad mF_\delta^* - mF_\delta' < \delta.$$

Hence $mF_\delta' \geqq m\Omega - 2\delta$, and consequently the function $f_0 = \lim\limits_{k\to\infty} f_k$ is continuous on the set F_δ' with measure as close as we please to $m\Omega$, i.e., it is measurable, as we had to prove.

Using Theorem 19, we can prove the following important lemma.

LEMMA 8. *If a non-decreasing sequence of functions* $f_1, f_2, ..., f_k, ...$ *which are summable in the bounded domain* Ω *has the property that*

$$\int_\Omega f_k \, dv \leqq A$$

i.e., if the integrals of its members are bounded, then the sequence $f_1, f_2, ..., f_k, ...$ *converges almost everywhere and has as its limit a summable function*

$$\lim_{k\to\infty} f_k = f_0$$

and moreover

$$\lim_{k\to\infty} \int_\Omega f_k \, dv = \int_\Omega f_0 \, dv.$$

The lemma will be proved if we establish the existence of closed sets F_δ^* such that all the f_k are continuous on them, the sequence f_k converges uniformly, and also $m(\Omega - F_\delta^*) < \delta$, where δ is any positive number.

For, in this case the limit function f_0 will be continuous on all the sets F_δ^* and the inequality

$$\int_{F_\delta^*} f_0 \, dv \leqq A$$

will hold.

By Lemma 4, f_0 will be a function measurable and summable on Ω. The closed sets F_δ^* form an exhaustive system for f_0, and therefore

$$\int_\Omega f_0 \, dv = \lim_{\delta\to 0} \int_{F_\delta^*} f_0 \, dv \leqq A.$$

Finally, given any $\varepsilon > 0$, we shall have, for a sufficiently large k,

$$\int_{F_\delta^*} f_k \, dv \leqq \int_{F_\delta^*} f_0 \, dv \leqq \int_{F_\delta^*} f_k \, dv + \varepsilon,$$

and consequently,

$$\int_\Omega f_k \, dv \leqq \int_\Omega f_0 \, dv \leqq \int_\Omega f_k \, dv + \varepsilon$$

which implies

$$\int_\Omega f_0 \, dv = \lim_{k \to \infty} \int_\Omega f_k \, dv,$$

as was to be shown.

We still have to show that sets exist having the required properties. Consider first the case when $f_k \geqq 0$.

We introduce the functions

$$\psi_k = \tan^{-1} f_k.$$

Clearly, ψ_k will be a non-decreasing bounded sequence. Hence ψ_k will converge everywhere on Ω.

We write

$$\lim_{k \to \infty} \psi_k = \psi_0.$$

By the arguments used in proving Egorov's theorem we see that closed sets F_δ exist for which $m\Omega - mF_\delta < \delta$, and for which the convergence is uniform and ψ_0 is continuous.

Let F_δ' be a closed set of those points of F_δ for which $\psi_0 = \pi/2$. On F_δ' the function ψ_k tends uniformly to $\pi/2$ and this implies that $f_k = \tan \psi_k$ tends uniformly to infinity. Consequently, $mF_\delta' = 0$, for otherwise we should have $\int_{F_\delta'} f_k \, dv \to \infty$, which is obviously impossible.

We enclose F_δ' in a domain Ω_δ such that $m\Omega_\delta < \delta$, and put

$$F_\delta^* = F_\delta - \Omega_\delta.$$

Then clearly,

$$mF_\delta^* > mF_\delta - m\Omega_\delta > m\Omega - 2\delta.$$

On the sets F_δ^* the sequence ψ_k converges uniformly to a limit different from $\pi/2$, i.e., f_k converges uniformly to a finite limit.

If f_k is a function of variable sign, it is sufficient to consider a sequence $\varphi_k = f_k - f_1$, and then the proof reduces to that just given. Hence the lemma is proved.

§ 7. Convergence in the Mean of Summable Functions

To illustrate the use of the previous theorems we now present some propositions dealing with the properties of measure.

THEOREM 20. *Suppose that on a bounded open set Ω a sequence is given of measurable sets $E_1, E_2, \ldots, E_k, \ldots$, no pair of which have any points in com-*

mon. Then the sum of these sets, $E_0 = E_1 + E_2 + \cdots + E_k + \cdots$ is measurable and

$$mE_0 = \sum_{i=1}^{\infty} mE_i.$$

Proof. Let $\xi_i(P)$ be the characteristic function of the set E_i. We put

$$\psi_s = \sum_{i=1}^{s} \xi_i(P).$$

By Theorem 13, each of the functions ψ_s is summable, since it is the sum of a finite number of summable functions. The sequence ψ_s is a non-decreasing sequence which converges to the characteristic function of the set E_0, *i.e.*, to $\xi_0(P)$. Moreover, $\psi_s \leqq 1$, since the sets have no common points and consequently there cannot be two function $\xi_i(P)$ different from zero at any point. By Lemma 8, the function $\xi_0(P) = \lim_{s \to \infty} \psi_s$ is summable, and

$$\int_{\Omega} \xi_0(P)\, dv = \lim_{s \to \infty} \int_{\Omega} \left(\sum_{k=1}^{s} \xi_k(P) \right) dv = \sum_{k=1}^{\infty} \int_{\Omega} \xi_k(P)\, dv.$$

Hence the theorem.

COROLLARY 1. *If $E_1, E_2, \ldots, E_k, \ldots$ is any sequence of measurable sets lying within a bounded domain Ω, and $E_0 = E_1 + E_2 + \cdots + E_k + \cdots$, then E_0 is measurable, and*

$$mE_0 \leqq \sum_{k=1}^{\infty} mE_k.$$

For, the set E_0 can be presented in the form

$$E_1 + [E_2 - E_1] + [E_3 - (E_1 + E_2)] + \cdots$$
$$+ [E_k - (E_1 + E_2 + \cdots + E_{k-1})] + \cdots.$$

Here all the addends are obviously measurable and no two of them have any points in common; also $m[E_k - (E_1 + E_2 + \cdots + E_{k-1})] \leqq mE_k$. Hence the proposition.

COROLLARY 2. *The sum of a sequence of sets each of zero measure*

$$E_0 = E_1 + E_2 + \cdots + E_k + \cdots, \quad mE_k = 0,$$

is also a set of zero measure.

This follows immediately from Corollary 1.

We showed earlier that if a function f vanishes almost everywhere, then its integral is zero. We now prove a theorem, which, in a certain sense, is the converse of this.

THEOREM 21. *If the integral of a non-negative summable function f on a bounded open set is zero, then the function f is zero almost everywhere, except perhaps at points of some set E of zero measure.*

Proof. Consider an expanding system of closed sets F_k on which the function f is continuous and such that $m(\Omega - F_k) < 1/2^k$. The set E_0 of points belonging to Ω which do not enter into any of the sets F_k has zero measure, since it can be included in the family of open sets $\Omega - F_k$ whose measure tends to zero as $k \to \infty$.

The set Φ_k of those points of F_k at which $f \geq 1/2^k$ is obviously a closed set. By hypothesis, $\int_{\Phi_k} f \, dv = 0$, and hence $m\Phi_k = 0$. Put $E = E_0 + \sum_{k=1}^{\infty} \Phi_k$. The sets Φ_k form an expanding sequence. Any point at which $f > 0$ must enter into either E_0 or one of the sets Φ_k, *i.e.*, it will belong to E. And from what we have already proved, $mE = 0$. Hence the theorem.

THEOREM 22. *If a function f is summable in an open set Ω, and if for any function ψ which is continuous in Ω the equality $\int_{\Omega} f\psi \, dv = 0$ holds, then f must satisfy the condition $\int_{\Omega} |f| \, dv = 0$ and consequently f is equal to zero almost everywhere.*

We prove this theorem by *reductio ad absurdum*.

If the integral $\int_{\Omega} |f| \, dv$ were different from zero, then at least one of the integrals

$$\int_{\Omega} f^+ \, dv, \quad \int_{\Omega} f^- \, dv$$

(whose sum is equal to $\int_{\Omega} |f| \, dv$) would be different from zero. The difference of these integrals is zero, since it is equal to the integral of f. Consequently, both the integrals must be different from zero. Hence there is a closed set F' on which $f^+ > 0$ and consequently $f = f^+$, and for this set

$$\int_{F'} f^+ \, dv = \int_{F'} f \, dv = h > 0.$$

This integral may be written in the form $\int_{\Omega} f(P) \, \xi_{F'}(P) \, dv$ where $\xi_{F'}(P)$ is the the characteristic function for the set F'. By Lemma 3 and by the definition of a measurable function (see p. 92), a continuous function $\chi_k(P)$ exists whose values lies between zero and unity

$$0 \leq \chi_k(P) \leq 1, \quad 0 \leq \chi_k(P) - \xi_{F'}(P) \leq 1,$$

and which can differ from $\xi_{F'}(P)$ only on a set Ω_k of measure as small as we please. Further,

$$\int_{\Omega} f(P) \, \xi_{F'}(P) \, dv = \int_{\Omega} f(P) \, \chi_k(P) \, dv + \int_{\Omega} f(P) \, [\xi_{F'}(P) - \chi_k(P)] \, dv.$$

But from the hypothesis, the first integral on the right-hand side is zero because of the continuity of $\chi_k(P)$. And using Theorem 15, we get

$$\left| \int_{\Omega} f(P) \left[\xi_{F'}(P) - \chi_k(P) \right] dv \right| \leq \int_{\Omega_k} |f(P)| \, dv.$$

Since the integral of $|f(P)|$ is absolutely convergent, the last integral, together with the measure of Ω_k, becomes as small as we please, and consequently cannot be equal to the constant number $h > 0$. We are thus led to a contradiction, and the theorem is proved.

LEMMA 9. *Any function $\psi(P)$ which is continuous in the open set Ω and which is different from zero only in an open set Ω_ψ lying together with its boundary inside Ω may be uniformly approximated on Ω as closely as we please by means of a function $\psi_h(P)$, whose derivatives of any order are continuous and which is different from zero only in an open set Ω_h consisting of points whose distance from Ω_ψ is less than h.*

Here h is a sufficiently small number, dependent on the desired degree of accuracy of approximation and on the configuration of Ω and Ω_ψ. We construct the function $\psi_h(P)$ by the formula

$$\psi_h(P) = \frac{\displaystyle\int_{r \leq h} e^{\frac{r^2}{r^2 - h^2}} \psi(P') \, dP'}{\displaystyle\int_{r \leq h} e^{\frac{r^2}{r^2 - h^2}} \, dP'}$$

where

$$r = \sqrt{(x_1 - x_1')^2 + (x_2 - x_2')^2 + \cdots + (x_n - x_n')^2}.$$

The function $\psi_h(P)$ is differentiable without restriction. To see this, we put

$$\int_{r \leq h} e^{\frac{r^2}{r^2 - h^2}} \, dP' = \alpha(h)$$

and define a function $\omega(P, P')$ by the relations

$$\omega(P, P') = \begin{cases} 0, & r > h \\ \dfrac{1}{\alpha(h)} e^{\frac{r^2}{r^2 - h^2}}, & r \leq h. \end{cases}$$

Then the function $\psi_h(P)$ can be written in the form

$$\psi_h(P) = \int_{\Omega} \psi(P') \, \omega(P, P') \, dP'.$$

In this form we can apply to it the theorem on differentiation with respect to a parameter under the sign of the integral.

For, the function $\omega(P, P')$ has continuous derivatives of any order with respect to the coordinates of the point P. Derivatives of any order of $\omega(P, P')$ obviously exist for $r < h$ and $r > h$. Their limiting values for $r > h$ and $r \to h$ will all be zero. We have to establish that the limiting values of all the derivatives of ω for $r < h$, $r \to h$ will also be zero. This follows from the fact that any derivative of ω of order s will have the form

$$\frac{\partial^s \omega}{\partial x_1^{a_1} \cdots \partial x_n^{a_n}} = \frac{e^{\frac{r^2}{r^2 - h^2}}}{(r^2 - h^2)^s} Q(x_1, \ldots, x_n, x_1', \ldots, x_n'),$$

where Q is a polynomial in its arguments. But $e^{\frac{r^2}{r^2 - h^2}}$ tends to zero faster than $1/(r^2 - h^2)^s$ tends to infinity, and this implies that

$$\lim_{r \to h} \frac{\partial^s \omega}{\partial x_1^{a_1} \cdots \partial x_n^{a_n}} = 0.$$

For the difference $\psi_h(P) - \psi(P)$ we have

$$\psi_h(P) - \psi(P) = \frac{\displaystyle\int_{r \leq h} e^{\frac{r^2}{r^2 - h^2}} [\psi(P') - \psi(P)]\, dP'}{\displaystyle\int_{r \leq h} e^{\frac{r^2}{r^2 - h^2}}\, dP'}.$$

For sufficiently small h,

$$\left| \psi(P') - \psi(P) \right| < \varepsilon,$$

whence, by the mean-value theorem,

$$\left| \psi_h(P) - \psi(P) \right| < \varepsilon.$$

Hence the lemma.

This lemma enables us to apply the integral criterion for the vanishing of the function in a rather weaker form than that given in Theorem 22. Let f be summable in Ω. Suppose that

$$\int_\Omega f(P)\psi(P)\, dP = 0 \tag{6.7}$$

for all functions $\psi(P)$ which firstly, have continuous derivatives of any order, and secondly, are each different from zero only in some inner subset Ω_ψ of the domain Ω.

Then this is sufficient to assert that

$$\int_\Omega \left| f(P) \right|\, dP = 0.$$

For, if the condition (6.7) holds good for all functions which are differentiable without limit, then it will also hold for all continuous functions. Suppose the contrary. Then a function $\psi_0(P)$ exists such that

$$\int_\Omega f(P)\,\psi_0(P)\,\mathrm{d}P > \varepsilon.$$

By the lemma, there is a function $\psi_h(P)$ as close as we please to ψ_0 which is differentiable without limit. We have:

$$\int_\Omega f(P)\psi_0(P)\,\mathrm{d}P = \int_\Omega f(P)\,\psi_h(P)\,\mathrm{d}P + \int_\Omega f(P)\,[\psi_0(P) - \psi_h(P)]\,\mathrm{d}P.$$

By the condition, the first integral on the right-hand side is zero. On the other hand,

$$\left|\int_\Omega f(P)\,[\psi_0(P) - \psi_k(P)]\,\mathrm{d}P\right| \leqq \max\left|\psi_0(P) - \psi_k(P)\right| \int_\Omega \left|f(P)\right|\,\mathrm{d}P < \varepsilon,$$

where ε is an arbitrary positive number. We are thus led to a contradiction, and this proves the assertion.

To prove the important Theorem 23 we require the following lemma.

Lemma on the Upper and Lower Limits of a Sequence. Let $x_1, x_2, \ldots, x_k, \ldots$ be any bounded sequence of real numbers. *Then a number Y is said to be the upper limit of the sequence $\{x_k\}$ if for any number $\varepsilon > 0$ there is among the numbers of $\{x_k\}$ only a finite set for which $x_k > Y + \varepsilon$, and an infinite set for which $x_k > Y - \varepsilon$.*

A number y is said to be the *lower limit of the sequence* if $x_k < y - \varepsilon$ holds only for a finite set of the numbers x_k and $x_k < y + \varepsilon$ holds for an infinite set of them.

If the upper and lower limits coincide, then the sequence obviously converges and has only a single limit point.

We usually write $\overline{\lim}\, x_k$ for Y and $\underline{\lim}\, x_k$ for y.

Our lemma asserts that

$$\overline{\lim}\, x_k = \lim_{j\to\infty}\left\{\lim_{s\to\infty}\max\,(x_j, \ldots, x_{j+s})\right\} \tag{6.8}$$

$$\underline{\lim}\, x_k = \lim_{j\to\infty}\left\{\lim_{s\to\infty}\min\,(x_j, \ldots, x_{j+s})\right\} \tag{6.9}$$

Both limits on the right-hand side of (6.8) and (6.9) are meaningful. For, as s increases, the numbers

$$z_s^{(j)} = \max\,(x_j, x_{j+1}, \ldots, x_{j+s})$$

do not decrease but remain bounded, and consequently tend to certain limits which we denote by $z_0^{(j)}$. The numbers $z_0^{(j)}$ are bounded from below and do not increase with increasing j, since $z_s^{(j)} \leqq z_{s+1}^{(j-1)}$, and consequently

$$z_0^{(j)} = \lim_{s \to \infty} z_s^{(j)} \leqq \lim_{s \to \infty} z_{s+1}^{(j-1)} = z_0^{(j-1)}.$$

Hence the limit of $z_0^{(j)}$ as $j \to \infty$ exists.

We shall now prove the relation (6.8). Given any $\varepsilon > 0$, an infinite set of numbers $z_s^{(j)}$ will satisfy the condition

$$z_s^{(j)} > \overline{\lim}\, x_k - \varepsilon$$

and hence it may be shown that, for all $j \geqq 1$,

$$z_0^{(j)} \geqq \overline{\lim}\, x_k, \quad \text{i.e.} \quad \lim_{j \to \infty} z_0^{(j)} \geqq \overline{\lim}\, x_k.$$

On the other hand, for a sufficiently large j all the $z_s^{(j)}$ will satisfy the inequality

$$z_s^{(j)} < \overline{\lim}\, x_k + \varepsilon,$$

from which it may be proved that

$$\lim_{j \to \infty} z_0^{(j)} \leqq \overline{\lim}\, x_k,$$

hence (6.8).

In an exactly similar way we can prove (6.9), which, incidentally, reduces to (6.8) if x_k is replaced by $-x_k$.

Suppose a sequence is given of functions $f_1, f_2, \ldots, f_k, \ldots$ which are summable in the open set Ω. If for any $\varepsilon > 0$ there is an integer $N = N(\varepsilon)$ such that

$$\int_\Omega |f_m - f_n|\, dv < \varepsilon$$

for all $m, n > N$, then the sequence $\{f_k\}$ is said to be *convergent in itself*.

THEOREM 23. *For any sequence of functions $f_1, f_2, \ldots, f_k, \ldots$ which is convergent in itself, a summable function f_0 exists which is called the generalized limit of the sequence $\{f_k\}$ and which has the property that*

$$\int_\Omega |f_k - f_0|\, dv \to 0 \text{ as } k \to \infty.$$

Before proving this theorem we make this observation. If f_1, f_2, are summable functions, then so is $f_3 = \max(f_1, f_2)$.

For, $\max(f_1, f_2) = f_1 + (f_2 - f_1)^+$. But if f_1, f_2 are summable, then so is their difference and also the positive part of this difference.

If $f_1, f_2, \ldots, f_m$ are summable functions, then clearly the function

$$\max(f_1, f_2, \ldots, f_m)$$

will also be summable.

We turn now to the proof of Theorem 23. We consider a sub-sequence $\{f_{N_k}\}$ chosen from the sequence $\{f_k\}$ so that

$$\int_\Omega |f_{N_k} - f_{N_{k_1}}|\, dv < \frac{1}{2^k} \quad \text{for} \quad k_1 > k.$$

We write ψ_k for f_{N_k}. We shall prove that the sequence ψ_k converges almost everywhere to a summable function f_0. To do this, we examine $\overline{\lim}\, \psi_k$ and $\underline{\lim}\, \psi_k$. Let

$$\chi_s^{(j)} = \max (\psi_1, \psi_2, \ldots, \psi_{j+s}).$$

We keep j fixed for the moment and consider the sequence $\chi_1^{(j)}, \chi_2^{(j)}, \ldots,$ $\chi_s^{(j)}, \ldots$ This is a non-decreasing sequence of summable functions. On the other hand,

$$\psi_j \leq \chi_s^{(j)} \leq \psi_j + |\psi_{j+1} - \psi_j| + |\psi_{j+2} - \psi_{j+1}| + \cdots + |\psi_{j+s} - \psi_{j+s-1}|$$

whence

$$0 \leq \int_\Omega \chi_s^{(j)}\, dv - \int_\Omega \psi_j\, dv \leq \int_\Omega \left\{ \sum_{k=j+1}^{j+s} |\psi_k - \psi_{k-1}|\, dv \right\} \leq \sum_{k=j}^{j+s-1} \frac{1}{2^k} < \frac{1}{2^{j-1}}.$$

Consequently, by Lemma 8, the sequence $\{\chi_s^{(j)}\}$ has almost everywhere a certain function $\chi_0^{(j)}$ as its limit, where $\chi_0^{(j)} \geq \psi_j$ and

$$0 \leq \int_\Omega \chi_0^{(j)}\, dv - \int_\Omega \psi_j\, dv \leq \frac{1}{2^{j-1}}.$$

We now consider the sequence $\chi_0^{(1)}, \chi_0^{(2)}, \ldots, \chi_0^{(k)}, \ldots$. It will be non-increasing, since $\chi_s^{(j)} \leq \chi_s^{(j-1)}$; hence $\chi_0^{(j)} \leq \chi_0^{(j-1)}$. The integrals of ψ_j are uniformly bounded from below, since

$$\int_\Omega \psi_j\, dv \geq \int_\Omega \psi_0\, dv - \sum_{k=1}^\infty \int_\Omega |\psi_k - \psi_{k-1}|\, dv \geq \int_\Omega \psi_0\, dv - \sum_{k=1}^\infty \frac{1}{2^k}$$

$$= \int_\Omega \psi_0\, dv - 1.$$

Consequently the integrals of $\chi_0^{(j)}$ are bounded from below. By the same Lemma 8 the sequence $\chi_0^{(j)}$ converges almost everywhere to a certain function f_0 where

$$\lim_{j \to \infty} \int_\Omega \chi_0^{(j)}\, dv - \int_\Omega f_0\, dv = \lim_{j \to \infty} \int |\chi_0^{(j)} - f_0|\, dv = 0.$$

By the lemma on the upper and lower limits of a sequence, $f_0 = \overline{\lim}\, \psi_k$.

We now show that f_0 satisfies the requirements of the theorem. We have

$$\int_\Omega \left| \psi_j - f_0 \right| dv \leqq \int_\Omega \left| \psi_j - \chi_0^{(s)} \right| dv + \int_\Omega \left| \chi_0^{(j)} - f_0 \right| dv.$$

Both integrals on the right-hand side become as small as we please for sufficiently large j. Hence

$$\int_\Omega \left| \psi_j - f_0 \right| dv < \varepsilon \quad \text{for} \quad j > J(\varepsilon).$$

Following exactly the same argument for the function $f_0^* = \underline{\lim} \, \psi_k$, we obtain

$$\int_\Omega \left| \psi_j - f_0^* \right| dv < \varepsilon$$

and this implies

$$\int_\Omega \left| f_0 - f_0^* \right| dv \leqq \int_\Omega \left| \psi_j - f_0 \right| dv + \int_\Omega \left| \psi_j - f_0^* \right| dv < 2\varepsilon.$$

Consequently,

$$\int_\Omega \left| f_0 - f_0^* \right| dv = 0,$$

and the functions f_0 and f_0^* coincide almost everywhere.

Finally, for any function of the original sequence f_k we shall have

$$\int_\Omega \left| f_k - f_0 \right| dv \leqq \int_\Omega \left| f_k - \psi_j \right| dv + \int_\Omega \left| \psi_j - f_0 \right| dv.$$

Consequently, given any $\varepsilon > 0$, we can find an integer $N(\varepsilon)$ such that, for $k > N(\varepsilon)$,

$$\int_\Omega \left| f_k - f_0 \right| dv < \varepsilon,$$

as we had to prove.

We shall say that a series $\sum_{k=1}^{\infty} u_k(P)$ \hfill (6.10)

converges in the Lebesgue sense if a summable function u exists such that

$$\lim_{N \to \infty} \int_\Omega \left| u - \sum_{k=1}^{N} u_k \right| dv = 0.$$

It is clear that convergence in the Lebesgue sense implies that it is possible to pass to the limit under the integral sign.

We shall say that the series (6.10) converges in itself in the Lebesgue sense if

$$\int_\Omega \left| \sum_{k=1}^{m+p} u_k \right| dv < \varepsilon \quad \text{for} \quad m > N(\varepsilon).$$

Theorem 23 can be reformulated for series thus:

A series which converges in itself in the Lebesgue sense converges in the Lebesgue sense to a certain limit function.

A series which converges in itself may be multiplied term by term by any measurable bounded function and will still remain convergent.

THEOREM 24. *The members of a sequence $f_1, f_2, \ldots, f_k, \ldots$ which converges in itself have Lebesgue integrals which are absolutely equicontinuous.*

In other words, given any $\varepsilon > 0$, a number δ can be found such that

$$\left| \int_{\Omega_\delta} f_k \, dv \right| < \varepsilon$$

provided only that the measure of the open set Ω_δ is less than δ.

Proof. Given $\varepsilon > 0$, we choose an N so that, for $k > N$, $m \geqq N$, we have

$$\int_\Omega |f_k - f_m| \, dv < \frac{\varepsilon}{2}.$$

For any f_k of the functions $f_1, f_2, \ldots, f_N$ we can, by virtue of the absolute continuity of the Lebesgue integral, find a δ_k such that

$$\int_\Omega |f_k| \, dv < \frac{\varepsilon}{2}$$

as soon as $m\Omega_1 < \delta_k$. We take as $\delta(\varepsilon)$ the smallest of the number $\delta_1, \delta_2, \ldots \delta_N$. Then for $k \leqq N$ the assertion made by the theorem is fulfilled because of the choice of $\delta(\varepsilon)$, and it is also true for k > N because

$$\left| \int_{\Omega_1} f_k \, dv \right| \leqq \int_{\Omega_1} |f_k - f_N| \, dv + \int_{\Omega_1} |f_N| \, dv < \frac{\varepsilon}{2} + \frac{\varepsilon}{2} = \varepsilon.$$

THEOREM 25. *In order that the sequence $\{f_k\}$ should be convergent in itself in the Lebesgue sense in an open interval, and consequently that it should have a limit in the Lebesgue sense, it is sufficient that two conditions be fulfilled:*

1. The sequence $\{f_k\}$ shall converge almost everywhere to a certain limit function f_0.

2. The members of the sequence $\{f_k\}$ shall have integrals which are absolutely equicontinuous.

Proof. By virtue of the lemma to Egorov's theorem, there are closed sets F_δ, with measure as close as we please to the measure of Ω, on which the convergence is uniform. We choose $\delta > 0$ so small that

$$\int_{\Omega_1} |f_k| \, dv < \frac{\varepsilon}{3}$$

as soon as $m\Omega_1 < \delta$. Further, we find an N so that on the set F_δ the inequality

$$\left|f_k - f_m\right| < \frac{\varepsilon}{3m\Omega}$$

holds as soon as $k, m > N$. Then, for $k > N$ and $m > N$, we shall have

$$\int_\Omega \left|f_k - f_m\right| dv = \int_{\Omega - F_\delta} \left|f_k - f_m\right| dv + \int_{F_\delta} \left|f_k - f_m\right| dv$$

$$\leqq \int_{\Omega - F_\delta} \left|f_k\right| dv + \int_{\Omega - F_\delta} \left|f_m\right| dv + \int_{F_\delta} \left|f_k - f_m\right| dv$$

$$< \frac{\varepsilon}{3} + \frac{\varepsilon}{3} + \frac{\varepsilon}{3} = \varepsilon.$$

This implies that the sequence $\{f_k\}$ converges in itself, as was to be shown.

COROLLARY. *If a sequence of summable functions* $f_1, f_2, \ldots, f_k, \ldots$ *converges almost everywhere in the open set* Ω *to a limit function* f_0, *and if the functions* $\left|f_k\right|$ *do not exceed a certain summable function* ψ, *then the integral* $\int_\Omega f_0 \, dv$ *exists and*

$$\int_\Omega f_0 \, dv = \lim_{k \to \infty} \int_\Omega f_k \, dv. \tag{6.11}$$

For, it follows from the inequality $\left|f_k\right| < \psi$ that the integrals of the sequence $\{f_k\}$ are absolutely equicontinuous and this implies that the sequence converges in the Lebesgue sense. It is clear that its limit in the Lebesgue sense must coincide with f_0 almost everywhere. The possibility of passing to the limit under the integral sign is, as we have already pointed out, a consequence of the convergence in the Lebesgue sense.

§ 8. The Lebesgue–Fubini Theorem

The Lebesgue–Fubini Theorem. Let $f(x_1, x_2, \ldots, x_n, y_1, y_2, \ldots, y_m)$ *be a summable function, of* $m + n$ *variables, given in a cube* Ω_0 $(0 < x_i < 1,$ $0 < y_j < 1),$ $i = 1, 2, \ldots, n, j = 1, 2, \ldots, m.$ *Consider a certain measurable set* E *of* Ω_0 *and let* $E(y_1, \ldots, y_m)$ *be the section of this set by the manifold*

$$y_1 = \text{const.}, \quad y_2 = \text{const.}, \quad \ldots, \quad y_m = \text{const.},$$

i.e., the set of points having the same coordinates

$$x_1, x_2, \ldots, x_n$$

as the points $(x_1, x_2, \ldots, x_n, y_1, y_2, \ldots, y_m)$ *of* E *for fixed* $y_1, y_2, \ldots, y_m.$ *Let* E_y *be the projection of* E *on to the manifold* $y_1, \ldots, y_m,$ *i.e., the set of all points whose coordinates* $y_1, \ldots, y_m$ *are the same as those of any points of* $E.$

Let $\psi^{(E)}(y_1, \ldots, y_m)$ be a function defined in the cube $0 < y_j < 1$ by

$$\psi^{(E)}(y_1, \ldots, y_m) = \begin{cases} 0 & \text{outside } E_y \\ \displaystyle\int_{E(y_1, \ldots, y_m)} f \, dx_1 \ldots dx_n & \text{inside } E_y \end{cases} \qquad (6.12)$$

Then:

(a) *The integrals on the right-hand side of (6.12) exist for almost all values of $y_1, \ldots, y_m$, (i.e., for all points except perhaps those of some set of zero measure).*

(b) *The function $\psi^{(E)}(y_1, \ldots, y_m)$ is a summable function in the variables $y_1, \ldots, y_m$.*

(c) $\displaystyle\int_E f \, dx_1 \ldots dx_n \, dy_1 \ldots dy_m = \int_{0 < y_j < 1} \psi^{(E)}(y_1, \ldots, y_m) \, dy_1 \ldots dy_m.$ (6.13)

Without loss of generality we may take f to be positive. We shall prove the theorem first for the case when E is an open set Ω, and the function f is continuous in Ω. We construct a system of net sets Φ_h exhaustive for Ω. Then it is clear that

$$\int_{\Phi_h} f \, dx_1 \ldots dx_n \, dy_1 \ldots dy_m = \int_{0 < y_j < 1} \psi^{(\Phi_h)}(y_1, \ldots, y_m) \, dy_1 \ldots dy_m. \quad (6.14)$$

The sequence $\psi^{(\Phi_h)}(y_1, \ldots, y_m)$ is a sequence, which does not decrease as $h \to 0$, of measurable functions such that

$$\int_{0 < y_j < 1} \psi^{(\Phi_h)}(y_1, \ldots, y_m) \, dy_1 \ldots dy_m \leqq \int_{\Omega} f \, dx_1 \ldots dx_n \, dy_1 \ldots dy_m.$$

By Lemma 8 this sequence has almost everywhere a limit $\psi'(y_1, \ldots, y_m)$ such that

$$\lim_{h \to 0} \psi^{(\Phi_h)}(y_1, \ldots, y_m) = \psi'(y_1, \ldots, y_m)$$

and

$$\lim_{h \to 0} \int_{0 < y_j < 1} \psi^{(\Phi_h)}(y_1, \ldots, y_m) \, dy_1 \ldots dy_m$$

$$= \int_{0 < y_j < 1} \psi'(y_1, \ldots, \ldots y_m) \, dy_1 \ldots dy_m.$$

But the $\Phi_h(y_1, \ldots, y_m)$ form an exhaustive system of polyhedra for $\Omega(y_1, \ldots, y_m)$, since any point of $\Omega(y_1, \ldots, y_m)$ falls into one of the $\Phi_h(y_1, \ldots, y_m)$, and consequently $\psi'(y_1, \ldots, y_m) = \psi(\Omega)(y_1, \ldots, y_m)$. Then passing to the limit in (6.14) we obtain the required result (6.13) for this case.

Remark. If we consider a system of domains $\Omega_1 \supset \Omega_2 \supset \cdots \supset \Omega_k \supset \cdots$ such that

$$\lim_{k \to \infty} m\Omega_k = 0,$$

then for almost all $y_1, \ldots, y_m$ we shall have $m\Omega_k(y_1, \ldots, y_m) \to 0$ as $k \to \infty$.

For, by what we have proved,

$$m\Omega_k = \int_{0 < y_j < 1} m\Omega_k(y_1, \ldots, y_m)\, dy_1 \ldots dy_m.$$

The sequence of functions $m\Omega_k(y_1, \ldots, y_m) = \psi_k(y_1, \ldots, y_m)$ is monotonic and bounded; hence at all points it has a limit

$$\lim_{k \to \infty} \psi_k = \psi_0(y_1, \ldots, y_m).$$

By Lemma 8, we can pass to the limit under the integral sign:

$$\lim_{k \to \infty} \int_{0 < y_j < 1} \psi_k\, dy_1 \ldots dy_m = \int_{0 < y_j < 1} \psi_0\, dy_1 \ldots dy_m = \lim_{k \to \infty} m\Omega_k = 0.$$

Consequently, $\psi_0 = 0$ almost everywhere, as was asserted.

We next consider the case when E is a closed set F on which f is continuous. This case reduces to the previous one if we extend f over the whole cube Ω_0 and consider the open set $\Omega = \Omega_0 - F$. Clearly,

$$\int_F f\, dx_1 \ldots dx_n\, dy_1 \ldots dy_m = \int_{\Omega_0} f\, dx_1 \ldots dx_n\, dy_1 \ldots dy_m$$

$$- \int_\Omega f\, dx_1 \ldots dx_n\, dy_1 \ldots dy_m. \tag{6.15}$$

On the other hand,

$$\psi^{(F)} + \psi^{(\Omega)} = \psi^{(\Omega_0)}$$

which implies

$$\int_{0 < y_j < 1} \psi^{(F)}\, dy_1 \ldots dy_m = \int_{0 < y_j < 1} \psi^{(\Omega_0)}\, dy_1 \ldots dy_m$$

$$- \int_{0 < y_j < 1} \psi^{(\Omega)}\, dy_1 \ldots dy_m \tag{6.16}$$

Substituting in the right-hand side of (6.15) for the $(m + n)$-tuple integrals their expressions in terms of the functions $\psi^{(\Omega_0)}$ and $\psi^{(\Omega)}$, and using (6.16), we obtain the required result for this case.

Finally we consider the case when E is an open set Ω and f is an arbitrary summable function. The general case of a measurable set E may easily be reduced to this one.

We construct an exhaustive system of closed sets F_k for the function f. By what has been proved

$$\int_{F_k} f \, dx_1 \dots dx_n \, dy_1 \dots dy_m = \int_{0 < y_j < 1} \psi^{(F_k)} \, dy_1 \dots dy_m. \qquad (6.17)$$

The sequence $\psi^{(F_k)}$ is an increasing sequence of measurable functions such that

$$\int_{0 < y_j < 1} \psi^{(F_k)} \, dy_1 \dots dy_m \leqq \int_{\Omega} f \, dx_1 \dots dx_n \, dy_1 \dots dy_m.$$

By Lemma 8 it will have almost everywhere a finite limit ψ' such that

$$\lim \psi^{(F_k)} = \psi', \quad \lim_{k \to \infty} \int_{0 < y_j < 1} \psi^{(F_k)} \, dy_1 \dots dy_m = \int_{0 < y_j < 1} \psi' \, dy_1 \dots dy_m.$$

We shall show that

$$\psi' = \psi^{(\Omega)}.$$

To do this, we shall establish that $F_k(y_1, y_2, \dots, y_m)$ forms an exhaustive system for $\Omega(y_1, y_2, \dots, y_m)$ for almost all $y_1, \dots, y_m$. For, the measure of $\Omega - F_k$ is as small as we please for sufficiently large k. Consequently, in view of the Remark on p. 123, for almost all $y_1, \dots, y_m$,

$$\lim_{k \to \infty} m[\Omega(y_1 \, y_2, \dots, y_m) - F_k(y_1, y_2, \dots, y_m)] = 0.$$

This implies that $F_k(y_1, y_2, \dots, y_m)$ does indeed form an exhaustive system for $\Omega(y_1, y_2, \dots, y_m)$ for almost all $y_1, y_2, \dots, y_m$, and consequently $\psi' = \psi^{(\Omega)}$. Then a passage to the limit in (6.17) proves the Lebesgue–Fubini theorem.

Remark. If the function $f(x_1, \dots, x_n, y_1, \dots, y_m)$ is measurable, if $f \geqq 0$, and if moreover the function $\psi^{(E)}$ exists and is summable, *i.e.*, if proposition (*a*) of the Lebesgue–Fubini theorem is satisfied and the right-hand side of (6.13) is meaningful, then the function f will be summable and, consequently, the Lebesgue–Fubini theorem in its entirety will be valid. That is, the left-hand side of (6.13) will be meaningful and will be equal to the right-hand side.

This is almost obvious. For, suppose it happened that the function f was not summable; then the integrals

$$\int_F f \, dx_1 \dots dx_n \, dy_1 \dots dy_m$$

taken over closed sets on which f is continuous could be made as large as we please. Suppose that the integral on the right-hand side of (6.13) is equal to A. We then choose a closed set F so as to have

$$\int_F f \, dx_1 \, \ldots \, dx_n \, dy_1 \, \ldots \, dy_m > A + 1.$$

We construct a function

$$f_1 = \xi(P) f(P),$$

where ξ is the characteristic function of the set F. Then $\psi_1^{(E)}$ exists and is summable and

$$\int_{0 < y_j < 1} \psi_1^{(E)} \, dy_1 \, dy_2 \, \ldots \, dy_m > A + 1.$$

On the other hand, it is clear that $\psi_1^{(E)} \leqq \psi^{(E)}$, which implies

$$\int_{0 < y_j < 1} \psi_1^{(E)} \, dy_1 \, dy_2 \, \ldots \, dy_m \leqq \int_{0 < y_j < 1} \psi^{(E)} \, dy_1 \, dy_2 \, \ldots \, dy_m = A.$$

We are thus led to a contradiction, showing that our supposition was false.

There is one further obvious consequence of the Lebesgue–Fubini theorem which is noteworthy:

In a repeated integral of a summable function the order of integration may be changed.

INTEGRALS DEPENDENT ON A PARAMETER

§ 1. Integrals which are Uniformly Convergent for a Given Value of Parameter

Suppose that in a closed domain D of the variables $x_1, \ldots, x_n$ a function $F(x_1, \ldots, x_n, \lambda)$ is given, which depends on the parameter λ whose values vary over the interval $a \leq \lambda \leq b$. Consider the integral

$$\psi(\lambda) = \int_D \cdots \int F(x_1, \ldots, x_n, \lambda) \, dx_1 \ldots dx_n. \tag{7.1}$$

We shall suppose that this integral exists in the Lebesgue sense. In other words, we assume that the function $F(x_1, \ldots, x_n, \lambda)$ considered as a function of the point $(x_1, \ldots, x_n)$ in n-dimensional space becomes continuous if we exclude from the closed domain a certain open set of points σ whose measure may be arbitrarily small (in applications we shall for the most part consider cases when $F(x_1, \ldots, x_n, \lambda)$ is discontinuous at points belonging to certain manifolds of fewer dimensions: $e.g.$, at isolated points, or on a finite number of surfaces, curves, $etc.$) and that

$$\int_D \cdots \int F(x_1, \ldots, x_n, \lambda) \, dx_1 \ldots dx_n = \lim \int_{D-\sigma} \cdots \int F(x_1, \ldots, x_n, \lambda) \, dx_1 \ldots dx_n.$$

However, the open set σ may decreas eprovided only that its measure tends to zero.

We shall say that the integral (7.1) *converges uniformly* for $\lambda = \lambda_0$ if, given any $\varepsilon > 0$, a number $h(\varepsilon)$ and a part $\sigma(\varepsilon)$ of the domain D can be found such that:

1. The function $F(x_1, \ldots, x_n, \lambda)$ is continuous in $x_1, x_2, \ldots, x_n, \lambda$ on the closed set $D - \sigma(\varepsilon)$ for $|\lambda - \lambda_0| \leq h(\varepsilon)$.

2. The integral

$$\int_{\sigma(\varepsilon)} \cdots \int \left| F(x_1, x_2, \ldots, x_n, \lambda) \right| \, dx_1 \, dx_2 \ldots dx_n$$

is less than ε for all values of λ in the interval $\lambda_0 - h(\varepsilon) \leq \lambda \leq \lambda_0 + h(\varepsilon)$.

3. The domain $D - \sigma(\varepsilon)$ is bounded (this condition is not needed if D is bounded).

LEMMA 1. *If the integral converges uniformly at $\lambda = \lambda_0$, then it is a continuous function of λ at this point.*

For,

$$\left| \int_D \cdots \int F(x_1, \ldots, x_n, \lambda_1)\, dx_1 \ldots dx_n - \int_D \cdots \int F(x_1, \ldots, x_n, \lambda_0)\, dx_1 \ldots dx_n \right|$$

$$\leq \left| \int_{D-\sigma} \cdots \int [F(x_1, \ldots, x_n, \lambda_1) - F(x_1, \ldots, x_n, \lambda_0)]\, dx_1 \ldots dx_n \right|$$

$$+ \left| \int_\sigma \cdots \int F(x_1, \ldots, x_n, \lambda_1)\, dx_1 \ldots dx_n \right|$$

$$+ \left| \int_\sigma \cdots \int F(x_1, \ldots, x_n, \lambda_0)\, dx_1 \ldots dx_n \right|$$

$$\leq \int_{D-\sigma} \cdots \int \left| F(x_1, \ldots, x_n, \lambda_1) - F(x_1, \ldots, x_n, \lambda_0) \right| dx_1 \ldots dx_n$$

$$+ \int_\sigma \cdots \int \left| F(x_1, \ldots, x_n, \lambda_1) \right| dx_1 \ldots dx_n$$

$$+ \int_\sigma \cdots \int \left| F(x_1, \ldots, x_n, \lambda_0) \right| dx_1 \ldots dx_n.$$

We choose $\sigma(\varepsilon)$ and $h(\varepsilon)$ so that the second and third addends are each less than $\varepsilon/3$. Then, since a function which is continuous on a bounded closed set is uniformly continuous, we can choose $|\lambda_1 - \lambda_0|$ so small that, for all points $x_1, \ldots, x_n$ belonging to the domain $D - \sigma$, we have

$$\left| F(x_1, \ldots, x_n, \lambda_1) - F(x_1, \ldots, x_n, \lambda_0) \right| < \frac{\varepsilon}{3(D - \sigma)}$$

where $D - \sigma$ denotes the measure (volume) of the domain $D - \sigma$. We then have

$$\left| \int_D \cdots \int F(x_1, \ldots, x_n, \lambda_1) - \int_D \cdots \int F(x_1, \ldots, x_n, \lambda)\, dx_1 \ldots dx_n \right| < \varepsilon$$

and the lemma is proved.

An integral (7.1) which converges uniformly for any λ in a certain interval will be a continuous function of λ in this interval.

If it is possible over the whole of this interval to choose the open set $\sigma(\varepsilon)$ to be the same and independent of the value of λ, then the integral will be uniformly convergent over the whole interval in the sense of the usual definition in analysis.

In some cases it is convenient to generalize the concept of uniform convergence of the integral for a given value of parameter to meet the situation where we take as parameter some point P with coordinates $(\tilde{x}_1, \tilde{x}_2, \dots, \tilde{x}_n)$ in n-dimensional space. Consider the integral

$$\int_D \cdots \int F(x_1, \dots, x_n, P)\, dx_1 \dots dx_n. \tag{7.2}$$

We shall say that this integral *converges uniformly at the point* P_0 if, for any $\varepsilon > 0$, we can find a neighbourhood $h(\varepsilon)$ about the point P_0 and a part $\sigma(\varepsilon)$ of the domain D such that:

1. The function F is continuous when the point P is in the neighbourhood $h(\varepsilon)$ of the point P_0 and the point Q with coordinates $(x_1, \dots, x_n)$ is on the closed set $D - \sigma(\varepsilon)$.
2. The integral $\int_\sigma \dots \int |F(x_1, \dots, x_n, P)|\, dx_1 \dots dx_n$ is less than ε for all P of $h(\varepsilon)$.
3. The domain $D - \sigma$ is bounded.

For this case the following lemma holds good.

LEMMA 2. *If the integral converges uniformly at* $P = P_0$, *then it is a continuous function of* P *at this point* P_0.

The proof is a repetition of the argument for Lemma 1.

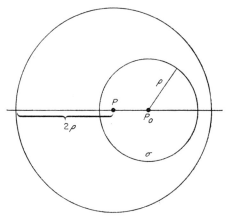

FIG. 10.

We shall often have to use the following sufficient criterion for uniform convergence. Let a function $F(x_1, x_2, \dots x_n, P)$ given in a bounded domain D satisfy the condition

$$\left| F(x_1, \dots, x_n, P) \right| < \frac{A}{r^{n-a}},$$

where $a > 0$, A is a certain constant, and r denotes the distance from the point $P(\tilde{x}_1, \tilde{x}_2, \ldots, \tilde{x}_n)$ to the point $Q(x_1, \ldots, x_n)$.

It is supposed that F depends continuously on the $x_1, \ldots, x_n, \tilde{x}_1, \ldots, \tilde{x}_n$ everywhere except at points where P coincides with Q.

It is easy to establish the uniform convergence of the integral

$$\int_D \cdots \int F(x_1, \ldots, x_n, P)\, dx_1 \ldots dx_n$$

at any point P_0 lying inside D. To do this, it is sufficient to take as σ a sphere, centre at P_0 (Fig. 10), and radius ϱ such that

$$\int_{r \leqq 2\varrho} \cdots \int \frac{A}{r^{n-a}}\, dx_1 \ldots dx_n < \varepsilon$$

and for $h(\varepsilon)$ a sphere of radius $\varrho_1 < \varrho$. For, we then have

$$\left| \int_\sigma \cdots \int F(x_1, \ldots, x_n, P)\, dx_1 \ldots dx_n \right| \leqq \int_\sigma \cdots \int |F|\, dx_1 \ldots dx_n$$

$$< \int_{r \leqq 2\varrho} \cdots \int \frac{A}{r^{n-a}}\, dx_1 \ldots dx_n < \varepsilon.\dagger$$

It is clear that in $D - \sigma$ and on its boundary, the function F will be a continuous function of its arguments for points P of the neighbourhood $h(\varepsilon)$.

§ 2. The Derivative of an Improper Integral with respect to a Parameter

Suppose we have an integral

$$\psi(\lambda) = \int_D \cdots \int F(x_1, x_2, \ldots, x_n, \lambda)\, dx_1\, dx_2 \ldots dx_n$$

over a domain D which does not depend on the parameter λ. In analysis we often encounter the formula

$$\frac{\partial}{\partial \lambda} \int_D \cdots \int F(x_1, x_2, \ldots, x_n, \lambda)\, dx_1\, dx_2 \ldots dx_n$$

$$= \int_D \cdots \int \frac{\partial F}{\partial \lambda}\, dx_1\, dx_2 \ldots dx_n = \varphi(\lambda)$$

† It is clear that a sphere of radius $r \leqslant 2\varrho$ with centre at the point P will contain the sphere P_0 (Fig. 10) if the point P is no further from P_0 than $\varrho_1 < \varrho$.

which is valid if the integral on the right-hand side is proper. It will be useful to us to get rid of this premise. We shall prove the following theorem.

THEOREM. *Suppose*

(a) *the function $F(x_1, \ldots, x_n, \lambda)$ is the improper integral with respect to λ of the function $f(x_1, \ldots, x_n, \lambda)$ in the domain D of the variables $x_1, x_2, \ldots, x_n$ for $a \leqq \lambda \leqq b$, i.e.,*

$$F(x_1, \ldots, x_n, \lambda) - F(x_1, \ldots, x_n, a) = \int_a^\lambda f(x_1, \ldots, x_n, \lambda) \, d\lambda$$

for all $x_1, x_2, \ldots, x_n$ of D except perhaps points of some set of measure zero. (For the most part this set will consist of separate points, curves, or surfaces.)

(b) *the function f is a summable function of the $(n + 1)$ variables $(x_1, \ldots, x_n, \lambda)$ i.e., the $(n + 1)$-tuple integral*

$$\int_a^\lambda \int_D \cdots \int f(x_1, x_2, \ldots, x_n, \lambda) \, dx_1 \ldots dx_n \, d\lambda$$

exists for $a \leqq \lambda \leqq b$.

(c) *the function $\varphi(\lambda) = \int_D \ldots \int f(x_1, x_2, \ldots, x_n, \lambda) \, dx_1 \ldots, dx_n$ is continuous at $\lambda = \lambda_0$.*

Then the following relation will hold:

$$\left[\frac{\partial \varphi(\lambda)}{\partial \lambda}\right]_{\lambda = \lambda_0} = \left[\frac{\partial}{\partial \lambda} \int_D \cdots \int F(x_1, \ldots, x_{n,\lambda}) \, dx_1 \ldots dx_n\right]_{\lambda = \lambda_0}$$

$$= \int_D \cdots \int f(x_1, \ldots, x_n, \lambda_0) \, dx_1 \ldots dx_n.$$

The proof of this theorem presents no difficulty. By definition,

$$\left[\frac{\partial \psi}{\partial \lambda}\right]_{\lambda = \lambda_0} = \lim_{h \to 0} \frac{1}{h} \int_D \cdots \int [F(x_1, \ldots, x_n, \lambda_0 + h)$$

$$- F(x_1, \ldots, x_n, \lambda_0)] \, dx_1 \ldots dx_n.$$

By the Lebesgue–Fubini theorem, on change of order of integration (Lecture 6), the integral on the right-hand side is equal to the $(n + 1)$-tuple integral

$$\int_{\substack{D \\ \{\lambda_0 < \lambda < \lambda_0 + h\}}} \cdots \int f \, dx_1 \, dx_2 \ldots dx_n \, d\lambda,$$

and this last integral can be expressed as

$$\int_{\lambda_0}^{\lambda+h} \left(\int_D \cdots \int f \, dx_1 \, dx_2 \, \ldots \, dx_n \right) d\lambda.$$

Using the property of indefinite Lebesgue integrals at a point of continuity of the function under the integral sign (see Lecture 6, Theorem 16), we have

$$\left[\frac{\partial \psi}{\partial \lambda} \right]_{\lambda=\lambda_0} = \varphi(\lambda_0),$$

which is what we had to prove.

We shall often have to make use of this theorem for some integral or another. Usually it is easy to verify that the condition (*a*) holds, but we shall give several simpler criteria to satisfy the condition (*b*).

Criterion 1. Conditions (*b*) and (*c*) of the theorem will be satisfied if the integral

$$\int_D \cdots \int f(x_1, x_2, \ldots, x_n, \lambda) \, dx_1 \, dx_2 \, \ldots \, dx_n$$

converges uniformly in the closed interval $a \leqq \lambda \leqq b$.

The proof is easy and we leave it to the reader.

Criterion 2. Condition (*b*) will be satisfied if the function $f(x_1, x_2, \ldots, x_n, \lambda)$ ($n \geqq 2$) is continuous in all its arguments everywhere with the exception of a smooth curve, continuous in λ, given by:

$$x_1 = a_1(\lambda), \quad x_2 = a_2(\lambda), \ldots, x_n = a_n(\lambda), \quad a \leqq \lambda \leqq b, \qquad (7.3)$$

and if in the neighbourhood of this curve the inequality

$$\left| f(x_1, x_2, \ldots, x_n, \lambda) \right| \leqq A \varrho^{-n+\varepsilon}$$

is satisfied, where

$$\varrho = \min_{a \leqq \lambda \leqq b} \sqrt{\sum_{i=1}^{n} [x_i - a_i(\lambda)]^2}, \quad \varepsilon > 0,$$

and if the domain D is bounded.

We separate the curve (7.3) from the $(n + 1)$-dimensional domain

$$\begin{cases} D \\ a \leqq \lambda \leqq b \end{cases}$$

by means of a domain σ_0 at the points of which $\varrho < \varrho_0$, where ϱ_0 is a constant. The integral

$$\int_{\left\{ \substack{D \\ a \leqq \lambda \leqq b} \right\}} \cdots \int \left| f(x_1, x_2, \ldots, x_n, \lambda) \right| \, dx_1 \, \ldots \, dx_n \, d\lambda \qquad (7.4)$$

will be convergent if the limit of the integral

$$\int_{\{a \leq \lambda \leq b\}}^{D} \cdots \int_{-\sigma_0} |f(x_1, x_2, \ldots, x_n, \lambda)| \, dx_1 \ldots dx_n \, d\lambda$$

exists when $\varrho_0 \to 0$. This limit will exist if, as $\varrho_0 \to 0$, the integral

$$\int_{\sigma_0} \cdots \int |f(x_1, x_2, \ldots, x_n, \lambda)| \, dx_1 \ldots dx_n \, d\lambda$$

tends to zero uniformly with respect to λ. We have

$$\int_{\sigma_0} \cdots \int |f(x_1, x_2, \ldots, x_n, \lambda)| \, dx_1 \, dx_2 \ldots dx_n \, d\lambda$$

$$\leq \int_a^b d\lambda \int_{0 < \varrho \leq \varrho_0} \cdots \int A\varrho^{-n+\varepsilon} \, dx_1 \ldots dx_n.$$

But

$$\int_{0 < \varrho \leq \varrho_0} \cdots \int \varrho^{-n+\varepsilon} \, dx_1 \ldots dx_n = c_n \varrho_0^\varepsilon:$$

from which it is clear that the integral investigated converges, and also that the integral

$$\int_D \cdots \int f(x_1, \ldots, x_n, \lambda) \, dx_1 \ldots dx_n$$

converges uniformly with respect to the parameter λ.

Criterion 3. If the domain D is not bounded, then we must examine the behaviour of the function at infinity as well as the singularity considered in Criterion 2.

The convergence of the integral (7.4) will be guaranteed if

$$|F(x_1, x_2, \ldots, x_n, \lambda)| \leq Ar^{-n-\varepsilon}$$

where

$$\varepsilon > 0 \quad \text{and} \quad r = \sqrt{x_1^2 + x_2^2 + \cdots + x_n^2}.$$

The reader will easily be able to prove the validity of this criterion.

THE EQUATION OF HEAT CONDUCTION

§ 1. Principal Solution

Having already considered a few problems concerning equations with two independent variables, we now pass on to equations in several independent variables.

We first consider the equation

$$\frac{\partial^2 u}{\partial x^2} + \frac{\partial^2 u}{\partial y^2} + \frac{\partial^2 u}{\partial z^2} = \frac{1}{a^2} \frac{\partial u}{\partial t} + f(x, y, z, t), \tag{8.1}$$

which determines the temperature at points of a certain homogeneous, isotropic body.

A change of variable using $t_1 = a^2 t$ enables us to get rid of the coefficient a^2. So for convenience we shall in the rest of this lecture take $a = 1$. And if we denote the sum of the second-order derivatives, as before, by $\nabla^2 u$, we can write the equation of heat conduction in the form

$$\nabla^2 u = \frac{\partial u}{\partial t} + f(x, y, z, t). \tag{8.1'}$$

It is this equation which we shall solve.

We are going to examine the problem of heat conduction in an unbounded medium. We shall solve Cauchy's problem for the equation (8.1′), *i.e.*, the problem of integrating the equation (8.1′) with the initial condition

$$[u]_{t=0} = \varphi(x, y, z). \tag{8.2}$$

As we pointed out in Lecture 2, this condition means that the distribution of temperature throughout the body is known at the initial moment of time.

A particular solution of the homogeneous equation of heat conduction (that is, equation (8.1′) with $f \equiv 0$) plays an essential part here; it is known as the *principal solution*.

Consider the function

$$v = \frac{1}{8\pi^{\frac{3}{2}}(t_0 - t)^{\frac{3}{2}}} e^{-\frac{r^2}{4(t_0 - t)}} \tag{8.3}$$

133

where

$$r = \sqrt{(x_0 - x)^2 + (y_0 - y)^2 + (z_0 - z)^2}.$$

We shall prove some lemmas about this function.

LEMMA 1. *The function v satisfies the equations*

$$\nabla_0^2 v - \frac{\partial v}{\partial t_0} = 0, \quad \nabla^2 v + \frac{\partial v}{\partial t} = 0 \qquad (8.4)$$

where

$$\nabla_0^2 v \equiv \frac{\partial^2 v}{\partial x_0^2} + \frac{\partial^2 v}{\partial y_0^2} + \frac{\partial^2 v}{\partial z_0^2}.$$

For, differentiation with respect to x_0 gives

$$\frac{\partial^2 v}{\partial x_0^2} = -\frac{1}{16\pi^{\frac{3}{2}}(t_0 - t)^{\frac{5}{2}}} e^{-\frac{r^2}{4(t_0 - t)}} + \frac{(x_0 - x)^2}{32\pi^{\frac{3}{2}}(t_0 - t)^{\frac{7}{2}}} e^{-\frac{r^2}{4(t_0 - t)}}$$

with similar terms for the derivatives with respect to y_0 and z_0. Hence

$$\nabla_0^2 v = \left(\frac{-3}{16\pi^{\frac{3}{2}}(t_0 - t)^{\frac{5}{2}}} + \frac{r^2}{32\pi^{\frac{3}{2}}(t_0 - t)^{\frac{7}{2}}} \right) e^{-\frac{r^2}{4(t_0 - t)}},$$

and moreover

$$\frac{\partial v}{\partial t_0} = \left(\frac{-3}{16\pi^{\frac{3}{2}}(t_0 - t)^{\frac{5}{2}}} + \frac{r^2}{32\pi^{\frac{3}{2}}(t_0 - t)^{\frac{7}{2}}} \right) e^{-\frac{r^2}{4(t_0 - t)}},$$

from which it is easily seen that the first of the equations (6.4) is satisfied. And the second of these equations may be deduced from the first, by noting that the function v depends only on the differences $(x_0 - x)$, $(y_0 - y)$, $(z_0 - z)$, $(t_0 - t)$, and consequently

$$\frac{\partial^2 v}{\partial x_0^2} = \frac{\partial^2 v}{\partial x^2}, \quad \frac{\partial v}{\partial t_0} = -\frac{\partial v}{\partial t}.$$

LEMMA 2. *If* $t_0 > t$, *then*

$$\int_{-\infty}^{+\infty} \int_{-\infty}^{+\infty} \int_{-\infty}^{+\infty} v \, dx \, dy \, dz = 1.$$

For, introducing polar coordinates r, θ, φ, making use of the symmetry of

v, and integrating by parts, we get

$$\int_{-\infty}^{+\infty}\int_{-\infty}^{+\infty}\int_{-\infty}^{+\infty} v \, dx \, dy \, dz = \int_0^{+\infty} r^2 \, dr \int_0^{2\pi}\int_0^{\pi} v \sin\theta \, d\theta \, d\varphi$$

$$= \int_0^{\infty} \frac{r^2}{2\pi^{\frac{1}{2}}(t_0 - t)^{\frac{3}{2}}} e^{-\frac{r^2}{4(t_0-t)}} \, dr = -\frac{1}{\pi^{\frac{1}{2}}(t_0 - t)^{\frac{1}{2}}} \int_0^{\infty} r \, d\left(e^{-\frac{r^2}{4(t_0-t)}}\right)$$

$$= \frac{1}{\pi^{\frac{1}{2}}(t_0 - t)^{\frac{1}{2}}} \int_0^{\infty} e^{-\frac{r^2}{4(t_0-t)}} \, dr = \frac{1}{\sqrt{\pi}} \int_{-\infty}^{+\infty} e^{-\frac{r^2}{4(t_0-t)}} \, d\left(\frac{r}{2(t_0 - t)^{\frac{1}{2}}}\right)$$

$$= \frac{1}{\sqrt{\pi}} \int_{-\infty}^{+\infty} e^{-z^2} \, dz.$$

Hence our lemma follows from the well-known result that

$$\int_{-\infty}^{+\infty} e^{-z^2} \, dz = \sqrt{\pi}. \tag{8.5}$$

Equation (8.5) may be proved, for instance, as follows:

$$\left(\int_{-\infty}^{+\infty} e^{-z^2} \, dx\right)^2 = \int_{-\infty}^{+\infty} e^{-x^2} \, dx \int_{-\infty}^{+\infty} e^{-y^2} \, dy = \int_{-\infty}^{+\infty}\int_{-\infty}^{+\infty} e^{-(x^2+y^2)} \, dx \, dy$$

$$= \int_0^{\infty} r e^{-r^2} \, dr \int_0^{2\pi} d\theta = \pi \int_0^{\infty} e^{-r^2} \, d(r^2) = -\pi[e^{-r^2}]_0^{\infty} = \pi.$$

Hence

$$\int_{-\infty}^{+\infty} e^{-z^2} \, dz = \sqrt{\pi}.$$

LEMMA 3. *Consider the integral $f(t_0 - t, x_0, y_0, z_0) = \iiint_{\Omega} v \, dx \, dy \, dz$ taken over some finite domain Ω. As $t_0 - t \to 0$, this integral tends to 1 if the point (x_z, y_0, z_0) lies within the domain Ω, and to 0 if this point lies outside the domain.*

We introduce new variables x_1, y_1, z_1 defined by

$$x - x_0 = x_1\sqrt{t_0 - t}, \quad y - y_0 = y_1\sqrt{t_0 - t}, \quad z - z_0 = z_1\sqrt{t_0 - t}.$$

Then

$$f(t_0 - t, x_0, y_0, z_0) = \frac{1}{8\pi^{\frac{3}{2}}} \int\int\int_{\Omega_1} e^{-\frac{1}{4}(x_1^2+y_1^2+z_1^2)} \, dx_1 \, dy_1 \, dz_1$$

where the domain of integration Ω_1 for the variables x_1, y_1, z_1 is obtained from the domain Ω of the variables x, y, z by a homothetic transformation with the coefficient of similitude $1/\sqrt{t_0 - t}$ and (x_0, y_0, z_0) as the centre of similitude, followed by a change of origin to the point (x_0, y_0, z_0). As $t_0 - t$ tends to zero, this domain will expand without limit.

If the point (x_0, y_0, z_0) lies inside the domain Ω, then the domain Ω_1 will expand in all directions and in the limit will occupy all space. By the previous lemma, we conclude that in this case the limit of the integral will be equal to 1.

If, on the other hand, the point (x_0, y_0, z_0) lies outside the domain Ω, then as $t_0 - t \to 0$ the domain Ω_1 will expand and in so doing will move away from the origin so that its distance from the origin in the coordinate system x_1, y_1, z_1 will increase without limit. By virtue of the convergence of the integral of v over an infinite space, which was established in the previous lemma, the limit of the integral $f(t_0 - t, x_0, y_0, z_0)$ in this case will be zero. Hence the lemma.

LEMMA 4. *Let $F(x, y, z)$ be any function which is continuous and bounded in a certain domain Ω (in particular, Ω may be the whole of space). Then*

$$\lim_{\xi \to 0} \frac{1}{8\pi^{\frac{3}{2}} \xi^{\frac{3}{2}}} \iiint_{\Omega} e^{-\frac{r^2}{4\xi}} F(x, y, y) \, dx \, dy \, dz = F(x_0, y_0, z_0) \quad (8.6)$$

if the point (x_0, y_0, z_0) belongs to the domain Ω. In this formula the passage to the limit is uniform with respect to (x_0, y_0, z_0) in any finite domain Ω_1 lying together with its boundary within Ω. It is here assumed that ξ remains positive as it tends to zero.
Proof. We form the difference

$$\frac{1}{8\pi^{\frac{3}{2}} \xi^{\frac{3}{2}}} \iiint_{\Omega} e^{-\frac{r^2}{4\xi}} F(x, y, z) \, dx \, dy \, dz$$

$$- F(x_0, y_0, z_0) \frac{1}{8\pi^{\frac{3}{2}} \xi^{\frac{3}{2}}} \iiint_{\Omega} e^{-\frac{r^2}{4\xi}} \, dx \, dy \, dz \quad (8.7)$$

$$= \frac{1}{8\pi^{\frac{3}{2}} \xi^{\frac{3}{2}}} \iiint_{\Omega} e^{-\frac{r^2}{4\xi}} [F(x, y, z) - F(x_0, y_0, z_0)] \, dx \, dy \, dz.$$

The second term of this difference tends to $F(x_0, y_0, z_0)$ by Lemma 3 and moreover, does so uniformly. To prove Lemma 4 it suffices to show that in any region $\Omega_1(x_0, y_0, z_0)$ this difference tends uniformly to zero as $\xi \to 0$.

We isolate the point (x_0, y_0, z_0) by a sphere of small radius δ. Let δ be so small that for $r \leqslant \delta$,

$$\left| F(x, y, z) - F(x_0, y_0, z_0) \right| < \frac{\varepsilon}{2},$$

this inequality being uniformly satisfied for all points (x_0, y_0, z_0) of the domain Ω_1. We shall evaluate the difference in (8.7) by breaking down the integral on the right-hand side into two integrals, one taken over the volume of the sphere $r \leqq \delta$ and the other over its exterior. We get

$$\left. \begin{aligned} & \iiint_{\Omega} v[F(x, y, z) - F(x_0, y_0, z_0)] \, dx \, dy \, dz \\ = {} & \iiint_{r \leqq \delta} v[F(x, y, z) - F(x_0, y_0, z_0)] \, dx \, dy \, dz \\ & + \iiint_{r \geqq \delta} v[F(x, y, z) - F(x_0, y_0, z_0)] \, dx \, dy \, dz. \end{aligned} \right\} \tag{8.8}$$

Estimating the first of these integrals, we have

$$\left| \iiint_{r \leqq \delta} v[F(x, y, z) - F(x_0, y_0, z_0)] \, dx \, dy \, dx \right|$$

$$\leqq \int_{-\infty}^{+\infty} \int_{-\infty}^{+\infty} \int_{-\infty}^{+\infty} \frac{\varepsilon}{2} v \, dx \, dy \, dz = \frac{\varepsilon}{2}.$$

Moreover, since F is bounded there is a constant M such that $|F| < M$, and consequently

$$\left| \frac{1}{8\pi^{\frac{3}{2}} \xi^{\frac{3}{2}}} \iiint_{r \geqq \delta} e^{-\frac{r^2}{4\xi}} [F(x, y, z) - F(x_0, y_0, z_0)] \, dx \, dy \, dz \right|$$

$$\leqq \frac{2M}{8\pi^{\frac{3}{2}} \xi^{\frac{3}{2}}} \iiint_{r \geqq \delta} e^{-\frac{r^2}{4\xi}} \, dx \, dy \, dz.$$

We change the variables in the last integral, putting

$$x = x_1 \sqrt{\xi}, \quad y = y_1 \sqrt{\xi}, \quad z = z_1 \sqrt{\xi}, \quad \text{so that} \quad r_1^2 = r^2/\xi,$$

and we have

$$\frac{2M}{8\pi^{\frac{3}{2}}} \iiint_{r_1 \geqq \delta/\sqrt{\xi}} e^{-\frac{1}{4} r_1^2} \, dx_1 \, dy_1 \, dz_1 = \frac{M}{\sqrt{\pi}} \int_{\delta/\sqrt{\xi}}^{\infty} r_1^2 \, e^{-\frac{1}{4} r_1^2} \, dr_1.$$

For sufficiently small ξ this integral is less than $\varepsilon/2$ because of the convergence of the integral

$$\int_0^\infty r_1^2\, e^{-\frac{1}{4}r_1^2}\, dr_1.$$

Hence the lemma.

Remark. We could consider, in place of the integral (8.6), the more general integral

$$\Phi(x_0, y_0, z_0, \xi) = \frac{1}{8\pi^{\frac{3}{2}}\xi^{\frac{3}{2}}} \int\int\int_\Omega e^{-\frac{r^2}{4\xi}} F(x, y, z, \xi)\, dx\, dy\, dz, \qquad (8.9)$$

in which the function $F(x, y, z, \xi)$ under the integral sign depends on the parameter ξ and is continuous and bounded for positive ξ:

$$|F(x, y, z, \xi)| < M.$$

Let

$$\lim_{\xi \to +0} F(x, y, z, \xi) = F(x, y, z), \qquad (8.10)$$

$F(x, y, z, \xi)$ tending to $F(x, y, z)$ uniformly in any bounded domain Ω_1 which, together with its boundary, lies within Ω. Then

$$\lim_{\xi \to +0} \Phi(x_0, y_0, z_0, \xi) = F(x_0, y_0, z_0)$$

at any interior point (x_0, y_0, z_0).

For, the difference

$$\Phi(x_0, y_0, z_0, \xi) - \frac{1}{8\pi^{\frac{3}{2}}\xi^{\frac{3}{2}}} \int\int\int_\Omega e^{-\frac{r^2}{4\xi}} F(x, y, z)\, dx\, dy\, dz$$

$$= \frac{1}{8\pi^{\frac{3}{2}}\xi^{\frac{3}{2}}} \int\int\int_\Omega e^{-\frac{r^2}{4\xi}} [F(x, y, z, \xi) - F(x, y, z)]\, dx\, dy\, dz.$$

We split the last integral up into two integrals, the first taken over the space inside a sphere σ defined by $(x - x_0)^2 + (y - y_0)^2 + (z - z_0)^2 \leqq \delta$, and the second taken over the space outside this sphere. Then, given any positive number ε, we get for sufficiently small ξ,

$$\left| \frac{1}{8\pi^{\frac{3}{2}}\xi^{\frac{3}{2}}} \int\int\int_\sigma e^{-\frac{r^2}{4\xi}} [F(x, y, z, \xi) - F(x, y, z)]\, dx\, dy\, dz \right|$$

$$\leqq \frac{\varepsilon}{8\pi^{\frac{3}{2}}\xi^{\frac{3}{2}}} \int\int\int_\sigma e^{-\frac{r^2}{4\xi}}\, dx\, dy\, dz < \varepsilon.$$

Further,

$$\left| \frac{1}{8\pi^{\frac{3}{2}} \xi^{\frac{3}{2}}} \int \int \int_{\Omega - \sigma} e^{-\frac{r^2}{4\xi}} [F(x, y, z, \xi) - F(x, y, z)] \, dx \, dy \, dz \right|$$

$$\leq \frac{M}{4\pi^{\frac{3}{2}} \xi^{\frac{3}{2}}} \int \int \int_{\Omega - \sigma} e^{-\frac{r^2}{4\xi}} \, dx \, dy \, dz.$$

The last expression tends to zero (see Lemma 4). Consequently,

$$\lim_{\xi \to +0} \Phi(x_0, y_0, z_0, \xi)$$

$$= \lim_{\xi \to +0} \left[\frac{1}{8\pi^{\frac{3}{2}} \xi^{\frac{3}{2}}} \int \int \int_{\Omega} e^{-\frac{r^2}{4\xi}} F(x, y, z) \, dx \, dy \, dz \right];$$

but by Lemma 4, the last limit is equal to $F(x_0, y_0, z_0)$, and so the assertion is true.

§ 2. The Solution of Cauchy's Problem

We now pass on to the solution of the equation (8.1′).

Suppose that for $t > 0$ a certain function u is given which has continuous derivatives with respect to the space coordinates up to the second order inclusive and a first-order derivative with respect to time. Suppose also that u and its first-order derivatives are bounded throughout space.

We apply Green's formula (5.16) to the functions u, v for values of t such that $0 \leq t < t_0$, taking as the domain of integration a sphere Ω bounded by the spherical surface S. We get

$$\int \int \int_{\Omega} (u \nabla^2 v - v \nabla^2 u) \, d\Omega = \int \int_S \left(v \frac{\partial u}{\partial n} - u \frac{\partial v}{\partial n} \right) dS. \qquad (8.11)$$

Further, it may easily be verified that

$$\int_0^{t_0} \left\{ \int \int \int_{\Omega} \left(u \frac{\partial v}{\partial t} + v \frac{\partial u}{\partial t} \right) d\Omega \right\} dt = \left[\int \int \int_{\Omega} uv \, d\Omega \right]_{+0}^{t_0 - 0}$$

for any two functions u, v such that the integrals

$$\int \int \int_{\Omega} uv \, d\Omega \quad \text{and} \quad \int \int \int_{\Omega} \left(u \frac{\partial v}{\partial t} + v \frac{\partial u}{\partial t} \right) d\Omega$$

are continuous functions of t in the interval $0 < t < t_0$. Combining this equation with (8.11), we get

$$\int_0^{t_0} \left\{ \iiint_\Omega \left[u \left(\nabla^2 v + \frac{\partial v}{\partial t} \right) - v \left(\nabla^2 u - \frac{\partial u}{\partial t} \right) \right] d\Omega \right\} dt$$

$$= \left[\iiint_\Omega uv \, d\Omega \right]_{+0}^{t_0 - 0} + \int_0^{t_0} \left\{ \iint_S \left(v \frac{\partial u}{\partial n} - u \frac{\partial v}{\partial n} \right) dS \right\} dt. \quad (8.12)$$

(8.12) holds identically for any functions satisfying the appropriate conditions. We now substitute for u the required solution of equation (8.1') and take v to be the particular solution of the equation of heat conduction which is given by (8.3). Then, using (8.2) and (8.4), we get

$$\int_0^{t_0} \left\{ \iint_S \left(v \frac{\partial u}{\partial n} - u \frac{\partial v}{\partial n} \right) dS \right\} dt + \left[\iiint_\Omega uv \, d\Omega \right]_{+0}^{t_0 - 0}$$

$$= - \int_0^{t_0} \left\{ \iiint_\Omega vf(x, y, z, t) \, d\Omega \right\} dt. \quad (8.13)$$

Now an integral of the form

$$\int_0^{t_0} \left\{ \int_{-\infty}^{+\infty} \int_{-\infty}^{+\infty} \int_{-\infty}^{+\infty} vF(x, y, z, t) \, d\Omega \right\} dt$$

will converge absolutely if $F(x, y, z, t)$ is a continuous, bounded function. For,

$$\int_0^{t_0} \left\{ \iiint_\Omega |vF(x, y, z, t)| \, d\Omega \right\} dt \leqq \int_0^{t_0} \left\{ \iiint_\Omega \max |F(x, y, z, t)| \, v d\Omega \right\} dt$$

$$\leqq \max |F(x, y, z, t)| \, t_0.$$

We shall assume at present that $f(x, y, z, t)$ and $\varphi(x, y, z)$ are continuous bounded functions. We shall further suppose that as $t \to t_0$, the solution $u(x, y, z, t)$ tends uniformly to its initial values in any bounded domain.

We now pass to the limit in equation (8.13), letting the radius of the sphere Ω tend to infinity. Then the integral over the surface S will tend to zero, since by hypothesis, both u and $\partial u/\partial n$ are bounded and $e^{-r^2/\xi}$ tends to zero more rapidly than any power of r as $r \to \infty$. We clearly get

$$\left[\int_{-\infty}^{+\infty} \int_{-\infty}^{+\infty} \int_{-\infty}^{+\infty} uv \, dx \, dy \, dz \right]_{+0}^{t_0 - 0}$$

$$= - \int_0^{t_0} \left\{ \int_{-\infty}^{+\infty} \int_{-\infty}^{+\infty} \int_{-\infty}^{+\infty} vf(x, y, z, t) \, dx \, dy \, dz \right\} dt.$$

By the remark made after Lemma 4,

$$\lim_{t \to t_0} \int_{-\infty}^{+\infty} \int_{-\infty}^{+\infty} \int_{-\infty}^{+\infty} uv \; dx \; dy \; dz = u(x_0, y_0, z_0, t_0);$$

it is clear that this integral is continuous for $t = 0$; hence

$$u(x_0, y_0, z_0, t_0) = \int_{-\infty}^{+\infty} \int_{-\infty}^{+\infty} \int_{-\infty}^{+\infty} \varphi(x, y, z) \; ([v]_{t=0}) \; dx \; dy \; dz$$

$$- \int_{0}^{t_0} \left\{ \int_{-\infty}^{+\infty} \int_{-\infty}^{+\infty} \int_{-\infty}^{+\infty} vf(x, y, z, t) \; dx \; dy \; dz \right\} \; dt. \quad (8.14)$$

Equation (8.14) gives an expression for the solution of our problem.

In deriving this formula, we have assumed that a solution to the problem does exist. Hence we should now verify whether the function u defined by (8.14) does in fact satisfy the equation (8.2) and the initial conditions.

It should be noted that the equation (8.2), which we have solved for $t > 0$ with initial conditions given for $t = 0$, may have no solution at all for $t < 0$. The solution (8.14) in any case will cease to have any significance in these circumstances.

Passing on now to the proof, we first show that the function u given by (8.14) has continuous derivatives up to some definite order. For, on changing the variables of integration by

$$\xi = x - x_0, \quad \eta = y - y_0, \quad \zeta = z - z_0, \quad \tau = t - t_0,$$

we can write (8.14) in the form

$$u(x_0, y_0, z_0, t_0)$$

$$= \int_{-\infty}^{+\infty} \int_{-\infty}^{+\infty} \int_{-\infty}^{+\infty} v(\xi, \eta, \zeta, t_0) \; \varphi(x_0 + \xi, y_0 + \eta, z_0 + \zeta) \; d\xi \; d\eta \; d\zeta$$

$$- \int_{0}^{t_0} \int_{-\infty}^{+\infty} \int_{-\infty}^{+\infty} \int_{-\infty}^{+\infty} v(\xi, \eta, \zeta, \tau) \times$$

$$\times f(x_0 + \xi, y_0 + \eta, z_0 + \zeta, t_0 - \tau) \; d\xi \; d\eta \; d\zeta \; d\tau,$$

where

$$v(\xi, \eta, \zeta, \tau) = \frac{1}{8\pi^{\frac{3}{2}} \tau^{\frac{3}{2}}} e^{-\frac{\xi^2 + \eta^2 + \zeta^2}{4\tau}}.$$

Now it is shown in calculus text-books that such a formula may be differentiated with respect to the parameters at least the same number of times as the order of derivatives of f and φ which remain bounded, since the integrals of derivatives of the functions under the integral sign converge uniformly. We shall assume that the functions f and φ have the required number of continuous, bounded derivatives.

When $f(x, y, z, t_0) = 0$ the formula (8.14) gives

$$u_1(x_0, y_0, z_0, t_0) = \frac{1}{8\pi^{\frac{3}{2}} t_0^{\frac{3}{2}}} \int_{-\infty}^{+\infty} \int_{-\infty}^{+\infty} \int_{-\infty}^{+\infty} e^{-\frac{r^2}{4t_0}} \varphi(x, y, z) dx \, dy \, dz$$

(8.15)

where u_1 denotes the solution of the equation

$$\nabla_0^2 u_1 - \frac{\partial u_1}{\partial t_0} = 0$$

which satisfies the condition

$$[u_1]_{t_0=0} = \varphi(x_0, y_0, z_0).$$

By Lemma 4, the right-hand side of (8.15) is a function which in any bounded domain tends uniformly to $\varphi(x_0, y_0, z_0)$ as $t_0 \to 0$. Moreover, it is not difficult to see that the integral may be differentiated with respect to the variables x_0, y_0, z_0, t_0 as often as we like provided only that $t_0 > 0$. We get

$$\nabla_0^2 u_1(x_0, y_0, z_0, t_0) - \frac{\partial u_1(x_0, y_0, z_0, t_0)}{\partial u_0}$$

(8.16)

$$= \int_{-\infty}^{+\infty} \int_{-\infty}^{+\infty} \int_{-\infty}^{+\infty} \left(\nabla_0^2 v - \frac{\partial v}{\partial t_0} \right) \varphi(x, y, z) \, dx \, dy \, dz.$$

By Lemma 1 we see that $u_1(x_0, y_0, z_0, t_0)$ satisfies the equation (8.1′) when $f(x_0, y_0, z_0, t_0) = 0$, as we had to show. The uniqueness of the bounded, smooth solution, *i.e.*, of a solution having the required number of continuous derivatives, follows immediately from the arguments which led us to the formula (8.14). Thus the formula (8.14), when $f = 0$, gives the solution of the problem.

It remains to show that the second addend in (8.14), *i.e.*, the function

$$u_2(x_0, y_0, z_0, t_0) = -\int_0^{t_0} \left\{ \int_{-\infty}^{+\infty} \int_{-\infty}^{+\infty} \int_{-\infty}^{+\infty} vf(x, y, z, t) \, dx \, dy \, dz \right\} dt,$$

satisfies the equation (8.1) and the condition $[u_2]_{t_0=0} = 0$. It will then follow that the function u satisfies equation (8.1) and the initial condition (8.2).

We introduce an auxiliary variable w defined for $t_0 > t$ by

$$w(x_0, y_0, z_0, t_0, t) = -\int_{-\infty}^{+\infty} \int_{-\infty}^{+\infty} \int_{-\infty}^{+\infty} vf(x, y, z, t) \, dx \, dy \, dz.$$

(8.17)

The function $w(x_0, y_0, z_0, t_0, t)$ satisfies the equation

$$\nabla_0^2 w - \frac{\partial w}{\partial t_0} = 0 \tag{8.18}$$

and the condition

$$[w]_{t_0=t} = -f(x_0, y_0, z_0, t).$$

The function $u_2(x_0, y_0, z_0, t_0)$ can be expressed in terms of w as

$$u_2(x_0, y_0, z_0, t_0) = \int_0^{t_0} w(x_0, y_0, z_0, t_0, t)\, \mathrm{d}t.$$

By its construction, it is clear that $u_2(x_0, y_0, z_0, 0) = 0$.

We now work out the expression for

$$\nabla_0^2 u_2 - \frac{\partial u_2}{\partial t_0}. \tag{8.19}$$

Differentiating u_2 under the integral sign with respect to x_0, y_0, z_0, we get

$$\nabla_0^2 u_2 = \int_0^{t_0} \nabla_0^2 w(x_0, y_0, z_0, t_0, t)\, \mathrm{d}t.$$

Also,

$$\frac{\partial u_2}{\partial t_0} = [w(x_0, y_0, z_0, t_0, t)]_{t=t_0} + \int_0^{t_0} \frac{\partial w(x_0, y_0, z_0, t_0, t)}{\partial t_0}\, \mathrm{d}t,$$

and hence

$$\nabla_0^2 u_2 - \frac{\partial u_2}{\partial t} = f(x, y, z, t),$$

as was to be shown.

The question whether the Cauchy problem which we have just considered is correctly formulated may be immediately resolved by analysing formula (8.14).

It is obvious that small changes in φ or f will have little influence on the solution.

There is continuous dependence of order $(0, 0)$ (see Lecture 2, § 2).

It is now time to mention another important circumstance.

If the free term in equation (8.1) is zero (this implies that there are no heat sources), then the solution of the Cauchy problem which we have just examined is a function which may be differentiated with respect to x_0, y_0, z_0 as often as we please, quite independently of whether the function φ has derivatives or not.

This smoothness of its solution distinguishes the equation of heat conduction in an essential way from, say, the equation for a vibrating string or from the wave equation.

For, the d'Alembert solution $u = \varphi(x - at)$ for the equation of a vibrating string cannot be differentiated with respect to x or t more than twice. This same function is also a solution of the wave equation in two or three independent variables, since, clearly,

$$\frac{\partial^2 u}{\partial y^2} = \frac{\partial^2 u}{\partial z^2} = 0.$$

Consequently the solutions of these equations are no smoother than the initial conditions are.

We have discussed the solution of the equation (8.1') in a space of three dimensions. In an exactly similar way the solution of the equation

$$\frac{\partial^2 u}{\partial x^2} + \frac{\partial^2 u}{\partial y^2} = \frac{\partial u}{\partial t} + f(x, y, t) \tag{8.20}$$

or of

$$\frac{\partial^2 u}{\partial x^2} = \frac{\partial u}{\partial t} + f(x, t) \tag{8.21}$$

may be examined. Without going into detailed consideration of these equations, we give the final results.

The principal solution of (8.20) will have the form

$$v = \frac{1}{4\pi(t_0 - t)} \, e^{-\frac{r^2}{t_0 - t}} \tag{8.22}$$

and the solution of (8.20) with the condition $[u]_{t=0} = \varphi(x, y)$ will have the form

$$u(x_0, y_0, t_0)$$

$$= \frac{1}{4\pi \, t_0} \int_{-\infty}^{+\infty} \int_{-\infty}^{+\infty} e^{-\frac{r^2}{4t_0}} \, \varphi(x, y) \, \mathrm{d}x \, \mathrm{d}y$$

$$- \frac{1}{4\pi} \int_0^{t_0} \left\{ \int_{-\infty}^{+\infty} \int_{-\infty}^{+\infty} \frac{1}{t_0 - t} \, e^{-\frac{r^2}{4(t_0 - t)}} \, f(x, y, t) \, \mathrm{d}x \, \mathrm{d}y \right\} \mathrm{d}t \tag{8.23}$$

In exactly the same way the principal solution of (8.21) will be

$$v = \frac{1}{2 \sqrt{\pi} \, \sqrt{t_0 - t}} \, e^{-\frac{(x - x_0)^2}{4(t_0 - t)}}$$

and the solution of (8.21) with the condition $[u]_{t=0} = \varphi(x)$ will take the form

$$u(x_0, t_0) = \frac{1}{2\sqrt{\pi}\sqrt{t_0}} \int_{-\infty}^{+\infty} e^{-\frac{(x-x_0)^2}{4t_0}} \varphi(x)\, dx$$

$$- \frac{1}{\sqrt{2\pi}} \int_{0}^{t_0} \frac{1}{\sqrt{t_0 - t}} \left\{ \int_{-\infty}^{+\infty} e^{-\frac{(x-x_0)^2}{4(t_0-t)}} f(x, t)\, dx \right\} dt. \quad (8.24)$$

We leave the reader to verify the correctness of these formulae.

LAPLACE'S EQUATION AND POISSON'S EQUATION

§ 1. The Theorem of the Maximum

We have seen that a whole series of questions in mathematical physics reduces to the solution of this or that equation of elliptical type. We shall deal next with the simplest of such equations: Laplace's equation

$$\nabla^2 u(x, y, z) = 0 \tag{9.1}$$

and Poisson's equation

$$\nabla^2 u(x, y, z) = -4\pi\varrho(x, y, z). \tag{9.2}$$

Any function which has continuous second-order derivatives and which satisfies Laplace's equation in a certain domain is called a *harmonic function* in this domain.

Before going on to the solution of problems related to these equations, we shall examine certain general properties of the solutions of these equations.

LEMMA 1. *If the function $\varrho(x, y, z)$ is positive at a point (x_0, y_0, z_0) lying within the domain where the equation (9.2) is defined, then the solution of this equation cannot attain its minimum value at this point.*

For, if the function $u(x, y, z)$ satisfying equation (9.2) attained a minimum at this point, then it would attain a minimum with respect to each variable separately at this point. But then all the first-order derivatives of u would have to be zero at this point, and the second-order derivative with respect to each variable would be non-negative. Consequently the sum of the second-order derivatives would have to be non-negative, and this contradicts the hypothesis that $\varrho(x_0, y_0, z_0)$ is positive.

Hence the lemma.

COROLLARY. *If $\varrho(x, y, z)$ is negative at the point (x_0, y_0, z_0), then $u(x, y, z)$ cannot attain its maximum at this point.*

This is proved by changing the signs of ϱ and u.

THEOREM 1. *A harmonic function which is given in a certain domain Ω and is continuous up to the boundary S cannot take anywhere within Ω values which*

146

are greater than its greatest value on the boundary nor values which are less than its least value on the boundary.

Suppose, if possible, that

$$u(x_0, y_0, z_0) > u_S + \varepsilon,$$

where u_S is the value of u at an arbitrary point on the boundary of the domain.

The function

$$v = u + \eta r^2,$$

where

$$r^2 = (x - x_0)^2 + (y - y_0)^2 + (z - z_0)^2,$$

and η is some positive constant, is such that

$$\nabla^2 v = \nabla^2 u + 6\eta = 6\eta$$

and is hence a solution of Poisson's equation (9.2) with positive ϱ. But v, for a sufficiently small η, will take a value at (x_0, y_0, z_0) which is still greater than max v_S.

For,

$$v(z_0, y_0, z_0) = u(x_0, y_0, z_0)$$

and by hypothesis

$$u(x_0, y_0, z_0) > u_S + \varepsilon = v_S + \varepsilon - \eta r^2.$$

Choosing η to be so small that, throughout Ω,

$$\varepsilon - \eta r^2 > \tfrac{1}{2}\varepsilon$$

we get

$$v(x_0, y_0, z_0) > v_S + \tfrac{1}{2}\varepsilon,$$

and consequently v will attain its maximum somewhere within the domain. This contradicts Lemma 1, so our supposition was false.

The second part of the theorem is proved by changing u to $-u$.

We shall show later that a harmonic function which takes a value inside the domain which is equal to its maximum or minimum value on the boundary is a constant.

COROLLARY 1. *A harmonic function which is equal to zero on the boundary of a finite domain is identically zero throughout this domain.*

Hence it follows that two harmonic functions which take equal values at points of the boundary of a domain will have equal values everywhere within their domain. For their difference is a harmonic function and is equal to zero on the boundary and consequently throughout the domain.

COROLLARY 2. *If a sequence of functions u_n, which are harmonic in a domain Ω and continuous up to the boundary, converges uniformly on the boundary S of this domain, then it will converge uniformly throughout this closed domain.*

This follows because $|u_{n_1} - u_{n_2}| < \varepsilon(N)$ for $n_1 > N$ and $n_2 > N$ for every point on the boundary and hence also throughout the domain. Consequently, by Cauchy's test, the sequence u_n converges uniformly throughout the closed domain.

§ 2. The Principal Solution. Green's Formula

In order to investigate the properties of harmonic functions more deeply, we shall prove some more lemmas.

LEMMA 2. *The function*

$$\frac{1}{r} = \frac{1}{\sqrt{(x - x_0)^2 + (y - y_0)^2 + (z - z_0)^2}}$$

satisfies Laplace's equation:

$$\nabla^2 \left(\frac{1}{r} \right) = 0 \qquad (9.3)$$

except at the point (x_0, y_0, z_0) where $1/r$ becomes infinite.
For,

$$\frac{\partial^2}{\partial x^2} \left(\frac{1}{r} \right) = -\frac{1}{r^3} + \frac{3(x - x_0)^2}{r^5}$$

$$\frac{\partial^2}{\partial y^2} \left(\frac{1}{r} \right) = -\frac{1}{r^3} + \frac{3(y - y_0)^2}{r^5}$$

$$\frac{\partial^2}{\partial z^2} \left(\frac{1}{r} \right) = -\frac{1}{r^3} + \frac{3(z - z_0)^2}{r^5}$$

Equation (9.3) follows by addition.

LEMMA 3. *If the function u is continuous, and has continuous first-order derivatives throughout a domain Ω and on its boundary S, and has second-order derivatives continuous within Ω, then*

$$\iiint_{\Omega} \frac{1}{r} \nabla^2 u \, d\Omega = \iint_{S} \left(u \frac{\partial}{\partial n} \left(\frac{1}{r} \right) - \frac{1}{r} \frac{\partial u}{\partial n} \right) dS - 4\pi u(x_0, y_0, z_0).$$

$$(9.4)$$

Here $\partial/\partial n$ denotes, as usual, the derivative along the inward normal to the surface S. The surface S of the domain Ω satisfies the conditions of § 1 in

Lecture 1, *i.e.*, it is piecewise smooth and is intersected by any straight line parallel to a coordinate axis either in a finite number of points or in a finite number of whole intervals.

We shall suppose at first that u has continuous second-order derivatives throughout Ω and on its boundary S. To prove the lemma, we cut out of the domain Ω a sphere ω with radius ε and centre at the point (x_0, y_0, z_0), and apply Green's formula to the remaining domain. Then

$$
\iiint_\Omega \frac{1}{r} \nabla^2 u \, d\Omega = \lim_{\varepsilon \to 0} \iiint_{\Omega - \omega} \frac{1}{r} \nabla^2 u \, d\Omega
$$

$$
= \lim_{\varepsilon \to 0} \iiint_{\Omega - \omega} \left\{ \frac{1}{r} \nabla^2 u - u \nabla^2 \left(\frac{1}{r} \right) \right\} d\Omega
$$

$$
= \iint_S \left\{ u \frac{\partial}{\partial n} \left(\frac{1}{r} \right) - \frac{1}{r} \frac{\partial u}{\partial n} \right\} dS
$$

$$
+ \lim_{\varepsilon \to 0} \iint_\sigma \left\{ u \frac{\partial}{\partial n} \left(\frac{1}{r} \right) - \frac{1}{r} \frac{\partial u}{\partial n} \right\} dS \qquad (9.5)
$$

where σ is the surface of the sphere ω.

It is not difficult to see that on the surface σ

$$
\frac{\partial}{\partial n} \left(\frac{1}{r} \right) = \frac{d}{dr} \left(\frac{1}{r} \right) = -\frac{1}{r^2} = -\frac{1}{\varepsilon^2},
$$

and consequently,

$$
\iint_\sigma u \frac{\partial}{\partial n} \left(\frac{1}{r} \right) dS = -\frac{1}{\varepsilon^2} \iint_\sigma u \, dS
$$

$$
= -\frac{1}{\varepsilon^2} \iint_\sigma u(x_0, y_0, z_0) \, dS - \frac{1}{\varepsilon^2} \iint_\sigma \eta \, dS,
$$

where η tends uniformly to zero in ω as $\varepsilon \to 0$:

$$
|\eta| < \delta(\varepsilon), \quad \text{and} \quad \delta(\varepsilon) \to 0 \quad \text{if} \quad \varepsilon \to 0;
$$

and

$$
\iint_\sigma u(x_0, y_0, z_0) \, dS = 4\pi \varepsilon^2 u(x_0, y_0, z_0); \quad \left| \iint_\sigma \eta \, dS \right| \leq \delta(\varepsilon) 4\pi \varepsilon^2.
$$

Since $\partial u / \partial n$ is bounded, $|\partial u / \partial n| < k$, and so

$$
\left| \iint_\sigma \frac{1}{r} \frac{\partial u}{\partial n} \, dS \right| \leq \frac{1}{\varepsilon} \iint_\sigma k \, dS \leq 4\pi k \varepsilon
$$

and

$$\lim_{\varepsilon \to 0} \int\int_\sigma \left\{ \frac{\partial}{\partial n}\left(\frac{1}{r}\right) u - \frac{1}{r}\frac{\partial u}{\partial n} \right\} dS = -4\pi u(x_0, y_0, z_0).$$

Substituting this expression in (9.5), we obtain the required result.

To get rid of the requirement that the second-order derivatives shall be continuous on the boundary of the domain, we replace the domain Ω by a domain Ω_1 which, together with its boundary, lies within Ω. Then by first applying (9.4) to the domain Ω_1 and then letting $\Omega_1 \to \Omega$, we obtain the required result.

Formula (9.4) was deduced on the assumption that the point (x_0, y_0, z_0) belongs to the domain Ω. If (x_0, y_0, z_0) lies outside the domain, then the formula becomes

$$\int\int\int_\Omega \frac{1}{r} \nabla^2 u \, d\Omega = \int\int_s \left\{ u \frac{\partial}{\partial n}\left(\frac{1}{r}\right) - \frac{1}{r}\frac{\partial u}{\partial n} \right\} dS. \qquad (9.6)$$

This may easily be established by repeating the whole argument, noting that there is now no need to introduce the integral over a domain σ.

The proof can also be extended without difficulty to cover the case when the derivatives have a finite number of points of discontinuity within the domain but otherwise satisfy the conditions of the theorem.

The name *Green's formulae* is often given to (9.4) and (9.6). The likeness will readily be seen between Green's formula (9.4) and the similar formula (8.14) which we obtained as the principal solution of the equation of heat conduction. Here then *the function* $1/r$ *is the principal solution of Laplace's equation.*

§ 3. The Potential due to a Volume, to a Single Layer, and to a Double Layer

If by some previous consideration we know the values of the u, $\nabla^2 u$ and $\partial u/\partial n$ which appear in Green's formula (9.4),

$$\nabla^2 u = -4\pi\varrho, \quad [u]_s = f_1, \quad \left[\frac{\partial u}{\partial n}\right]_s = f_2,$$

then the formula gives us the unknown function u explicitly:

$$u(x_0, y_0, z_0) = \int\int\int_\Omega \frac{\varrho}{r} \, dx \, dy \, dz + \frac{1}{4\pi}\int\int_s \frac{\partial}{\partial u}\left(\frac{1}{r}\right) f_1 \, dS$$

$$- \frac{1}{4\pi}\int\int_s \frac{1}{r} f_2 \, dS. \qquad (9.7)$$

But we cannot specify f_1 and f_2 completely arbitrarily. Hence (9.7) does not allow us to construct a solution of equation (9.2) which shall itself have arbitrary values on the boundary and also have arbitrary values for its normal derivative thereon, as we did for the problem of a vibrating string.

We give special names to the integrals which appear in (9.7). We call the integral $\iiint_\Omega \varrho/r \, dx \, dy \, dz$ *the Newtonian potential* and the function ϱ the *density* of this potential. Similarly, we call the integral $\iint_S f_2/r \, dS$ the *potential of the single layer* with density f_2, and the integral $\iint_S f_1(\partial/\partial n) \,[(1/r)] \, dS$ the *potential of the double layer* with density f_1.

The Newtonian potential has a very simple physical interpretation. Imagine that mass is distributed through the volume Ω with density ϱ. We calculate the attraction of this mass on a material particle. By Newton's law, a mass m at the point (x, y, z) attracts a unit mass at (x_0, y_0, z_0) with a force F of magnitude m/r^2 directed towards the point (x, y, z). In other words,

$$F = \text{grad}_0 \, U$$

where $U = \dfrac{m}{r} = \dfrac{m}{\sqrt{(x - x_0)^2 + (y - y_0)^2 + (z - z_0)^2}}.$

The function U whose gradient is equal to the attractive force is called the potential of the gravitational force. Hence we can call m/r the Newtonian potential of the mass m at the point (x, y, z).

By dividing the whole volume Ω into small elements and replacing the attraction of the mass $\varrho \Delta\Omega$ in the element $\Delta\Omega$ by that of an equal mass concentrated at some internal point this element, we obtain an expression for the force F acting on a unit mass concentrated at the point (x_0, y_0, z_0), *viz.*,

$$F = \sum \text{grad}_0 \, \frac{1}{r_\xi} \, \varrho \Delta\Omega$$

where $\text{grad}_0 \, (1/r_\xi)$ is the value of $\text{grad}_0 \, 1/r$ at some internal point of $\Delta\Omega$. Passing to the limit, we get

$$F = \iiint_\Omega \varrho \, \text{grad}_0 \, \frac{1}{r} \, d\Omega, \quad F_x = \iiint_\Omega \varrho \, \frac{x - x_0}{r^3} \, d\Omega,$$

$$F_y = \iiint_\Omega \varrho \, \frac{y - y_0}{r^3} \, d\Omega, \quad F_z = \iiint_\Omega \varrho \, \frac{z - z_0}{r^3} \, d\Omega.$$

Suppose $\varrho(x, y, z)$ is a continuous function. Then the quantities F_x, F_y, F_z will be uniformly convergent integrals and, by the theorem in §2 of Lecture 7, they are the derivatives with respect to x_0, y_0, z_0 respectively of the function

$$U = \iiint_\Omega \frac{\varrho}{r} \, d\Omega$$

and consequently

$$F = \mathrm{grad}_0\ U.$$

Thus U is the gravitational potential of a mass distributed with density ϱ over the volume Ω.

The same interpretation can be given to the potential of a single layer. This is the potential due to a gravitating mass distributed over the surface S at points outside the surface. If f_2 is the density of this mass, then, by replacing the effect of each surface element ΔS at the point (x_0, y_0, z_0) by $\mathrm{grad}_0\ (1/r_\xi) f_2\ \Delta S$ where $\mathrm{grad}_0\ (1/r_\xi)$ denotes the value of $\mathrm{grad}_0\ (1/r)$ at some mean point of the surface element, and by summing over all the elements ΔS and passing to the limit, we obtain for the total attractive force at the point (x_0, y_0, z_0) the expression

$$F = \mathrm{grad}_0\ U \quad \text{where } U = \iint_S \frac{f_2}{r}\ \mathrm{d}S$$

as we had to show.

We have considered ϱ and f_2 to be the density of masses gravitating according to Newton's law, but we could equally well have derived the same expression as an electric potential due to charges attracting or repelling according to Coulomb's law, or as a magnetic potential.

We now pass on to explain the geometrical significance of the potential due to a double layer, to which end we put the integral into a rather different form. We consider a surface, which, in general, is not closed. We select some point O and construct round it a sphere C of unit radius.

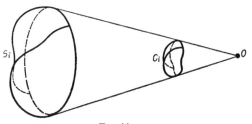

Fig. 11.

We suppose that the surface is two-sided, and we designate one side as the inner side, the other as the outer side, and in so doing we define the direction of the inward normal at all points of the surface. We suppose that the surface S can be divided into a finite number of parts $S_1, S_2, \ldots, S_k$ so that each part either meets each radius vector drawn from the point O in a single point or else forms part of a conical surface with vertex at the point O. For each such piece the sign of the cosine of the angle between the internal and a radius vector drawn from O will be constant. The radii vectores drawn to all points of the piece S_i will intersect the unit sphere C and form on it a certain domain C_i (see Fig. 11).

We shall call the area of this domain C_i the solid angle which the inner surface of the piece S_i subtends at the point O and we denote it by ω_{S_i}. We shall call this solid angle positive if the radius vector directed from O to a point of S_i forms an obtuse angle with the inward normal, *i.e.*, if an observer situated at O would actually see the inner surface of S_i; and we shall call the solid angle negative if the radius vector makes an acute angle with the inward normal so that an observer at O would actually see the outer surface of S_i. For pieces of a conical surface with vertex at O, we shall take the solid angle to be zero.

Using polar coordinates with pole at O, the solid angle subtended by the piece S_i at O may be written as

$$\omega_{S_i} = -\iint_{S_i} \text{sign} \, [\cos (r, n)] \sin \theta \, d\theta \, d\varphi \tag{9.8}$$

where the notation sign $[\cos(r, n)]$ denotes the sign of the cosine of the angle between the radius vector r and the inward normal n.

For the whole surface S we define the solid angle which it subtends at the origin as the sum of the solid angles for the pieces S_i, and it will be expressed by the integral

$$\omega_S = -\iint_{S} \text{sign} \, [\cos (r, n)] \sin \theta \, d\theta \, d\varphi.$$

We can now formulate a lemma.

LEMMA 4.

$$\omega_S = \iint_{S} \frac{\partial}{\partial n} \left(\frac{1}{r} \right) dS \tag{9.9}$$

Proof. The solid angle ω_S depends only on the boundary l of the surface S and does not change, no matter how the surface S may be deformed, provided that the boundary l stays the same and that the surface S does not pass through the point O during the deformation.

The integral (9.9) also does not depend on the position of the surface S during such a deformation. For if S_1 and S_2 are two surfaces having the same boundary, the difference of the integrals

$$\iint_{S_1} \frac{\partial}{\partial n} \left(\frac{1}{r} \right) dS - \iint_{S_2} \frac{\partial}{\partial n} \left(\frac{1}{r} \right) dS$$

is the integral over the closed surface S' formed by S_1 and S_2. And since by supposition the point O is outside the domain enclosed by S', we have by

Green's formula (9.6), putting $u \equiv 1$,

$$\iint_{S'} \frac{\partial}{\partial n} \left(\frac{1}{r} \right) dS = 0.$$

We can now prove (9.9) by showing that ω_S and the integral in (9.9) coincide for any one particular type of surface S bounded by the boundary l. We may take in particular the surface consisting of part of the sphere C and a part L of the conical surface formed by the radii vectores through the origin and points on the boundary l. But for both these pieces (9.9) is evidently true: for on L, $\partial(1/r)/\partial n = 0$ and the solid angle is zero; and on C,

$$\frac{\partial}{\partial n} \left(\frac{1}{r} \right) = + \frac{d}{dr} \left(\frac{1}{r} \right) \text{sign} \, [\cos(r, n)]$$

$$= - \text{sign} \, [\cos(r, n)] \text{ as } r = 1 \text{ on } C$$

and therefore

$$\frac{\partial}{\partial n} \left(\frac{1}{r} \right) dS = - \text{sign} \, [\cos(r, n)] \sin \theta \, d\theta \, d\varphi.$$

(9.9) could also have been proved by direct calculation.
From (9.8) and (9.9) we see that for any surface

$$- \text{sign} \, [\cos(r, n)] \sin \theta \, d\theta \, d\varphi = \frac{\partial}{\partial n} \left(\frac{1}{r} \right) dS.$$

We shall sometimes denote the expression $\partial(1/r)/\partial n \, dS$ by $d\omega$ and call it an elementary solid angle:

$$d\omega = \frac{\partial}{\partial n} \left(\frac{1}{r} \right) dS. \qquad (9.10)$$

Using the notation (9.10) we can put the expression for the potential of a double layer in the form

$$\iint_S f_1 \, d\omega. \qquad (9.11)$$

Hence it follows that if the density f_1 of the potential of a double layer is equal to unity, then this potential expresses the solid angle subtended by the surface S at the point (x_0, y_0, z_0).

SOME GENERAL CONSEQUENCES OF GREEN'S FORMULA

§ 1. The Mean-Value Theorem for a Harmonic Function

When u is a harmonic function, Green's formula (9.4) takes the form

$$u(x_0, y_0, z_0) = \frac{1}{4\pi} \int\int_S \left\{ u \frac{\partial}{\partial n} \left(\frac{1}{r} \right) - \frac{1}{r} \frac{\partial}{\partial n} \right\} dS \qquad (10.1)$$

As we have seen, this formula is always valid if the function u has first-order derivatives continuous throughout and on the boundary of the domain and second-order derivatives continuous within the domain. The right-hand side of (10.1) consists of the sum of the potentials of a single layer and of a double layer. We shall examine the properties of these potentials in detail later, but we now show that each of them is a harmonic function everywhere outside the surface S.

For, if the point (x_0, y_0, z_0) does not lie on this surface, then the potentials of the single and double layers can be differentiated under the integral sign with respect to the variables x_0, y_0, z_0 as many times as we please. We have already seen that $\nabla_0^2(1/r) = 0$, and so $1/r$ is a harmonic function of these variables; consequently the potentials of the single and double layers will also be harmonic functions outside S.

We next show that these potentials will be analytical functions. Near a point x_0^*, y_0^*, z_0^*, the function $1/r$ can be expanded in a series of powers of $x_0 - x_0^*, y_0 - y_0^*, z_0 - z_0^*$ which converges uniformly for sufficiently small values of the absolute values of these differences.

For,

$$\frac{1}{r} = \frac{1}{\sqrt{(x - x_0)^2 + (y - y_0)^2 + (z - z_0)^2}}$$

$$= \{[(x - x_0^*) - (x_0 - x_0^*)]^2 + [(y - y_0^*) - (y_0 - y_0^*)]^2$$

$$+ [(z - z_0^*) - (z_0 - z_0^*)]^2\}^{-\frac{1}{2}}$$

$$= [(x - x_0^*)^2 + (y - y_0^*)^2 + (z - z_0)^2]^{-\frac{1}{2}} \times$$

$$\times \left[1 - 2\frac{(x - x_0^*)(x_0 - x_0^*) + (y - y_0^*)(y_0 - y_0^*) + (z - z_0^*)(z_0 - z_0^*)}{(x - x_0^*)^2 + (y - y_0^*)^2 + (z - z_0^*)^2} \right.$$

$$\left. + \frac{(x_0 - x_0^*)^2 + (y_0 - y_0^*)^2 + (z_0 - z_0^*)^2}{(x - x_0^*)^2 + (y - y_0^*)^2 + (z - z_0^*)^2} \right]^{-\frac{1}{2}}.$$

On expanding this expression in a series using Newton's binomial theorem, it is easy to see that the series will converge uniformly for values of $(x_0 - x_0^*)^2 + (y_0 - y_0^*)^2 + (z_0 - z_0^*)^2$ which are sufficiently small in comparison with the minimum value of $(x - x_0^*)^2 + (y - y_0^*)^2 + (z - z_0^*)^2$ as the point (x, y, z) moves over the surface S.

By integrating this series term by term, we see that the functions

$$v = \frac{1}{4\pi} \iint_S \frac{\partial u}{\partial n} \frac{1}{r} \, dS \quad \text{and} \quad w = \frac{1}{4\pi} \iint_S u \frac{\partial}{\partial n} \left(\frac{1}{r} \right) dS$$

also admit of similar expansions, and are therefore analytical functions of x_0^*, y_0^*, z_0^* everywhere except on the surface S.

Let u be a function which is harmonic inside a sphere and which, together with its first-order derivatives, is continuous throughout and on the surface of the sphere. We apply Green's formula (10.1) to a sphere of radius h drawn about the point (x_0, y_0, z_0); S is now the surface of this sphere. We get

$$u(x_0, y_0, z_0) = \frac{1}{4\pi} \iint_S \left(\frac{1}{h^2} u - \frac{1}{h} \frac{\partial u}{\partial n} \right) dS.$$

But for any closed surface S_1

$$\iint_{S_1} \frac{\partial u}{\partial n} \, dS = -\iiint_{\Omega_1} \Delta u \, dx \, dy \, dz = 0,$$

where Ω_1 is the volume bounded by the surface S_1, and the function u is harmonic within Ω_1 and has continuous derivatives up to and on the boundary. The first equality follows from (1.2) since $\partial u/\partial n = (\text{grad } u)_n$ and $\Delta u = \text{div} (\text{grad } u)$; and the second equality is obvious. Consequently,

$$u(x_0, y_0, z_0) = \frac{1}{4\pi h^2} \iint_S u \, dS = \frac{1}{4\pi} \int_0^\pi d\theta \int_0^{2\pi} u(R, \theta, \varphi) \sin \theta \, d\varphi \quad (10.2)$$

where R, θ and φ are spherical coordinates.

Hence we have:

THEOREM 1. *The value of a harmonic function at the centre of any sphere is equal to the arithmetic mean of its values on the surface of this sphere.*

This property of harmonic functions completely characterizes this class of functions. In other words, the converse theorem holds:

THEOREM 2. *If a function $u(x, y, z)$ which is continuous in a domain Ω has the property that its value at the centre of any sphere lying wholly within Ω is equal to the arithmetic mean of its values on the surface of this sphere, then the function has continuous second-order derivatives and $\nabla^2 u = 0$, i.e., $u(x, y, z)$ is a harmonic function.*

We postpone the proof of Theorem 2 until a little later, and meanwhile point out certain properties of functions which satisfy the conditions of Theorem 2.

LEMMA 1. *A continuous function $u(x, y, z)$ having the property that its value at the centre of any sphere included within the domain of its definition is equal to the arithmetic mean of its values on the surface of this sphere cannot have either a maximum or a minimum value within this domain unless it simply reduces to a constant.*

For if at some point P_0 within the domain Ω the function u attained a maximum value, then the arithmetic mean of its values on the surface of any sufficiently small sphere σ with its centre at P_0 could equal $u(P_0)$ only if the function u were equal to its maximum value everywhere on the surface of σ. This would imply that the function was constant within the sphere σ. But if the function u is constant within any sphere surrounding any point where it attains a maximum value, then, as may easily be seen, it must be constant everywhere within the domain Ω. For we can join the point P_0 to an arbitrary internal point P by some smooth curve l lying wholly within the domain Ω. Then by Weierstrass's theorem there will be a minimum distance from an arbitrary point of the curve l to points of the boundary of the domain, and this minimum distance will be positive. Hence there exists a positive number η such that a sphere of radius η described about any point of the curve l will lie entirely within Ω. And we find on the curve l a finite number of points $P_0, P_1, P_2, \ldots, P_n = P$ such that each successive point lies within a sphere of radius η described about its predecessor. Using the demonstrated fact that u is constant in any internal sphere circumscribing a point at which u attains a maximum, and passing in sequence from one vertex of the polygon to the next, we get

$$u(P) = u(P_0).$$

LEMMA 2. *If a function $u(x, y, z)$ which satisfies the conditions of Lemma 1 vanishes on the boundary of some region Ω, then it will be identically zero throughout this domain.*

For, if the function u took values within the domain which were different from zero, then it would have a maximum or minimum value within the domain and would not be constant, contrary to Lemma 1.

LEMMA 3. *If a function $u(x, y, z)$ which satisfies the conditions of Lemma 1 coincides with a harmonic function u_0 on any closed surface S, then it will coin-*

cide with this harmonic function everywhere within the domain bounded by S, and so will be itself a harmonic function.

For, the difference $u - u_0$ will then also satisfy the conditions of Lemma 1, since u and u_0 separately do. But this difference is zero on S. Therefore, by Lemma 2, $u - u_0 = 0$ everywhere within the domain bounded by S.

We shall prove later that, for any continuous function $f(S)$ which is defined on the surface of a sphere, there exists a harmonic function u which is continuous within the closed sphere and is equal to $f(S)$ at points of its boundary. From this, Theorem 2 will follow immediately. In fact, for any internal sphere σ in the domain Ω, a function u_0 exists which is harmonic within the sphere σ and which takes the same values as u on its surface. Then, by Lemma 3, u coincides with u_0. This implies that u is harmonic in any internal sphere, and consequently at any internal point of Ω, as Theorem 2 asserts.

The formula (10.2) implies the following theorem.

THEOREM 3. *If a sequence of functions $\{v_n\}$ which are harmonic in the domain Ω converges uniformly in this domain to a limit function v, then this limit function v is also a harmonic function.*

For, if each of the functions v_n satisfies

$$v_n(x_0, y_0, z_0) = \frac{1}{4\pi h^2} \int\int_S v_n \, \mathrm{d}S.$$

where S is a sphere of radius $h = \sqrt{(x - x_0)^2 + (y - y_0)^2 + (z - z_0)^2}$, then the limit function will satisfy the same relation. Consequently, v is a harmonic function in Ω, as we had to show. Hence it follows, incidentally, that if a harmonic function $v(x, y, z, \lambda)$ depends continuously on the parameter λ in a finite interval $a \leq \lambda \leq b$, then the integral of v taken over this interval will also be a harmonic function in the domain Ω.

§ 2. Behaviour of a Harmonic Function near a Singular Point

Suppose a function u has a singular point in the domain Ω, but is harmonic in the neighbourhood of this singular point. We shall examine its behaviour near the singular point. For simplicity we shall take this point as the coordinate origin, and we shall transform the function u into polar coordinates, so that $u = u(R, \theta, \varphi)$.

LEMMA 4. *If a function $u(x, y, z)$ satisfies the inequalities*

$$|u| \leq \frac{A}{R^n}, \qquad \left|\frac{\partial u}{\partial R}\right| \leq \frac{A}{R^{n+1}}$$

where A is a constant, and $R = \sqrt{x^2 + y^2 + z^2}$, and if u is a harmonic func-

tion everywhere in Ω except at the origin, then u can be represented in Ω by

$$u(x_0, y_0, z_0) = \sum_{m=0}^{n-1} \sum_{i+j+k=m} a_{ijk} \frac{\partial^m}{\partial x_0^i \, \partial y_0^j \, \partial z_0^k} \left(\frac{1}{R_0} \right) + u^*(x_0, y_0, z_0)$$

$$(10.3)$$

where u^* is a function which is harmonic everywhere in Ω including the origin.
(If $n = 0$, the first addend on the right-hand side of (10.3) is absent.)

The proof follows immediately from Green's formula (10.1). On surrounding the origin with a sufficiently small sphere with centre at the origin and such that the point (x_0, y_0, z_0) lies within it, we get

$$u(x_0, y_0, z_0) = \frac{1}{4\pi} \iint_S \left(u \frac{\partial}{\partial n} \left(\frac{1}{r} \right) - \frac{1}{r} \frac{\partial u}{\partial n} \right) dS$$

$$+ \frac{1}{4\pi} \iint_\sigma \left(u \frac{\partial}{\partial n} \left(\frac{1}{r} \right) - \frac{1}{r} \frac{\partial u}{\partial n} \right) dS.$$

Using Maclaurin's formula, we can express the function $1/r$ near the origin by

$$\frac{1}{r} = \frac{1}{R_0} + \sum_{m=1}^{n-1} \sum_{i+j+k=m} \frac{1}{i! \, j! \, k!} \, x^i \, y^j \, z^k \left[\frac{\partial^m}{\partial x^i \, \partial y^j \, \partial z^k} \left(\frac{1}{r} \right) \right]_0 + R_n,$$

where the derivatives are evaluated at the origin, and $R_0 = \sqrt{x_0^2 + y_0^2 + z_0^2}$ and $|R_n| < CR^n$ where C is a constant. Using the fact that

$$\left[\frac{\partial^m}{\partial x^i \, \partial y^j \, \partial z^k} \left(\frac{1}{r} \right) \right]_0 = (-1)^m \frac{\partial^m}{\partial x_0^i \, \partial y_0^j \, \partial z_0^k} \left(\frac{1}{R_0} \right),$$

we get

$$\frac{1}{r} = \frac{1}{R_0} + \sum_{m=1}^{n-1} \frac{(-1)^m}{i! \, j! \, k!} \, x^i \, y^j \, z^k \frac{\partial^m}{\partial x_0^i \, \partial y_0^j \, \partial z_0^k} \frac{1}{R_0} + R_n. \quad (10.4)$$

Similarly, $\partial(1/r)/\partial n$ can also be expressed as a linear combination of derivatives of $1/R_0$ with respect to x_0, y_0, z_0 of up to the $(n-1)$th order with coefficients independent of x_0, y_0, z_0, and a remainder R_n' satisfying the inequality $|R_n'| < C_1 R^{n-1}$. This follows from the fact that the derivatives of $1/r$ with respect to x, y, z differ only in sign from the derivatives with respect to x_0, y_0, z_0, and for these derivatives the required expression is clearly valid. But

$$\frac{\partial}{\partial n} = \cos (n, x) \frac{\partial}{\partial x} + \cos (n, y) \frac{\partial}{\partial y} + \cos (n, z) \frac{\partial}{\partial z}.$$

where $\cos (n, x)$, $\cos (n, y)$, $\cos (n, z)$ do not depend on x_0, y_0, z_0, and are bounded.

Hence we get

$$u(x_0, y_0, z_0) = \frac{1}{4\pi} \iint_S \left\{ u \frac{\partial}{\partial n} \left(\frac{1}{r} \right) - \frac{1}{r} \frac{\partial u}{\partial n} \right\} dS + \frac{a_0^{(\sigma)}}{R_0}$$

$$+ \sum_{m=1}^{n-1} a_{ijk}^{(\sigma)} \frac{\partial^m}{\partial x_0^i \, \partial y_0^j \, \partial z_0^k} \frac{1}{R_0}$$

$$+ \frac{1}{4\pi} \iint_\sigma \left(R_n' u - R_n \frac{\partial u}{\partial n} \right) dS,$$

where $a_{ijk}^{(\sigma)}$ are certain numbers.

We pass to the limit as σ contracts into the points $x = y = z = 0$. The limit of the integral taken over σ will be zero, by the conditions of the lemma. The first integral on the right-hand side is a harmonic function of x_0, y_0, z_0 and is independent of σ (it is the sum of the potentials of a single layer and a double layer). It follows that the sum

$$\frac{a_0^{(\sigma)}}{R_0} + \sum_{m=1}^{n-1} \sum_{i+j+k=m} a_{ijk}^{(\sigma)} \frac{\partial^m}{\partial x_0^i \, \partial y_0^j \, \partial z_0^k} \left(\frac{1}{R_0} \right)$$

must also tend to a certain limit $\Phi(x_0, y_0, z_0) = \Phi(P_0)$. We shall prove that this limit can be expressed in the form

$$\frac{a_0}{R_0} + \sum_{m=1}^{n-1} \sum_{i+j+k=m} a_{ijk} \frac{\partial^m}{\partial x_0^i \, \partial y_0^j \, \partial z_0^k} \left(\frac{1}{R_0} \right).$$

To do this it is sufficient to prove the following lemma.

LEMMA 5. *Let $\varphi_1, \varphi_2, \ldots, \varphi_k$ be certain functions of $x_1, x_2, \ldots, x_n$ (k is a finite number). If a linear combination of these functions having variable coefficients which do not depend on the $x_1, x_2, \ldots, x_n$,*

$$\varphi^{(m)} = \sum_{i=1}^k y_i^{(m)} \varphi_i(x_1, \ldots, x_n),$$

converges as $m \to 0$, then its limit $\varphi^{(0)}$ is also a linear combination of the same functions φ_i.

For, the function $\varphi^{(m)}$ obviously cannot take independent arbitrary values throughout space, since on assigning a value to it at some point $P_0^{(s)}$ we obtain a certain linear equation which the coefficients $y_i^{(m)}$ must satisfy. The number of such linearly independent equations is not greater than k. Hence we can find N points $P_0^{(1)}, \ldots, P_0^{(N)}$ ($N \leq k$) such that the value of the function $\varphi^{(m)}$ at any point P can be defined completely in terms of its values at these points:

$$\varphi^{(m)}(P) = \sum_{s=1}^N \varphi^{(m)}(P_0^{(s)}) \psi_s(P) \tag{10.5}$$

where $\psi_s(P)$ are linear combinations of the φ_i with numerical coefficients.

Conversely, any function satisfying (10.5) is expressible in the form of a linear combination of the φ_i.

Passing now to the limit in (10.5), we see that $\varphi^{(0)}$ must satisfy the equation

$$\varphi^{(0)}(P) = \sum_{s=1}^{N} \varphi^{(0)}(P_0^{(s)}) \, \psi_s(P),$$

i.e., $\varphi^{(0)}$ is a linear combination of the $\varphi_i(P)$, as was to be shown.

Hence Lemma 5 and with it Lemma 4 are proved.

THEOREM 4. *The representation* (10.3) *is valid if* $|u| \leq A/R^n$.

We introduce in place of $u(R, \theta, \varphi)$ an auxiliary function

$$v(R, \theta, \varphi) = \frac{1}{R^{n+1}} \int_0^R u(R_1, \theta, \varphi) \, R_1^n \, dR_1: \tag{10.6}$$

the integral on the right-hand side converges. It can also be put into the form

$$v(R, \theta, \varphi) = \frac{1}{R^{n+1}} \int_0^1 u(R\xi, \theta, \varphi) \, R^n \xi^n \, d(R\xi) = \int_0^1 u(R\xi, \theta, \varphi) \, \xi^n \, d\xi$$

$$= \int_0^1 u(\xi x, \xi y, \xi z) \, \xi^n \, d\xi.$$

The integrand of the last integral, as may easily be seen by direct differentiation, is a harmonic function of the variables x, y, z; hence, by Theorem 3 of § 1, v is also a harmonic function.

Differentiation with respect to R gives

$$\frac{\partial v(R, \theta, \varphi)}{\partial R} = -\frac{n+1}{R^{n+2}} \int_0^R u(R_1, \theta, \varphi) \, R_1^n \, dR_1 + \frac{u(R, \theta, \varphi)}{R} \tag{10.7}$$

(10.6), (10.7) imply

$$|v| \leq \frac{1}{R^{n+1}} \int_0^R A \, dR_1 = \frac{A}{R^n}$$

$$\left| \frac{\partial v}{\partial R} \right| \leq \frac{n+1}{R^{n+2}} \int_0^R A \, dR_1 + \frac{A}{R^{n+1}} = \frac{A(n+2)}{R^{n+1}}.$$

Consequently the function v satisfies all the conditions of Lemma 4 and can be expressed in the form

$$v = \frac{a_0}{R_0} + \sum_{m=1}^{n-1} a_{ijk} \frac{\partial^m}{\partial x_0^i \, \partial y_0^j \, \partial z_0^k} \left(\frac{1}{R_0} \right) + w(R_0, \theta, \varphi),$$

where w is a function which is harmonic near the origin.

The function u can be simply calculated from v:

$$u(R, \theta, \varphi) = \frac{1}{R^n} \frac{\partial}{\partial R} [R^{n+1} v(R, \theta, \varphi)]$$

$$= (n + 1) v(R, \theta, \varphi) + x \frac{\partial v}{\partial x} + y \frac{\partial v}{\partial y} + z \frac{\partial v}{\partial z}.$$

Since $\psi_{ijk} = \dfrac{\partial^m}{\partial x^i \, \partial y^j \, \partial z^k} \left(\dfrac{1}{R}\right)$ is a homogeneous function of degree $-(m + 1)$, we have, by Euler's theorem on homogeneous functions,

$$x \frac{\partial \psi_{ijk}}{\partial x} + y \frac{\partial \psi_{ijk}}{\partial y} + z \frac{\partial \psi_{ijk}}{\partial z} = -(m + 1) \psi_{ijk},$$

and on noting that if w is a harmonic function, then so is $x \dfrac{\partial \psi}{\partial x} + y \dfrac{\partial \psi}{\partial y} + z \dfrac{\partial \psi}{\partial z}$ (as may easily be verified by direct differentiation), we see that our theorem is valid.

COROLLARY. *If in the neighbourhood of the origin the function $u(x, y, z)$ is everywhere (except possibly at the origin itself) bounded and harmonic, then it can be made harmonic everywhere, including the origin, by a suitable choice of the value of $u(0, 0, 0)$.*

The function u coincides in the neighbourhood of the origin with a harmonic function w. Consequently it can only fail to be harmonic everywhere because $u(0, 0, 0)$ is not equal to $w(0, 0, 0)$. Hence our proposition.

It is clear that the singularity need not be at the origin but could be at any point.

§ 3. Behaviour of a Harmonic Function at Infinity. Inverse Points

The function $1/r$ can be transformed into a certain form which will be extremely useful later. We have

$$\frac{1}{r} = \frac{1}{\sqrt{x^2 + y^2 + z^2 - 2(xx_0 + yy_0 + zz_0) + x_0^2 + y_0^2 + z_0^2}}$$

$$= \frac{1}{RR_0 \sqrt{\dfrac{1}{R_0^2} - 2 \dfrac{xx_0 + yy_0 + zz_0}{R^2 R_0^2} + \dfrac{1}{R^2}}}$$

where

$$R^2 = x^2 + y^2 + z^2, \quad R_0^2 = x_0^2 + y_0^2 + z_0^2.$$

Hence

$$\frac{1}{r} = \frac{1}{RR_0} \frac{1}{\sqrt{\left(\dfrac{x}{R^2} - \dfrac{x_0}{R_0^2}\right)^2 + \left(\dfrac{y}{R^2} - \dfrac{y_0}{R_0^2}\right)^2 + \left(\dfrac{z}{R^2} - \dfrac{z_0}{R_0^2}\right)^2}}$$

We write

$$\frac{x}{R^2} = \xi, \quad \frac{y}{R^2} = \eta, \quad \frac{z}{R^2} = \zeta, \quad \frac{x_0}{R_0^2} = \xi_0, \quad \frac{y_0}{R_0^2} = \eta_0, \quad \frac{z_0}{R_0^2} = \zeta_0.$$

We call (ξ, η, ζ) the *inverse point* to (x, y, z) with respect to the unit sphere. It is not difficult to see that

$$P^2 = \xi^2 + \eta^2 + \zeta^2 = \frac{1}{R^2},$$

$$x = \frac{\xi}{P^2}, \quad y = \frac{\eta}{P^2}, \quad z = \frac{\zeta}{P^2}.$$

Thus the relation is reciprocal: if (ξ, η, ζ) is inverse to (x, y, z), then (x, y, z) is inverse to (ξ, η, ζ). The point inverse to a given point lies on the radius vector from the origin to the given point at a distance which is the reciprocal of that of the given point.

Writing

$$\varrho = \sqrt{(\xi - \xi_0)^2 + (\eta - \eta_0)^2 + (\zeta - \zeta_0)^2},$$

$$P = \sqrt{\xi^2 + \eta^2 + \zeta^2}, \quad P_0 = \sqrt{\xi_0^2 + \eta_0^2 + \zeta_0^2}$$

we have

$$\frac{1}{r} = \frac{1}{RR_0} \frac{1}{\varrho} = \frac{PP_0}{\varrho}. \tag{10.8}$$

From this formula we get a theorem which is important for its applications:

THEOREM 5. *If the function $u(R, \theta, \varphi)$ is harmonic in a certain domain Ω of the variables R, θ, φ, then the function*

$$\frac{1}{P} u\left(\frac{1}{P}, \theta, \varphi\right) = v(P, \theta, \varphi)$$

is also a harmonic function in the corresponding domain Ω_1 of the variables P, θ, φ which is obtained from the domain Ω by the substitution $P = 1/R$. In other words, $\dfrac{1}{P} u\left(\dfrac{1}{P}, \theta, \varphi\right)$ is a harmonic function of ξ, η, ζ.

For, by Green's formula (10.1) any harmonic function $u(x, y, z)$ can be expressed as the sum of the potentials of a single layer and of a double layer

$$u(R_0, \theta_0, \varphi_0) = \iint_S \mu \frac{\partial}{\partial n}\left(\frac{1}{r}\right) dS + \iint_S \nu \frac{1}{r} dS,$$

where

$$\mu = \left[\frac{1}{4\pi} u\right]_S, \quad \text{and} \quad \nu = \left[-\frac{1}{4\pi} \frac{\partial u}{\partial n}\right]_S.$$

Consequently,

$$\frac{1}{P_0} u = R_0 u = \iint_S \mu \frac{\partial}{\partial n}\left(\frac{R_0}{r}\right) dS + \iint_S \nu \frac{R_0}{r} dS; \qquad (10.9)$$

but $R_0/r = 1/R\varrho$ is a harmonic function of ξ_0, η_0, ζ_0 and therefore the integrals in (10.9) are also harmonic functions of these variables, as we had to show.

COROLLARY 1. *If a function which is harmonic outside a certain sphere satisfies the inequality*

$$|u| < AR^{n-1}, \qquad (10.10)$$

then this function can be expressed in the form

$$u = a_0 + \sum_{m=1}^{n-1} a_{ijk} R^{2m+1} \frac{\partial^m}{\partial x^i \, \partial y^j \, \partial z^k}\left(\frac{1}{R}\right) + u^*, \qquad (10.11)$$

$$i + j + k = m$$

where u^* is a harmonic function tending to zero at infinity.

For, if the inequality (10.10) holds for u, then we shall have $|v| < A/P^n$ for the function $v(P, \theta, \varphi) = \frac{1}{P} u\left(\frac{1}{P}, \theta, \varphi\right)$. But by Theorem 4 of this lecture

$$v(P, \theta, \varphi) = \frac{a_0}{P} + \sum_{m=1}^{n-1} a_{ijk} \frac{\partial^m}{\partial \xi^i \, \partial \eta^j \, \partial \zeta^k}\left(\frac{1}{P}\right) + v^*(P, \theta, \varphi). \quad (10.12)$$

The function $\dfrac{\partial^m}{\partial \xi^i \, \partial \eta^j \, \partial \zeta^k}\left(\dfrac{1}{P}\right)$ is a homogeneous function of degree $-(m+1)$ in ξ, η, ζ.

Consequently, $P^{m+1} \dfrac{\partial^m}{\partial \xi^i \, \partial \eta^j \, \partial \zeta^k} \dfrac{1}{P}$ is a homogeneous function of zero degree in ξ, η, ζ and depends only on the ratios of these variables. But ξ, η, ζ are proportional to x, y, z. Hence

$$P^{m+1} \frac{\partial^m}{\partial \xi^i \, \partial \eta^j \, \partial \zeta^k}\left(\frac{1}{P}\right) = R^{m+1} \frac{\partial^m}{\partial x^i \, \partial y^j \, \partial z^k}\left(\frac{1}{R}\right),$$

so that

$$\frac{\partial^m}{\partial \xi^i \, \partial \eta^j \, \partial \zeta^k} \left(\frac{1}{P} \right) = R^{2m+2} \frac{\partial^m}{\partial x^i \, \partial y^j \, \partial z^k} \left(\frac{1}{R} \right). \qquad (10.13)$$

Substituting (10.13) in (10.12), and using the fact that $u = (1/R)\, v$, we get (10.11). Hence the proposition.

Another important corollary follows from (10.11).

COROLLARY 2. *A function $u(x, y, z)$ which is harmonic and bounded outside a sphere will tend to a definite limit at infinity.*

POISSON'S EQUATION IN AN UNBOUNDED MEDIUM. NEWTONIAN POTENTIAL

WE SHALL next investigate Poisson's equation for an unbounded medium.

LEMMA 1. *A function which is harmonic throughout space and which tends to zero at infinity is identically zero.*

For, applying Theorem 1 of Lecture 9 to an arbitrarily large sphere, we see that the value of our harmonic function at any point of space is as small as we please. Hence the lemma.

COROLLARY 1. *The solution of Poisson's equation*

$$\nabla^2 v = -4\pi\varrho(x, y, z) \tag{11.1}$$

in infinite space which tends to zero at infinity is unique.

For, if v_1 and v_2 are two such solutions, then their difference $v_1 - v_2$ is a harmonic function which tends to zero at infinity. Hence by Lemma 1, this difference is identically zero.

COROLLARY 2. *(Liouville's Theorem) A harmonic function which is bounded throughout space is constant.*

For, by Corollary 2 to Theorem 5 of Lecture 10, this function tends to a certain limit c at infinity. The difference $u - c$ is itself a harmonic function and tends to zero at infinity. Hence by Lemma 1, this difference is everywhere zero, i.e., $u = c$ everywhere, as we had to show.

We now pass on to the solution of Poisson's equation (11.1) in infinite space.

Suppose that the function $\varrho(x, y, z)$ is integrable and that it satisfies the inequalities

$$\left.\begin{array}{ll} \left|\varrho(x, y, z)\right| < \dfrac{A}{R^{2+a}} & \text{if } R \geqq 1 \\[2ex] \left|\varrho(x, y, z)\right| < A & \text{if } R < 1 \end{array}\right\} \tag{11.2}$$

where

$$R = \sqrt{x^2 + y^2 + z^2} \quad \text{and} \quad a > 0.$$

Without loss of generality we can suppose that $a < 1$, for otherwise the replacement of a by $a_1 < a$ would only weaken the inequality (11.2).

The solution of equation (11.1) subject to the conditions (11.2) can easily be constructed using Green's formula (9.4). Let $u(x_0, y_0, z_0)$ be a solution of (11.1) which tends to zero at infinity. For any volume Ω bounded by a surface S we have by Green's formula,

$$u(x_0, y_0, z_0)$$

$$= -\frac{1}{4\pi} \iiint_\Omega (-4\pi\varrho) \frac{1}{r} \, dx \, dy \, dz + \frac{1}{4\pi} \iint_S u \frac{\partial}{\partial n} \left(\frac{1}{r}\right) dS - \frac{1}{4\pi} \iint_S \frac{1}{r} \frac{\partial u}{\partial n} dS$$

$$= \iiint_\Omega \frac{\varrho}{r} dx \, dy \, dz + \frac{1}{4\pi} \iint_S u \frac{\partial}{\partial n} \left(\frac{1}{r}\right) dS - \frac{1}{4\pi} \iint_S \frac{1}{r} \frac{\partial u}{\partial n} \, dS, \quad (11.3)$$

where r is, as usual, the distance between the points (x, y, z) and (x_0, y_0, z_0). We take as the volume Ω a sphere of radius R with centre at the origin and let $R \to \infty$. Then the first integral on the right-hand side of (11.3) will tend to a definite limit, since by virtue of the conditions (11.2) the volume integral converges. The sum of the other two integrals is a harmonic function. We shall show in a moment that the limit of the first integral gives a solution of the problem posed. Then because of the proved uniqueness of solution it will follow that the sum of the second and third integrals tends to zero.

We now show that the function

$$u(x_0, y_0, z_0) = \int_{-\infty}^{+\infty} \int_{-\infty}^{+\infty} \int_{-\infty}^{+\infty} \frac{\varrho}{r} \, dx \, dy \, dz \quad (11.4)$$

where the triple integral is taken over all space, does really satisfy equation (11.1) and the stipulated conditions.

The function (11.4) is called the Newtonian potential, and $\varrho(x, y, z)$ its density as already defined in § 3, Lecture 9.

We investigate first how $u(x_0, y_0, z_0)$ behaves at infinity. We have

$$|u(x_0, y_0, z_0)| \leqq A \int_{-\infty}^{+\infty} \int_{-\infty}^{+\infty} \int_{-\infty}^{+\infty} \frac{1}{rR^{2+a}} \, dx \, dy \, dz.$$

The magnitude of the last integral depends only on $R_0 = \sqrt{x_0^2 + y_0^2 + z_0^2}$, and if we put $x_0 = R_0$, $y_0 = 0$, $z_0 = 0$, its value is unchanged. For this integral obviously does not change when the coordinate axes are rotated, and we may choose the axes so that the axis OX goes through the point (x_0, y_0, z_0). Changing to new variables ξ, η, ζ given by

$$x = R_0\xi, \quad y = R_0\eta, \quad z = R_0\zeta,$$

the integral becomes

$$A \int_{-\infty}^{+\infty} \int_{-\infty}^{+\infty} \int_{-\infty}^{+\infty} \frac{R_0^3 \, d\xi \, d\eta \, d\zeta}{R_0^{3+a} \, P^{2+a} \, P_1} = \frac{A}{R_0^a} \int_{-\infty}^{+\infty} \int_{-\infty}^{+\infty} \int_{-\infty}^{+\infty} \frac{d\xi \, d\eta \, d\zeta}{P_1 \, P^{2+a}},$$

where

$$P = \sqrt{\xi^2 + \eta^2 + \zeta^2}, \quad P_1 = \sqrt{(\xi - 1)^2 + \eta^2 + \zeta^2}.$$

The last integral converges: since (i) when $P \to \infty$, the integrand decreases like P^{-3-a}; (ii) near $P = 0$ the singularity of order P^{-2-a} is integrable; and (iii) near $P_1 = 0$ the singularity of order P_1^{-1} is integrable. Writing

$$\int_{-\infty}^{+\infty} \int_{-\infty}^{+\infty} \int_{-\infty}^{+\infty} \frac{d\xi \, d\eta \, d\zeta}{P_1 \, P^{2+a}} = k,$$

we get

$$|u(x_0, y_0, z_0,)| \leqq \frac{A_k}{R_0^a},$$

which shows that $u \to 0$ at infinity.

We next show that u has continuous derivatives, obtained by differentiating under the integral sign. For example,

$$\frac{\partial u}{\partial x_0} = \int_{-\infty}^{+\infty} \int_{-\infty}^{+\infty} \int_{-\infty}^{+\infty} \varrho \frac{\partial}{\partial x_0} \left(\frac{1}{r}\right) dx \, dy \, dz.$$

Differentiating under the integral sign is permissible because the integral so obtained converges uniformly with respect to x_0. For,

$$\frac{\partial}{\partial x_0} \left(\frac{1}{r}\right) = \frac{x - x_0}{r^3} \quad \text{and} \quad \left|\frac{\partial}{\partial x_0} \left(\frac{1}{r}\right)\right| \leqq \frac{1}{r^2}$$

and this guarantees the convergence (see Criterion 2 in Lecture 7). Simultaneously we have proved that continuous first-order derivatives of the Newtonian potential exist. In order to prove the existence and continuity of the second-order derivatives, certain new restrictions must be laid on the function $\varrho(x, y, z)$; we shall in fact assume that this function has continuous derivatives of the first order. This restriction is not really essential, but to replace it by a weaker one would make the investigation longer.

The function $\varrho(x, y, z)$ can always be split into two parts ϱ_1 and ϱ_2 such that, in the neighbourhood of the given point (x_0, y_0, z_0) ϱ_2 will be identically zero, while ϱ_1 will be identically zero in a certain neighbourhood of infinity, i.e., outside a certain domain D; and at the same time, both ϱ_1 and ϱ_2 will have continuous first-order derivatives. Then

$$u(x_0, y_0, z_0) = \int_{-\infty}^{+\infty} \int_{-\infty}^{+\infty} \int_{-\infty}^{+\infty} \frac{\varrho_1}{r} dx \, dy \, dz + \int_{-\infty}^{+\infty} \int_{-\infty}^{+\infty} \int_{-\infty}^{+\infty} \frac{\varrho_2}{r} dx \, dy \, dz.$$

Since $\varrho_2 \equiv 0$ in the neighbourhood of (x_0, y_0, z_0), we can exclude this neighbourhood from the second integral; and then we can differentiate it twice with respect to the parameters and obtain uniformly convergent integrals. Considering now the first integral, we shall have, for example,

$$\frac{\partial}{\partial x_0} \int_{-\infty}^{+\infty} \int_{-\infty}^{+\infty} \int_{-\infty}^{+\infty} \frac{\varrho_1}{r} \, dx \, dy \, dz = \int_{-\infty}^{+\infty} \int_{-\infty}^{+\infty} \int_{-\infty}^{+\infty} \varrho_1 \, \frac{x - x_0}{r^3} \, dx \, dy \, dz.$$

Introducing new variables $x = x_0 + \xi$, $y_0 = y_0 + \eta$, $z = z_0 + \zeta$, we get

$$\frac{\partial}{\partial x_0} \int_{-\infty}^{+\infty} \int_{-\infty}^{+\infty} \int_{-\infty}^{+\infty} \frac{\varrho_1}{r} \, dx \, dy \, dz$$

$$= \int_{-\infty}^{+\infty} \int_{-\infty}^{+\infty} \int_{-\infty}^{+\infty} \frac{\xi \varrho_1(x_0 + \xi, y_0 + \eta, z_0 + \zeta) \, d\xi \, d\eta \, d\zeta}{\sqrt{(\xi^2 + \eta^2 + \zeta^2)}}$$

and this last integral can obviously be differentiated with respect to the parameters x_0, y_0, z_0, since the integrals so obtained will converge uniformly.

It remains to show that the Newtonian potential satisfies Poisson's equation.

We take a function $\psi(x_0, y_0, z_0)$ which is equal to zero everywhere except in a certain sphere C with centre at the point (x_0, y_0, z_0) and which has continuous derivatives of certain orders. Then using Green's formula we find, noting that outside the sphere C the functions ψ and $\partial \psi / \partial n$ vanish,

$$\psi(x_0, y_0, z_0) = -\frac{1}{4\pi} \int_{-\infty}^{+\infty} \int_{-\infty}^{+\infty} \int_{-\infty}^{+\infty} \frac{\nabla^2 \psi(x, y, z)}{r} \, dx \, dy \, dz.$$

Multiplying throughout by $\varrho(x_0, y_0, z_0)$ and integrating with respect to x_0, y_0, z_0, we get

$$\int_{-\infty}^{+\infty} \int_{-\infty}^{+\infty} \int_{-\infty}^{+\infty} \psi(x_0, y_0, z_0) \, \varrho(x_0, y_0, z_0) \, dx_0 \, dy_0 dz_0$$

$$= -\frac{1}{4\pi} \underbrace{\int_{-\infty}^{+\infty} \int_{-\infty}^{+\infty} \cdots \int_{-\infty}^{+\infty}}_{6 \text{ times}} \frac{\nabla^2 \psi(x, y, z) \, \varrho(x_0, y_0, z_0)}{r} \, dx \, dy \, dz \, dx_0 \, dy_0 \, dz_0$$

$$= -\frac{1}{4\pi} \int_{-\infty}^{+\infty} \int_{-\infty}^{+\infty} \int_{-\infty}^{+\infty} \nabla^2 \psi(x, y, z)$$

$$\times \left\{ \int_{-\infty}^{+\infty} \int_{-\infty}^{+\infty} \int_{-\infty}^{+\infty} \frac{\varrho}{r} \, dx_0 \, dy_0 \, dz_0 \right\} dx \, dy \, dz$$

$$= -\frac{1}{4\pi} \int_{-\infty}^{+\infty} \int_{-\infty}^{+\infty} \int_{-\infty}^{+\infty} u(x, y, z) \, \nabla^2 \psi(x, y, z) \, dx \, dy \, dz. \tag{11.5}$$

The last integral can be transformed by using the fact that for a sufficiently large domain D

$$\iiint_D u\,\nabla^2\psi\,dx\,dy\,dz = \iiint_D \psi\nabla^2 u\,dx\,dy\,dz.$$

Comparing this with (11.5) we conclude that

$$\int_{-\infty}^{+\infty}\int_{-\infty}^{+\infty}\int_{-\infty}^{+\infty} \psi(x,\,y,\,z)\,[\varDelta u + 4\pi\varrho]\,dx\,dy\,dz = 0,$$

and since $\psi(x, y, z)$ is arbitrary, it follows that

$$\nabla^2 u = -4\pi\varrho,$$

as was to be shown.

THE SOLUTION OF THE DIRICHLET PROBLEM FOR A SPHERE

IN LECTURE 2 we met the Dirichlet problem for Laplace's equation: this was the problem of determining a function harmonic within a domain when its values on the boundary are prescribed. The Dirichlet problem can also be posed for other equations of elliptic type besides Laplace's equation; the problem then is to find a solution of the given equation within the given domain which shall take specified values on the boundary of the domain. In this lecture we shall investigate the solution of the Dirichlet problem posed for Poisson's equation $\nabla^2 u = \varrho$ for the case of a sphere.

We take a point (x_0, y_0, z_0) within a domain Ω bounded by a surface S. We have already seen that the function $1/r$, where

$$r = \sqrt{(x - x_0)^2 + (y - y_0)^2 + (z - z_0)^2},$$

is a solution of Laplace's equation.

Applying Green's formula (9.4) to $1/r$ and some solution u of Poisson's equation $\nabla^2 u = \varrho$, we get

$$u(x_0, y_0, z_0) = \frac{1}{4\pi} \iint_S \left\{ u \frac{\partial}{\partial n} \left(\frac{1}{r} \right) - \frac{1}{r} \frac{\partial u}{\partial n} \right\} dS$$

$$- \frac{1}{4\pi} \iiint_\Omega \frac{\varrho}{r} \, dx \, dy \, dz. \qquad (12.1)$$

If we were to construct a function $g(x, y, z, x_0, y_0, z_0)$ harmonic throughout and such that $[g]_S = [1/r]_S$, then, applying Green's formula to u and g, we should have

$$\frac{1}{4\pi} \iiint_\Omega \{ g\nabla^2 u - u\nabla^2 g \} \, dx \, dy \, dz = \frac{1}{4\pi} \iiint_\Omega g\varrho \, dx \, dy \, dz$$

$$= \frac{1}{4\pi} \iint_S \left\{ u \frac{\partial g}{\partial n} - g \frac{\partial u}{\partial n} \right\} dS. \qquad (12.2)$$

Combining (12.1) and the second equality (12.2) and taking into account the value of g on S, we get

$$u(x_0, y_0, z_0) = -\iiint_\Omega \left(\frac{1}{4\pi r} - \frac{g}{4\pi}\right) \varrho \, dx \, dy \, dz$$

$$+ \iint_S u \frac{\partial}{\partial n} \left(\frac{1}{4\pi r} - \frac{g}{4\pi}\right) dS. \qquad (12.3)$$

Writing

$$\frac{1}{4\pi r} - \frac{g}{4\pi} = G(x, y, z, x_0, y_0, z_0)$$

we have

$$u(x_0, y_0, z_0) = -\iiint_\Omega G\varrho \, d\Omega + \iint_S \frac{\partial G}{\partial n} f(S) \, dS.$$

The function G is called *Green's function*. Thus, on the assumption that a solution of the formulated Dirichlet problem exists, the formula

$$u(x_0, y_0, z_0) = -\iiint_\Omega G\varrho \, d\Omega + \iint_S \frac{\partial G}{\partial n} f(S) \, dS \qquad (12.5)$$

will give this solution in explicit form if Green's function is known.

Green's function takes the value zero on the boundary and it is the sum of the function $1/r$ and of the function g which is harmonic everywhere within the domain. It is clear that it is determined uniquely.

We now pass on to the solution of the Dirichlet problem for a sphere. In this case Green's function can be constructed in explicit form. We put

$$r_1 = \sqrt{(x - \xi_0)^2 + (y - \eta_0)^2 + (z - \zeta_0)^2}$$

where $\xi_0 = x_0/R_0^2$, $\eta_0 = y_0/R_0^2$, $\zeta_0 = z_0/R_0^2$, $R_0^2 = x_0^2 + y_0^2 + z_0^2$, as on p. 163. If the point (x, y, z) lies on a sphere of unit radius, then $x = \xi$, $y = \eta$, $z = \zeta$. By virtue of (10.8), on the sphere $R = 1$, we shall have

$$\left[\frac{1}{R_0 r_1}\right]_{R=1} = \frac{1}{r}.$$

The function $1/R_0 r_1$ is obviously a harmonic function of the variables x, y, z inside the sphere $R < 1$. Consequently $g = 1/R_0 r_1$.

Hence

$$G = \frac{1}{4\pi r} - \frac{1}{4\pi R_0 r_1}$$

where

$$R_0 = \sqrt{x_0^2 + y_0^2 + z_0^2}, \quad r = \sqrt{(x - x_0)^2 + (y - y_0)^2 + (z - z_0)^2}.$$

$$r_1 = \sqrt{\left(x - \frac{x_0}{R_0^2}\right)^2 + \left(y - \frac{y_0}{R_0^2}\right)^2 + \left(z - \frac{z_0}{R_0^2}\right)^2}$$

$$= \frac{1}{R_0} \sqrt{R^2 R_0^2 - 2(xx_0 + yy_0 + zz_0) + 1}.$$

Hence

$$G(x, y, z, x_0, y_0, z_0) = \frac{1}{4\pi}\left[\frac{1}{r} - \frac{1}{\sqrt{R^2 R_0^2 - 2(xx_0 + yy_0 + zz_0) + 1}}\right].$$

As we see, the function G turns out to be a symmetrical function as regards the arguments (x, y, z) and (x_0, y_0, z_0). Consequently it is also a harmonic function of the variables (x_0, y_0, z_0) if $r \neq 0$.

We next verify that the formula (12.5) really does give a solution of the Dirichlet problem for the sphere. Our proof will consist of two parts. We shall show separately that the first term in (12.5) is a solution of Poisson's equation and vanishes on the boundary S of Ω, and that the second term is a solution of Laplace's equation and takes the specified values $f(S)$ on S. We begin with the proof of the first assertion.

We shall prove that the function

$$u_1(x_0, y_0, z_0) = -\iiint_\Omega G\varrho \, dx \, dy \, dz \tag{12.6}$$

vanishes on the boundary and satisfies Poisson's equation. To do this we must estimate the magnitude of Green's function $G(x, y, z, x_0, y_0, z_0)$. We shall show that

$$0 < G < \frac{1}{r}. \tag{12.7}$$

We first establish that G is positive. We surround an internal point (x_0, y_0, z_0) by a small sphere σ, and we consider the function G in the domain Ω' included between the spheres σ and S. In this domain G is a harmonic function. If the sphere σ is sufficiently small, then G will be positive on it, since the first term is as large as we please and the second term is bounded. On the sphere S the function G is, by definition, zero. Hence G is non-negative everywhere on the boundary of Ω' and is positive on part of this boundary. Since it cannot have a minimum value, it must be positive everywhere within the domain Ω'.

To prove the second inequality in (12.7), it is sufficient to show that the function g is positive. This follows from the fact that it takes positive values on the boundary of Ω and is harmonic within Ω.

By Criterion 2 of Lecture 7, it follows from the inequalities (12.7) that the integral converges uniformly at the point (x_0, y_0, z_0); consequently the integral is a continuous function. Its value is zero if (x_0, y_0, z_0) is a point on the boundary. Hence the integral (12.6) tends to zero if the point (x_0, y_0, z_0) tends towards a point on the boundary.

Suppose that the point (x_0, y_0, z_0) lies within the sphere. We write the integral (12.6) in the form

$$u_1(x_0, y_0, z_0) = -\frac{1}{4\pi} \iiint_\Omega \frac{\varrho}{r} \, dx \, dy \, dz + \frac{1}{4\pi} \iiint_\Omega g\varrho \, dx \, dy \, dz.$$

The first term is the Newtonian potential and so the Laplacian operator applied to it gives ϱ. The second term is a harmonic function, since

$$\nabla_0^2 \left[\frac{1}{4\pi} \iiint_\Omega g\varrho \, dx \, dy \, dz \right] = \frac{1}{4\pi} \iiint_\Omega \varrho \nabla_0^2 g \, dx \, dy \, dz = 0.$$

(We denote the Laplacian operator by ∇_2^0 here to stress that the derivatives are taken with respect to the arguments x_0, y_0, z_0.) Hence formula (12.6) gives the required solution of Poisson's equation.

Passing on now to the second part of the proof, it is useful to transform formula (12.5). Let γ be the angle between the radii vectores of the points (x, y, z) and (x_0, y_0, z_0). Then the distance r, as the side of the triangle opposite the angle γ, can be expressed in the form

$$r = \sqrt{R^2 + R_0^2 - 2RR_0 \cos \gamma} :$$

similarly, r_1 can be expressed in the form

$$r_1 = \sqrt{R^2 + \frac{1}{R_0^2} - 2\frac{R}{R_0} \cos \gamma} ,$$

and then Green's function will take the form

$$G = \frac{1}{4\pi} \left[\frac{1}{\sqrt{R^2 + R_0^2 - 2RR_0 \cos \gamma}} - \frac{1}{\sqrt{R^2 R_0^2 - 2RR_0 \cos \gamma + 1}} \right],$$

and

$$\left[\frac{\partial G}{\partial n} \right]_{R=1} = \left[-\frac{\partial G}{\partial R} \right]_{R=1}$$

$$= \frac{1}{4\pi} \left[\frac{R - R_0 \cos \gamma}{(R^2 + R_0^2 - 2RR_0 \cos \gamma)^{\frac{3}{2}}} - \frac{RR_0^2 - R_0 \cos \gamma}{(R^2 R_0^2 - 2RR_0 \cos \gamma + 1)^{\frac{3}{2}}} \right]_{R=1}$$

$$= \frac{1}{4\pi} \frac{1 - R_0^2}{(1 - 2R_0 \cos \gamma + R_0^2)^{\frac{3}{2}}} .$$

To investigate the second term on the right-hand side of (12.5), we apply this formula (12.5) to the case of Laplace's equation ($\varrho \equiv 0$). Then the first term will vanish. The solution of the Dirichlet problem, if it exists, may be written in the form

$$u_2(x_0, y_0, z_0) = \frac{1}{4\pi} \int\int_S \frac{1 - R_0^2}{(1 - 2R_0 \cos \gamma + R_0^2)^{\frac{3}{2}}} f(S) \, dS \quad (12.8)$$

where $f(S)$ are the specified values of $u(x_0, y_0, z_0)$ on the unit sphere. (12.8) is known as *Poisson's formula.*

We have obtained in explicit form the solution of the Dirichlet problem for a sphere of unit radius. The solution for a sphere of arbitrary radius P can also easily be obtained from (12.8). We introduce a function

$$v(R_0, \theta_0, \varphi_0,) = u(R_0/P, \theta_0, \varphi_0)$$

$$= \frac{1}{4\pi} \int\int_S \frac{1 - (R_0/P)^2}{\left[1 - 2\dfrac{R_0}{P} \cos \gamma + \left(\dfrac{R_0}{P}\right)^2\right]^{\frac{3}{2}}} f(S) \sin \theta \, d\theta \, d\varphi.$$

Replacing $P^2 \sin \theta \, d\theta \, d\varphi$ by dS, we finally obtain

$$v(x_0, y_0, z_0) = \frac{1}{4\pi} \int\int_S \frac{P^2 - R_0^2}{(P^2 - 2PR_0 \cos \gamma + R_0^2)^{\frac{3}{2}} \, P} f(S) \, dS. \quad (12.9)$$

We have to show that the function $v(x_0, y_0, z_0)$ takes on the boundary the values $f(S)$ and that it is a harmonic function. From the very method of obtaining (12.9), this formula is valid in the particular case when $f(S) \equiv 1$ and when a solution of the Dirichlet problem obviously exists (it is identically equal to 1). Thus we have

$$\frac{1}{4\pi} \int\int_S \frac{P^2 - R_0^2}{P(P^2 - 2PR_0 \cos \gamma + R_0^2)^{\frac{3}{2}}} \, dS = 1.$$

Let S_0 be some point of the sphere S. We form the difference

$$v(x_0, y_0, z_0) - f(S_0)$$

$$= \frac{1}{4\pi} \int\int_S \frac{P^2 - R_0^2}{P(P^2 - 2PR_0 \cos \gamma + R_0^2)^{\frac{3}{2}}} [f(S) - f(S_0)] \, dS.$$

We shall prove that this difference tends to zero when the point (x_0, y_0, z_0) tends to S_0. This proof is almost the same as that which we have already given when examining the equation for heat conduction.

We surround the point S_0 by a small sphere of radius η, choosing η to be so small that, at all points of the surface which fall within the sphere, by virtue of the continuity of f, we have

$$\left| f(S) - f(S_0) \right| < \varepsilon,$$

where ε is a given positive number. Let σ be the part of the surface S which is included within the sphere of radius η with centre at S_0. Then the above difference can be written in the form

$$v(x_0, y_0, z_0) - f(S_0)$$

$$= \frac{1}{4\pi} \iint_{S-\sigma} \frac{P^2 - R_0^2}{P(P^2 - 2PR_0 \cos \gamma + R_0^2)^{\frac{3}{2}}} [f(S) - f(S_0)] \, dS$$

$$+ \frac{1}{4\pi} \iint_{\sigma} \frac{P^2 - R_0^2}{P(P^2 - 2PR_0 \cos \gamma + R_0^2)^{\frac{3}{2}}} [f(S) - f(S_0)] \, dS = I_1 + I_2.$$

Being continuous, the function $f(S)$ is bounded on S, $i.e.$, $|f(S)| < M$, where M is a certain number. The last integral, which we have denoted by I_2, may be estimated thus:

$$|I_2| = \frac{1}{4\pi} \left| \iint_{\sigma} \frac{P^2 - R_0^2}{P(P^2 - 2PR_0 \cos \gamma + R_0^2)^{\frac{3}{2}}} [f(S) - f(S_0)] \, dS \right|$$

$$\leqq \frac{1}{4\pi} \cdot \frac{\varepsilon}{2} \iint_{\sigma} \frac{P^2 - R_0^2}{P(P^2 - 2PR_0 \cos \gamma + R_0^2)^{\frac{3}{2}}} \, dS$$

$$< \frac{\varepsilon}{2} \cdot \frac{1}{4\pi} \iint_{S} \frac{P^2 - R_0^2}{P(P^2 - 2PR_0 \cos \gamma + R_0^2)^{\frac{3}{2}}} \, dS = \frac{\varepsilon}{2}.$$

The first integral I_1 may be estimated as follows:

$$|I_1| \leqq 2M \cdot \frac{1}{4\pi} \iint_{S-\sigma} \frac{P^2 - R_0^2}{P(P^2 - 2PR_0 \cos \gamma + R_0^2)^{\frac{3}{2}}} \, dS$$

$$= \frac{M}{2\pi} \iint_{S-\sigma} \frac{(P - R_0)(P + R_0)}{P[(P - R_0)^2 + 2PR_0(1 - \cos \gamma)]^{\frac{3}{2}}} \, dS.$$

Thus, $|I_1|$ may be made as small as we please by taking the point (x_0, y_0, z_0) sufficiently close to S_0. For then $(1 - \cos \gamma)$ will be greater than a certain positive number everywhere outside σ, and this implies that the denominator of the integrand is bounded from below; the numerator can be made as small as we please. Hence $|I_1| < \varepsilon/2$ and $|v(x_0, y_0, z_0) - f(S_0)| < \varepsilon$, as we had to show.

That the function $v(x_0, y_0, z_0)$ is harmonic follows from the fact that, for $R_0 < P$, the function G, and therefore also $\partial G/\partial n$, is a harmonic function of the variables x_0, y_0, z_0.

We now formulate one more problem, the so-called *exterior Dirichlet problem* for a sphere: it is required to determine a function u which shall satisfy the equation $\nabla^2 u = \varrho$ outside the sphere, take specified values on the sphere, and tend to zero at infinity. The solution of such a problem is obviously unique.

As before, let r denote the distance from the point (x, y, z) to the point (x_0, y_0, z_0) lying outside the sphere, and let r_1 be the distance to the inverse point.

Then, similarly to the previous problem, a function g which is harmonic outside the sphere and takes on it the values $[1/r]_s$ will have the form

$$\frac{1}{R_0 r_1}.$$

If we suppose at first that at infinity the function u decreases like $1/R^a$, where $a > 0$, so that its first derivatives decrease like $1/R^{a+1}$ and its second derivatives decrease like $1/R^{a+2}$, then it is not difficult to obtain in the same way as before the formula

$$u(x_0, y_0, z_0) = -\iiint_\Omega G\varrho \, dx \, dy \, dz + \iint_S \frac{\partial G}{\partial n} u \, dS$$

where

$$G = \frac{1}{4\pi r} - \frac{1}{4\pi R_0 r_1}.$$

In the above formula the derivative function under the integral sign is taken along the external normal to the sphere; this is the internal normal to the domain in which the problem is being considered.

Restricting ourselves again, first of all, to the case when $[u]_s = 0$, we see that the solution of the equation $\nabla^2 u = \varrho$ with homogeneous conditions has the form

$$u = -\iiint_\Omega G\varrho \, dx \, dy \, dz. \tag{12.10}$$

We shall have, on again carrying out a change of variables,

$$G = \frac{1}{4\pi} \left[\frac{1}{\sqrt{R^2 + R_0^2 - 2RR_0 \cos \gamma}} \right.$$

$$\left. - \frac{1}{\sqrt{R^2 R_0^2 - 2RR_0 \cos \gamma + 1}} \right],$$

$$\left[\frac{\partial G}{\partial n} \right]_S = \left[\frac{\partial G}{\partial R} \right]_{R=1} = \frac{R_0^2 - 1}{(R_0^2 - 2R_0 \cos \gamma + 1)^{\frac{3}{2}}}.$$

We shall verify that the formula (12.10) gives the solution of the problem formulated if ϱ is identically equal to zero outside a certain bounded domain. The analysis of the general case, when, for example, $|\varrho| < A/R^{2+a}$, is left to the reader.

We now go on with the proof. As before, the formula which expresses the value of $u(x_0, y_0, z_0)$ falls into two parts, the first of which is a solution of Poisson's equation and satisfies the condition of vanishing on the boundary of the domain, and the second is a solution of Laplace's equation and takes on the boundary the values specified for u. We shall establish the truth of this assertion only for the first part, since the proof for the second part is exactly the same as in the previous case.

But this assertion is evidently true, since, for sufficiently large R_0, we have:

$$|G| \le \frac{M}{R_0}, \quad \left| \frac{\partial G}{\partial n} \right| \le \frac{M}{R_0}, \quad \left| \frac{\partial}{\partial x_0} \left(\frac{\partial G}{\partial n} \right) \right| \le \frac{M}{R_0^2}. \qquad (12.11)$$

The first of these inequalities gives at once for u

$$|u| \le \iiint_{\Omega} \max |\varrho| \frac{M}{R_0} \, dS = 4\pi \frac{\max |\varrho| M}{R_0}.$$

The proof that u satisfies the equation $\nabla^2 u = \varrho$ is the same as before, as is the proof that u vanishes on the boundary.

From the same inequalities (12.11) immediately follows another important consequence.

If a function u is harmonic outside a sphere of unit radius and tends to zero at infinity, then there is a constant M such that

$$|u| \le \frac{M}{R_0}, \quad \left| \frac{\partial u}{\partial x} \right| \le \frac{M}{R_0^2}, \quad \left| \frac{\partial u}{\partial y} \right| \le \frac{M}{R_0^2}, \quad \left| \frac{\partial u}{\partial z} \right| \le \frac{M}{R_0^2}. \qquad (12.12)$$

For, by the uniqueness theorem, such a function must coincide with the function

$$u_1 = \int\int_S \frac{\partial G}{\partial n} [u]_S \, dS,$$

which, as we have just seen, is a harmonic function which vanishes at infinity and takes on S the same values as u. This implies that

$$|u| \leqq \int\int_S \left|\frac{\partial G}{\partial n}\right| |u| \, dS \leqq \max |u|_S \frac{M}{R_0} 4\pi,$$

$$\left|\frac{\partial u}{\partial x_0}\right| \leqq \int\int_S \left|\frac{\partial}{\partial x_0}\left(\frac{\partial G}{\partial n}\right)\right| |u| \, dS \leqq \max |u|_S \frac{M}{R_0^2} 4\pi; \quad \text{and so on.}$$

This proves the proposition. The sphere of radius unity may easily be replaced by one of arbitrary radius by means of a transformation of the variables. We then obtain the following theorem:

THEOREM 1. *For any function u which is harmonic in the neighbourhood of an infinitely distant point and which tends to zero as $R \to \infty$, there is a number M such that the inequalities (12.12) hold good.*

THE DIRICHLET PROBLEM
AND THE NEUMANN PROBLEM
FOR A HALF-SPACE

THE two main types of boundary-value problem for Laplace's equation are the Dirichlet problem and the Neumann problem, and these have already been formulated on p. 25.

We recall that the Dirichlet problem for Laplace's equation consists in determining a function u in the domain Ω with the boundary S to satisfy the equation

$$\nabla^2 u = 0 \tag{13.1}$$

and the boundary conditions

$$[u]_S = f_1(S). \tag{13.2}$$

The Neumann problem consists in finding a solution of the equation (13.1) to satisfy the boundary conditions

$$\left[\frac{\partial u}{\partial n}\right]_S = f_2(S). \tag{13.3}$$

We assume that the functions $f_1(S)$ and $f_2(S)$ are continuous.

We now take as Ω the domain $z > 0$; the plane XOY will serve as the surface S. We shall prove that *in such a domain the solution of the Dirichlet problem, bounded everywhere, is unique; and that the solution of the Neumann problem is determined to within an additive constant.*

To make the solution of the Neumann problem also unique, it is sufficient to impose, for example, the further requirement that $u(x, y, z)$ shall tend to zero when the point (x, y, z) tends to infinity, *i.e.*, that $|u(x, y, z)| < \varepsilon$ if $x^2 + y^2 + z^2 > R(\varepsilon)$.

Suppose, for example, that the Dirichlet problem had two solutions u_1 and u_2. Then their difference $v = u_1 - u_2$ would be a harmonic function vanishing for $z = 0$. We define v for negative values of z so that it is an odd function:

$$v(x, y, z) = -v(x, y, -z).$$

We shall now prove that this function v will be harmonic throughout space including the plane $z = 0$.

We construct a sphere σ of arbitrary radius with its centre on the plane $z = 0$, and define a function v_1 harmonic within this sphere and taking on its surface the values

$$[v_1]_\sigma = [v]_\sigma. \tag{13.4}$$

It is easy to see that v_1 will be zero for $z = 0$. For, the function

$$w_1(x, y, z) = \tfrac{1}{2}[v_1(x, y, z) + v_1(x, y, -z)]$$

will be harmonic and will vanish on the sphere σ; hence $w_1(x, y, 0) = 0$. But $w_1(x, y, 0) = v_1(x, y, 0)$.

The plane $z = 0$ divides our sphere into two half-spheres. The function v_1 coincides with v on the boundary of each half-sphere; on the surface σ this follows from (13.4), and on the part of the plane $z = 0$ both functions are zero. Hence $v = v_1$, and this implies that everywhere within the sphere σ the function v will be differentiable any number of times and will be harmonic within this sphere. Since the position of the centre of the sphere σ is arbitrary, it follows that v will be harmonic throughout space.

Since by hypothesis v is bounded throughout space, Liouville's theorem (Lecture 11) implies that it must be identically equal to some constant. And this constant can only be zero, since $v = 0$ when $z = 0$.

We next prove the uniqueness of solution of the second problem under consideration.

Let u_1 and u_2 be two solutions of the Neumann problem for the half-space. Then the function $v = u_1 - u_2$ satisfies the conditions:

(1) $\nabla^2 v = 0$ for $z > 0$,

(2) $[\partial v/\partial z]_{z=0} = 0$.

Moreover, the function v is bounded throughout the upper half-space, since u_1 and u_2 are.

We define v for negative values of z by means of the formula

$$v(x, y, z) = v(x, y, -z).$$

We shall prove that the function v so defined will be harmonic everywhere including the plane $z = 0$.

Consider the derivative $\partial v/\partial z = w(x, y, z)$. This will be a function which is harmonic in the upper and in the lower half-spaces and which satisfies the conditions:

$$w(x, y, z) = -w(x, y, -z), \quad w(x, y, 0) = 0,$$

and consequently, as we have just proved, it will be harmonic throughout space.

The function

$$\omega(x, y, z) = \int_{z}^{z+1} \frac{\partial v(x, y, z)}{\partial z} = v(x, y, z + 1) - v(x, y, z)$$

will also be harmonic throughout space, as may easily be seen by an immediate differentiation.

Hence it follows that the function $v(x, y, z)$ is also harmonic throughout space. For, the possibility that v might not be harmonic on the plane $z = 0$ is ruled out because

$$v(x, y, z) = v(x, y, z + 1) - \omega(x, y, z), \qquad (13.5)$$

and since the right-hand side of (13.5) is harmonic on this plane, the left-hand side must also be.

Now, by Liouville's theorem, since v is bounded throught space, we have

$$v = \text{constant}.$$

Hence the solution of Neumann's problem is unique to within an additive constant, as we had to show.

We now pass on to the explicit solution of the Dirichlet problem and the Neumann problem.

We shall assume that the harmonic functions considered satisfy the conditions:

$$|u| \leq \frac{\mu}{R^a} : \left|\frac{\partial u}{\partial x}\right| \leq \frac{\mu}{R^{1+a}} : \left|\frac{\partial u}{\partial y}\right| \leq \frac{\mu}{R^{1+a}} : \left|\frac{\partial u}{\partial z}\right| \leq \frac{\mu}{R^{1+a}},$$

where $R = \sqrt{x^2 + y^2 + z^2}$, $a > 0$, and μ is a constant. After we have obtained the explicit solutions, the necessity for this assumption will drop out.

We apply Green's formula (9.4) to the function u, taking for the volume Ω a half-sphere with centre at the origin: $R \leq A$, $z \geq 0$. Since $\nabla^2 u = 0$,

$$u(x_0, y_0, z_0) = \frac{1}{4\pi} \iint_S \left(u \frac{\partial}{\partial n}\left(\frac{1}{r}\right) - \frac{1}{r} \frac{\partial u}{\partial n}\right) dS,$$

where

$$r = \sqrt{(x - x_0)^2 + (y - y_0)^2 + (z - z_0)^2}.$$

The surface S consists of a part S_1 of the plane $z = 0$ and of S_2, a hemispherical surface $R = A$. Letting A tend to infinity, we find

$$\lim_{R \to \infty} \iint_{S_2} \left\{u \frac{\partial}{\partial n}\left(\frac{1}{r}\right) - \frac{1}{r} \frac{\partial u}{\partial n}\right\} dS = 0:$$

for

$$\left| \iint_{S_2} u \frac{\partial}{\partial n} \left(\frac{1}{r} \right) dS \right| \leq \frac{1}{(A - R_0)^2} \iint_{S_2} |u| \, dS \leq \frac{4\pi\mu A^2}{(A - R_0)^2 \, A^a},$$

and

$$\left| \iint_{S_2} \frac{1}{r} \frac{\partial u}{\partial n} dS \right| \leq \frac{1}{A - R_0} \iint_{S_2} \left| \frac{\partial u}{\partial n} \right| dS \leq \frac{4\pi\mu A}{(A - R_0) \, A^a},$$

where

$$R_0 = \sqrt{x_0^2 + y_0^2 + z_0^2}.$$

Hence

$$u(x_0, y_0, z_0) = \lim_{A \to \infty} \frac{1}{4\pi} \iint_{S_1} \left\{ u \frac{\partial}{\partial n} \left(\frac{1}{r} \right) - \frac{1}{r} \frac{\partial u}{\partial n} \right\} dS$$

$$= \frac{1}{4\pi} \iint_{z=0} \left\{ u \frac{\partial}{\partial n} \left(\frac{1}{r} \right) - \frac{1}{r} \frac{\partial u}{\partial n} \right\} dS. \qquad (13.6)$$

We consider together with (x_0, y_0, z_0) its image-point $(x_0, y_0, -z_0)$ in the plane $z = 0$, and let $r_1 = \sqrt{(x - x_0)^2 + (y - y_0)^2 + (z + z_0)^2}$. In the upper half-space $1/r_1$ is a harmonic function, as well as u. Hence

$$\iiint_\Omega \left\{ \frac{1}{r_1} \nabla^2 u - u \nabla^2 \frac{1}{r_1} \right\} dx \, dy \, dz = 0,$$

and consequently

$$\iint_{S_1 + S_2} \left\{ u \frac{\partial}{\partial n} \left(\frac{1}{r_1} \right) - \frac{1}{r_1} \frac{\partial u}{\partial n} \right\} dS = 0.$$

Passing to the limit as A tends to infinity, and using the same inequalities as were used in deriving (13.6), we get

$$\iint_{z=0} \left\{ u \frac{\partial}{\partial n} \left(\frac{1}{r_1} \right) - \frac{1}{r_1} \frac{\partial u}{\partial n} \right\} dS = 0.$$

We now note that on the plane $z = 0$, $r_1 = r$ and $\dfrac{\partial}{\partial u} \left(\dfrac{1}{r_1} \right) = -\dfrac{\partial}{\partial n} \left(\dfrac{1}{r} \right)$ (the radii vectores r_1 and r are symmetrical relative to the plane $z = 0$), and hence

$$\iint_{z=0} \left\{ u \frac{\partial}{\partial n} \left(\frac{1}{r} \right) + \frac{1}{r} \frac{\partial u}{\partial n} \right\} dS = 0. \qquad (13.7)$$

Adding (13.6) and (13.7), we get

$$u(x_0, y_0, z_0) = \frac{1}{2\pi} \iint_{z=0} u \frac{\partial}{\partial n} \left(\frac{1}{r}\right) dS$$

$$= \frac{1}{2\pi} \iint_{z=0} f_1(S) \frac{\partial}{\partial n} \left(\frac{1}{r}\right) dS, \qquad (13.8)$$

and subtracting,

$$u(x_0, y_0, z_0) = -\frac{1}{2\pi} \iint_{z=0} \frac{1}{r} \frac{\partial u}{\partial n} dS$$

$$= -\frac{1}{2\pi} \iint_{z=0} \frac{1}{r} f_2(S) \, dS. \qquad (13.9)$$

We shall show next that (13.8) and (13.9) do actually give the solutions of the Dirichlet and Neumann problems.

We shall assume that $f_1(S) = f_1(x, y)$ and $f_2(S) = f_2(x, y)$ are continuous functions satisfying the inequalities

$$|f_1(x, y)| \leq M, \quad |f_2(x, y)| \leq \frac{M}{\varrho^{1+a}} \qquad (13.10)$$

where $\varrho = \sqrt{x^2 + y^2}$, $a > 0$, and M is a constant. Without loss of generality we may suppose $a < 1$.

We first verify that the integrals on the right-hand side of (13.8) and (13.9) satisfy Laplace's equation. This follows from the fact that, for $z_0 > 0$, we may everywhere differentiate with respect to x_0, y_0, z_0 under the integral sign. Thus, for example,

$$\nabla_0^2 \iint_{z=0} \frac{\partial}{\partial n} \left(\frac{1}{r}\right) f_1(S) \, dS = \iint_{z=0} f_1(S) \frac{\partial}{\partial n} \left[\nabla_0^2 \left(\frac{1}{r}\right)\right] dS = 0.$$

In exactly the same way we can show that the right-hand side of (13.9) satisfies Laplace's equation.

We next examine the behaviour of the right-hand side of (13.8) and (13.9) when $R_0 = \sqrt{x_0^2 + y_0^2 + z_0^2} \to \infty$. We begin with the integral (13.8). We show that this integral is bounded. We have

$$\int_{-\infty}^{+\infty} \int_{-\infty}^{+\infty} \frac{\partial}{\partial n} \left(\frac{1}{r}\right) f_1(x, y) \, dx \, dy = -\int_{-\infty}^{+\infty} \int_{-\infty}^{+\infty} \frac{z_0}{r^3} f_1(x, y) \, dx \, dy.$$

Subject to the conditions (13.10) the integral on the right-hand side clearly converges. Further, we have

$$\left| -\int_{-\infty}^{+\infty}\int_{-\infty}^{+\infty} \frac{z_0}{r^3} f_1(x, y)\, dx\, dy \right| \leq M \int_{-\infty}^{+\infty}\int_{-\infty}^{+\infty} \frac{z_0}{r^3}\, dx\, dy$$

$$= M \int_{-\infty}^{+\infty}\int_{-\infty}^{+\infty} \frac{\partial}{\partial n}\left(\frac{1}{r}\right) dx\, dy.$$

But the last integral is equal to the solid angle which the plane $z = 0$ subtend at the point (x_0, y_0, z_0) and consequently is equal to 2π.

Thus we get

$$\left| \int_{-\infty}^{+\infty}\int_{-\infty}^{+\infty} \frac{\partial}{\partial n}\left(\frac{1}{r}\right) f_1(x, y)\, dx\, dy \right| \leq 2\pi M.$$

Consequently the integral (13.8) is a bounded function of the variables x_0, y_0, z_0.

Considering next the integral (13.9), we first prove the following lemma.

LEMMA. *If* $0 \leq \theta \leq 1$, *then* $x^2 - 2\theta x + 1 \geq \frac{1}{2}(x - 1)^2$.

For, $2\theta x \leq 2x$ if $x \geq 0$, and $2\theta x \leq 0$ if $x \leq 0$. Hence $2\theta x \leq x + |x|$ and $x^2 - 2\theta x + 1 \geq x^2 - x + 1 - |x|$.

But $|x| \leq \frac{1}{2} + \frac{1}{2}x^2$ (this follows from the inequality $|ab| \leq (a^2 + b^2)/2$), and so

$$x^2 - 2\theta x + 1 \geq \frac{1}{2}x^2 - x + \frac{1}{2} = \frac{1}{2}(x - 1)^2.$$

In order to estimate the integral (13.9) at the point $x_0 = \varrho_0$, $y_0 = 0$, z_0 (and, thanks to symmetry, an estimate at this point will give all that is needed), we put

$$\sqrt{\varrho_0^2 + z_0^2} = R_0, \quad \frac{x}{R_0} = \xi, \quad \frac{y}{R_0} = \eta, \quad \frac{\varrho_0}{R_0} = \theta.$$

Then

$$\left| \int_{-\infty}^{+\infty}\int_{-\infty}^{+\infty} \frac{1}{r} f_2(x, y)\, dx\, dy \right|$$

$$\leq \int_{-\infty}^{+\infty}\int_{-\infty}^{+\infty} \frac{1}{\sqrt{x^2 - 2x\varrho_0 + y^2 + z_0^2 + \varrho_0^2}} \frac{M}{\sqrt{(x^2 + y^2)^{1+a}}}\, dx\, dy$$

$$= \frac{M}{R_0^a} \int_{-\infty}^{+\infty}\int_{-\infty}^{+\infty} \frac{1}{\sqrt{\xi^2 - 2\xi\theta + 1 + \eta^2}} \frac{1}{\sqrt{(\xi^2 + \eta^2)^{1+a}}}\, d\xi\, d\eta.$$

By the lemma we have

$$\frac{1}{\sqrt{\xi^2 - 2\xi\theta + 1 + \eta^2}} \leq \frac{1}{\sqrt{\dfrac{(\xi - 1)^2}{2} + \dfrac{\eta^2}{2}}} = \frac{\sqrt{2}}{\sqrt{(\xi - 1)^2 + \eta^2}}$$

and hence

$$\left| \int_{-\infty}^{+\infty} \int_{-\infty}^{+\infty} \frac{1}{r} f_2(x, y) \, dx \, dy \right|$$

$$\leq \frac{M\sqrt{2}}{R_0^a} \int_{-\infty}^{+\infty} \int_{-\infty}^{+\infty} \frac{1}{\sqrt{(\xi - 1)^2 + \eta^2}} \frac{1}{\sqrt{(\xi^2 + \eta^2)^{1+a}}} \, d\xi \, d\eta.$$

The last integral converges absolutely near its three singularities: (a) $\xi = 1$, $\eta = 0$; (b) $\xi = 0$, $\eta = 0$; (c) $\xi = \infty$, $\eta = \infty$. Denoting it by N, we have

$$\left| \int_{-\infty}^{+\infty} \int_{-\infty}^{+\infty} \frac{1}{r} f_2(x, y) \, dx \, dy \right| \leq \frac{MN\sqrt{2}}{R_0^a}.$$

Consequently our function $u(x_0, y_0, z_0)$ vanishes at infinity, as we had to show.

Finally, in order to be certain that we have obtained solutions of the Dirichlet problem and the Neumann problem, we have still to verify that the conditions on $z = 0$ are also satisfied. To do this, it is sufficient to establish that

$$\lim_{z_0 \to 0} \frac{1}{2\pi} \int_{-\infty}^{+\infty} \int_{-\infty}^{+\infty} \frac{z_0}{r^3} f_i(x, y) \, dx \, dy = f_i(x_0, y_0), \quad (i = 1, 2),$$

since the boundary values of the solution of the Dirichlet problem and the boundary values of the normal derivative of the solution of the Neumann problem can both be expressed in this form.

We surround the point (x_0, y_0) with a circle c so that within c

$$\left| f_i(x, y) - f_i(x_0, y_0) \right| < \frac{\varepsilon}{6\pi};$$

the remaining part of the xy-plane outside c we denote by c'. Since the functions $f_i(x, y)$ are bounded, let $|f_i(x, y)| \leq L$.

We take z_0 to be so small that the solid angle subtended by the circle c at the point (x_0, y_0, z_0) is greater than $2\pi - (\varepsilon/3L)$, and consequently the solid angle subtended by the rest of the plane c' will be less than $\varepsilon/3L$. (The sum of these two angles must be 2π.) Then

$$\int_{-\infty}^{+\infty}\int_{-\infty}^{+\infty} \frac{z_0}{r^3} f_i(x, y) \,dx\,dy = \int_{-\infty}^{+\infty}\int_{-\infty}^{+\infty} \frac{\partial}{\partial n}\left(\frac{1}{r}\right) f_i(x, y) \,dx\,dy$$

$$= \int_{-\infty}^{+\infty}\int_{-\infty}^{+\infty} f_i(x, y) \,d\omega$$

$$= \iint_C f_i(x_0, y_0) \,d\omega + \iint_C [f_i(x, y) - f_i(x_0, y_0)] \,d\omega + \iint_{C'} f_i(x, y) \,d\omega.$$

Now

$$\left| 2\pi f_i(x_0, y_0) - \iint_C f_i(x_0, y_0) \,d\omega \right| \leq \left| f_i(x_0, y_0)\left(2\pi - 2\pi + \frac{\varepsilon}{3L}\right) \right| \leq \frac{\varepsilon}{3};$$

$$\left| \iint_C [f_i(x, y) - f_i(x_0, y_0)] \,d\omega \right| \leq \iint_C \frac{\varepsilon}{6\pi} \,d\omega \leq \frac{\varepsilon}{3};$$

$$\left| \iint_{C'} f_i(x, y) \,d\omega \right| \leq L \iint_{C'} \,d\omega \leq \frac{\varepsilon}{3};$$

hence, as was to be shown,

$$\left| \int_{-\infty}^{+\infty}\int_{-\infty}^{+\infty} \frac{z_0}{r^3} f_i(x, y) \,dx\,dy - 2\pi f_i(x_0, y_0) \right| \leq \varepsilon.$$

THE WAVE EQUATION
AND THE RETARDED POTENTIAL

§ 1. The Characteristics of the Wave Equation

We shall consider the *wave equation* in four variables

$$\frac{\partial^2 u}{\partial x^2} + \frac{\partial^2 u}{\partial y^2} + \frac{\partial^2 u}{\partial z^2} - \frac{1}{a^2}\frac{\partial^2 u}{\partial t^2} = F(x, y, z, t) \qquad (14.1)$$

and first of all we shall deal with its characteristics.

The equation of the characteristic surfaces will have the form

$$\left(\frac{\partial \varphi}{\partial x}\right)^2 + \left(\frac{\partial \varphi}{\partial y}\right)^2 + \left(\frac{\partial \varphi}{\partial z}\right)^2 - \frac{1}{a^2}\left(\frac{\partial \varphi}{\partial t}\right)^2 = 0 \qquad (14.2)$$

or

$$\left(\frac{\partial t}{\partial x}\right)^2 + \left(\frac{\partial t}{\partial y}\right)^2 + \left(\frac{\partial t}{\partial z}\right)^2 = \frac{1}{a^2}. \qquad (14.2')$$

We consider in the four-dimensional space of the variables x, y, z, t the surface defined by

$$(x - x_0)^2 + (y - y_0)^2 + (z - z_0)^2 - a^2(t - t_0)^2 = 0. \qquad (14.3)$$

This surface is called the *characteristic conoid*. It is easily verified directly that the implicit function defined by (14.3) satisfies the equation (14.2'). The point (x_0, y_0, z_0, t_0) is a singular point of the surface (14.3). At this point the surface has no tangent plane, since the ratios

$$\cos (n, x): \cos (n, y): \cos (n, z): \cos (n, t)$$

become indeterminate there. By analogy with three-dimensional surfaces this point is called the vertex.

The characteristic surface for the equation of a vibrating membrane will be given by

$$(x - x_0)^2 + (y - y_0)^2 - a^2(t - t_0)^2 = 0.$$

This is the equation of a cone with vertex at the point (x_0, y_0, t_0) in the three-dimensional space of x, y, t.

We can put the equation (14.3) into the form

$$t = t_0 \pm \frac{r}{a} \qquad (14.4)$$

where

$$r = \sqrt{(x - x_0)^2 + (y - y_0)^2 + (z - z_0)^2} \, .$$

Depending on the choice of sign in (14.4) we obtain either the upper or the lower half of the conoid. We shall use the lower half, *i.e.*, the part defined by the equation

$$t = t_0 - \frac{r}{a}.$$

§ 2. Kirchhoff's Method of Solution of Cauchy's Problem

The idea underlying Kirchhoff's method of solution of Cauchy's problem for the wave equation is the same as that of the solution of a boundary-value problem of the first kind for a hyperbolic equation by the method of successive approximations, which we have already discussed in Lecture 5, § 1. A characteristic conoid is constructed with vertex at the given point (x_0, y_0, z_0). As was explained in Lecture 3, the values of the function u and its derivatives on this conoid are related by certain partial differential equations in three variables which can be derived from (14.2) by virtue of the wave equation. We shall see after a detailed examination that this circumstance enables the value of the unknown function at the vertex of the conoid to be expressed simply in terms of the known data, in the same way as in Lecture 4 the value of the function $\partial u/\partial \eta$ at a certain point was expressed in terms of its value at some other point on the same characteristic by means of quadrature.

Let us begin the examination of the method. In order to obtain the required relation on the characteristic conoid, we start off in the same way as in Lecture 3, § 4. We transform equation (14.1) by introducing new independent variables

$$x_1 = x, \quad y_1 = y, \quad z_1 = z, \quad t_1 = t - t_0 + \frac{r}{a}.$$

We write

$$u\left(x_1, y_1, z_1, t_1 + t_0 - \frac{r}{a}\right) = u_1(x_1, y_1, z_1, t_1)$$

and similarly for other functions.

Then

$$\frac{\partial u}{\partial x} = \frac{\partial u_1}{\partial x_1} + \frac{\partial u_1}{\partial t_1} \frac{x_1 - x_0}{ra}, \quad \frac{\partial u}{\partial t} = \frac{\partial u_1}{\partial t_1}, \quad \frac{\partial^2 u}{\partial t^2} = \frac{\partial^2 u_1}{\partial t_1^2},$$

$$\frac{\partial^2 u}{\partial x^2} = \frac{\partial^2 u_1}{\partial x_1^2} + 2 \frac{\partial^2 u_1}{\partial x_1 \partial t_1} \frac{x_1 - x_0}{ra} + \frac{\partial^2 u_1}{\partial t_1^2} \frac{(x_1 - x_0)^2}{r^2 a^2}$$

$$+ \left(\frac{1}{ra} - \frac{(x_1 - x_0)^2}{r^3 a} \right) \frac{\partial u_1}{\partial t_1}.$$

(It is clear that, with our substitution, $r_1 = r$ and we shall always write r.)
Equations (14.1) becomes

$$\frac{\partial^2 u_1}{\partial x_1^2} + \frac{\partial^2 u_1}{\partial y_1^2} + \frac{\partial^2 u_1}{\partial z_1^2} + \frac{2}{a} \frac{\partial}{\partial r} \left(\frac{\partial u_1}{\partial t_1} \right) + \frac{2}{ra} \frac{\partial u_1}{\partial t_1}$$

$$+ \left[\frac{(x_1 - x_0)^2}{r^2 a^2} + \frac{(y_1 - y_0)^2}{r^2 a^2} + \frac{(z_1 - z_0)^2}{r^2 a^2} - \frac{1}{a^2} \right] \frac{\partial^2 u_1}{\partial t_1^2}$$

$$= F_1(x_1, y_1, z_1, t_1)$$

where $\partial/\partial r$ stands for

$$\frac{x_1 - x_0}{r} \frac{\partial}{\partial x_1} + \frac{y_1 - y_0}{r} \frac{\partial}{\partial y_1} + \frac{z_1 - z_0}{r} \frac{\partial}{\partial z_1}.$$

Since

$$\frac{x_1 - x_0}{r} = \cos(x, r), \quad \frac{y_1 - y_0}{r} = \cos(y, r), \quad \frac{z_1 - z_0}{r} = \cos(z, r),$$

the operator $\partial/\partial r$ denotes the derivative taken in the direction of the radius r from the point (x_0, y_0, z_0) to the given point (x_1, y_1, z_1).
Since

$$\frac{(x_1 - x_0)^2}{a^2 r^2} + \frac{(y_1 - y_0)^2}{a^2 r^2} + \frac{(z_1 - z_0)^2}{a^2 r^2} - \frac{1}{a^2} = 0,$$

our equation may be written as

$$\nabla^2 u_1 + \frac{2}{ar} \frac{\partial}{\partial r} \left(r \frac{\partial u_1}{\partial t_1} \right) = F_1(x_1, y_1, z_1, t_1),$$

or

$$\frac{1}{r} \nabla^2 u_1 + \frac{2}{ar^2} \frac{\partial}{\partial r} \left(r \frac{\partial u_1}{\partial t_1} \right) = \frac{1}{r} F_1(x_1, y_1, z_1, t_1). \quad (14.5)$$

Equation (14.5) enables certain important particular solutions of the wave equation to be constructed immediately. Suppose $F_1 = 0$. Then the function $u = \dfrac{\Phi(t_1)}{r}$, where Φ is an arbitrary, twice-differentiable function, gives us a solution of the equation (14.5), since $\dfrac{\partial u_1}{\partial t_1} = \dfrac{1}{r} \Phi'(t_1)$ and therefore both terms on the left-hand side of (14.5) vanish. Replacing t_1 by its expression in terms of x, y, z, t, we have

$$u = \frac{1}{r} \Phi\left(t - t_0 + \frac{r}{a}\right).$$

In the given case the parameter t_0 is not essential and we can take the solution in the form

$$u = \frac{1}{r} \Phi_1\left(t + \frac{r}{a}\right).$$

It may easily be verified that

$$u = \frac{1}{r} \Phi_3\left(-t + \frac{r}{a}\right)$$

will also be a solution of the wave equation, since the latter is unchanged when t is replaced by $-t$. Putting

$$\Phi_3\left(-t + \frac{r}{a}\right) = \Phi_2\left(t - \frac{r}{a}\right)$$

and adding both particular solutions, we get

$$u = \frac{1}{r}\left[\Phi_1\left(t + \frac{r}{a}\right) + \Phi_2\left(t - \frac{r}{a}\right)\right]. \tag{14.6}$$

In appearance (14.6) is reminiscent of d'Alembert's solution of the equation for a vibrating string. The solution (14.6) represents *spherical waves*. The first term represents a spherical wave of constant form moving towards the point $r = 0$, its amplitude increasing as it approaches the centre. The second term represents a spherical wave of constant form moving away from the point $r = 0$ towards infinity, its amplitude decreasing as it moves away like $1/r$.

If Φ_1 and Φ_2 are different from zero only over a finite interval in the range of variation of their arguments, then at any point in space the function u will be zero, *i.e.*, there will be a state of rest, both before the arrival of the spherical waves and after their passage.

As we shall see later, these solutions for the wave equation play the same sort of role as the function $1/r$ played for Laplace's equation.

We shall integrate both sides of equation (14.5) over a certain domain Ω of the space x_1, y_1, z_1, the point (x_0, y_0, z_0) being included in Ω. S is the boundary of Ω. For convenience we shall separate the point (x_0, y_0, z_0) from this domain by means of a small sphere σ of radius ε and volume τ. Finally we shall take the limit as $\varepsilon \to 0$. We get

$$\lim_{\varepsilon \to 0} \iiint_{\Omega - \tau} \left[\frac{1}{r} \nabla^2 u_1 + \frac{2}{ar^2} \frac{\partial}{\partial r} \left(r \frac{\partial u_1}{\partial t_1} \right) \right] dx_1 \, dy_1 \, dz_1$$

$$= \iiint_{\Omega} \frac{1}{r} F_1(x_1, y_1, z_1, t_1) \, dx_1 \, dy_1 \, dz_1 . \tag{14.7}$$

Before proceeding to the limit with the left-hand side, we transform it somewhat. By Green's formula (9.4) we have

$$\lim_{\varepsilon \to 0} \iiint_{\Omega - \tau} \frac{1}{r} \nabla^2 u_1 \, dx_1 \, dy_1 \, dz_1 = \iiint_{\Omega} \frac{1}{r} \nabla^2 u_1 \, dx_1 \, dy_1 \, dz_1$$

$$= -4\pi u_1(x_0, y_0, z_0, t - t_0) + \iint_S \left\{ u_1 \frac{\partial}{\partial n} \left(\frac{1}{r} \right) - \frac{1}{r} \frac{\partial u_1}{\partial n} \, dS \right\}, \tag{14.8}$$

since $t_1 = t - t_0$ when $x_1 = x_0, y_1 = y_0, z_1 = z_0$. Further, introducing polar coordinates, we have

$$dx_1 \, dy_1 \, dz_1 = r^2 \sin \theta \, d\theta \, d\varphi \, dr$$

and

$$I_\sigma \equiv \iiint_{\Omega - \tau} \frac{2}{ar^2} \frac{\partial}{\partial r} \left(r \frac{\partial u_1}{\partial t_1} \right) dx_1 \, dy_1 \, dz_1$$

$$= \frac{2}{a} \iiint_{\Omega - \tau} \frac{\partial}{\partial r} \left(r \frac{\partial u_1}{\partial t_1} \right) \sin \theta \, d\theta \, d\varphi \, dr$$

$$= \frac{2}{a} \iint_{S + \sigma} r \frac{\partial u_1}{\partial t_1} \left[- \text{sign} \cos (r, n) \right] \sin \theta \, d\theta \, d\varphi = \frac{2}{a} \iint_{S + \sigma} r \frac{\partial u_1}{\partial t_1} \, d\omega$$

where ω denotes a solid angle [see (9.8)]. Continuing the transformation, we have

$$I_\sigma = \frac{2}{a} \iint_{S + \sigma} r \frac{\partial u_1}{\partial t_1} \frac{\partial}{\partial n} \left(\frac{1}{r} \right) dS = -\frac{2}{a} \iint_{S + \sigma} \frac{1}{r} \frac{\partial r}{\partial n} \frac{\partial u_1}{\partial t_1} \, dS .$$

The limit of $\iint_\sigma \frac{1}{r} \frac{\partial r}{\partial n} \frac{\partial u_1}{\partial t_1} \cdot dS$ is obviously zero if $\partial u_1 / \partial t_1$ is bounded, since $\partial r / \partial n$ is bounded and the integral is less than $\frac{M}{\varepsilon} \iint_\sigma dS = 4\pi\varepsilon M$, where M is a certain constant. Consequently

$$\lim_{\varepsilon \to 0} I_\sigma = -\frac{2}{a} \iint_S \frac{1}{r} \frac{\partial r}{\partial n} \frac{\partial u_1}{\partial t_1} \, dS. \tag{14.9}$$

The formulae (14.7), (14.8), (14.9) give

$$-4\pi u_1(x_0, y_0, z_0, t - t_0) + \iint_S \left\{ u_1 \frac{\partial}{\partial n}\left(\frac{1}{r}\right) - \frac{1}{r}\frac{\partial u_1}{\partial n} - \frac{2}{ra}\frac{\partial r}{\partial n}\frac{\partial u_1}{\partial t_1} \right\} dS$$

$$= \iiint_\Omega \frac{1}{r} F_1(x_1, y_1, z_1, t_1) \, dx_1 \, dy_1 \, dz_1.$$

We now put $t_1 = 0$; then $t = t_0 - (r/a)$. If further, $x_1 = x_0, y_1 = y_0, z_1 = z_0$, then $t = t_0$. Therefore

$$u_1(x_0, y_0, z_0, 0) = u(x_0, y_0, z_0, t_0).$$

$$F_1(x_1, y_1, z_1, 0) = F\left(x, y, z, t_0 - \frac{r}{a}\right)$$

and our formula can be written as

$$u(x_0, y_0, z_0, t_0) = \frac{1}{4\pi} \iint_S \left[u_1 \frac{\partial}{\partial n}\left(\frac{1}{r}\right) - \frac{1}{r}\frac{\partial u_1}{\partial n} - \frac{2}{ra}\frac{\partial r}{\partial n}\frac{\partial u_1}{\partial t_1} \right]_{t_1 = 0} dS$$

$$- \frac{1}{4\pi} \iiint_\Omega \frac{1}{r} F\left(x, y, z, t_0 - \frac{r}{a}\right) dx \, dy \, dz. \tag{14.10}$$

(14.10) is called *Kirchhoff's formula;* as we shall soon see, it enables a solution of Cauchy's problem for the wave equation to be found.

This formula is very similar in form to Green's formula, which we derived earlier. If u_1, $\partial u_1/\partial n$, and $\partial u_1/\partial t_1$ are specified on the surface S, then the right-hand side of (14.10) will be a known function. The integrals occurring on the right-hand side of (14.10) are commonly called the *retarded potentials*. We may explain this nomenclature by taking as an example the last integral

$$\frac{1}{4\pi} \iiint_\Omega \frac{1}{r} F\left(x, y, z, t_0 - \frac{r}{a}\right) dx \, dy \, dz. \tag{14.11}$$

This integral differs from the Newtonian potential only in that the function F enters into it not with the argument t_0 but with the *retarded* argument $t_0 - (r/a)$.

We now pass on to the solution of Cauchy's problem, *i.e.*, to the solution of equation (14.1) subject to the conditions

$$[u]_{t=0} = \varphi_0(x, y, z), \qquad \left[\frac{\partial u}{\partial t}\right]_{t=0} = \varphi_1(x, y, z). \tag{14.12}$$

As the surface S we take the surface defined by $t_0 - (r/a) = 0$; then on it $t = 0$ when $t_1 = 0$.

The domain bounded by S is a sphere of radius at_0 about the point (x_0, y_0, z_0), and consequently formula (14.10) may be applied to it. For $t = 0$ the conditions (14.12) define all the first-order derivatives of u and so of u_1 also. We shall have:

$$\left[\frac{\partial u}{\partial t}\right]_{t=0} = \left[\frac{\partial u_1}{\partial t_1}\right]_{t=0} = \varphi_1(x, y, z),$$

$$\frac{\partial u_1}{\partial} = \frac{\partial u_1}{\partial x_1} \cos(n, x) + \frac{\partial u_1}{\partial y_1} \cos(n, y) + \frac{\partial u_1}{\partial z_1} \cos(n, z)$$

$$= \left(\frac{\partial u}{\partial x} - \frac{1}{a}\frac{\partial u}{\partial t}\frac{\partial r}{\partial x}\right)\cos(n, x) + \left(\frac{\partial u}{\partial y} - \frac{1}{a}\frac{\partial u}{\partial t}\frac{\partial r}{\partial y}\right)\cos(n, y)$$

$$+ \left(\frac{\partial u}{\partial z} - \frac{1}{a}\frac{\partial u}{\partial t}\frac{\partial r}{\partial z}\right)\cos(n, z) = \frac{\partial u}{\partial n} - \frac{1}{a}\frac{\partial u}{\partial t}\frac{\partial r}{\partial n}$$

and consequently

$$\left[\frac{\partial u_1}{\partial n}\right]_{t=0} = \frac{\partial \varphi_0}{\partial n} - \frac{1}{a}\varphi_1(x, y, z)\frac{\partial r}{\partial n},$$

and finally

$$u(x_0, y_0, z_0, t_0) = \frac{1}{4\pi}\iint_{r=at_0}\left\{\varphi_0\frac{\partial}{\partial n}\left(\frac{1}{r}\right) - \frac{1}{r}\frac{\partial \varphi_0}{\partial n} - \frac{1}{ar}\frac{\partial r}{\partial n}\varphi_1\right\}dS$$

$$- \frac{1}{4\pi}\iiint_{r\le at_0}\frac{1}{r}F\left(x, y, z, t_0 - \frac{r}{a}\right)dx\,dy\,dz \quad (14.13)$$

The formula (14.13) gives an explicit expression for the value of the unknown function at any point (x_0, y_0, z_0) at an arbitrary moment of time $t_0 > 0$. It also shows that the solution of Cauchy's problem for the wave equation, if it exists, is unique. We shall show later that the function which we have obtained does satisfy the wave equation and the initial conditions, and we shall also establish that the Cauchy problem was correctly formulated.

When $F = 0$, the formula (14.13) can be put into another form known as *Poisson's formula*, which is often encountered in the literature.

Let $T_\varrho\{\psi\}$ denote the arithmetic mean of the values of the function ψ on a sphere of radius ϱ drawn about the point (x_0, y_0, z_0):

$$T_\varrho\{\psi\} = \frac{1}{4\pi}\int_0^{2\pi}\int_0^\pi \psi(\varrho, \theta, \varphi)\sin\theta\,d\theta\,d\varphi,$$

where ϱ, θ, φ are the polar coordinates with origin at (x_0, y_0, z_0) and are related to the rectangular Cartesian coordinates by

$$x = x_0 + \varrho \sin \theta \cos \varphi$$

$$y = y_0 + \varrho \sin \theta \sin \varphi$$

$$z = z_0 + \varrho \cos \theta$$

$$0 \leq \varphi \leq 2\pi, \quad 0 \leq \theta \leq \pi.$$

With this notation, (14.13) may be written in the following form (Poisson's formula):

$$u(x_0, y_0, z_0, t_0) = t_0 T_{at_0} \{\varphi_1\} + \frac{\partial}{\partial t_0} [t_0 T_{at_0}\{\varphi_0\}].$$

Proof. On the surface $\varrho = at_0$, we have

$$\frac{\partial \varrho}{\partial n} = -\frac{\partial \varrho}{\partial \varrho} = -1, \quad dS = a^2 t_0^2 \sin \theta \, d\theta \, d\varphi$$

and consequently

$$-\frac{1}{4\pi} \iint_{\varrho = at_0} \frac{1}{a\varrho} \frac{\partial \varrho}{\partial n} \varphi_1 \, dS = \frac{t_0}{4\pi} \iint_{\varrho = at_0} \varphi_1(\varrho, \theta, \varphi) \sin \theta \, d\theta \, d\varphi.$$

Further,

$$\frac{\partial \varphi_0}{\partial n} = -\frac{\partial \varphi_0}{\partial \varrho}, \quad \frac{\partial}{\partial n}\left(\frac{1}{\varrho}\right) = \frac{1}{\varrho^2} \frac{\partial \varrho}{\partial \varrho} = \frac{1}{a^2 t_0^2},$$

whence, after some reduction, we get

$$\frac{1}{4\pi} \iint_{\varrho = at_0} \left\{ \varphi_0 \frac{\partial}{\partial n}\left(\frac{1}{\varrho}\right) - \frac{1}{\varrho} \frac{\partial \varphi_0}{\partial n} \right\} dS$$

$$= \frac{1}{4\pi} \iint_{\varrho = at_0} \varphi_0 \sin \theta \, d\theta \, d\varphi + t_0 \frac{\partial}{\partial t_0} \frac{1}{4\pi} \iint_{\varrho = at_0} \varphi_0 \sin \theta \, d\theta \, d\varphi,$$

as was to be shown.

We now note some important consequences of formula (14.13).

Suppose there are no external disturbing forces, so that $F = 0$, and suppose that the initial disturbance at $t = 0$ is concentrated in a certain bounded domain ω. We shall investigate the behaviour of the solution at some point (x_0, y_0, z_0) lying outside the domain ω. Let δ be the distance from the domain ω to the point (x_0, y_0, z_0). For $t_0 < \delta/a$, the sphere S whose equation is $r = at_0$ will lie entirely outside the domain ω, and so the result of substituting such a value of t_0 in the right-hand side of (14.13) will be zero. At $t_0 = \delta/a$, u will begin to vary and will do so while S intersects the domain ω. Then at $t_0 = D/a$, where D is the greatest distance of a point of ω from (x_0, y_0, z_0), u will again become zero and will be zero thereafter. The further the point (x_0, y_0, z_0) is from ω, the later the disturbance will arrive there and the later the disturbance will pass this point. At any given instant t_0, we can construct a surface S_1 separating the points which the disturbance has not yet reached from those at which it has already arrived. This surface is called the leading wave-front.

A second surface S_2 separates the points at which the disturbance is still occurring from those points at which oscillation has ceased. This surface is called the rear wave-front.

The existence of the rear wave-front is to be explained by the fact that a sound emitted by a source does not die away gradually at a given point in

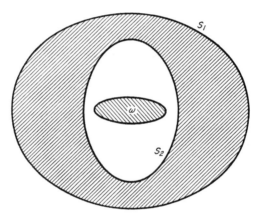

Fig. 12.

space but ceases at once after the sound wave has passed. If this were not so, sounds would merge into one another, like the sound of the notes of a piano when the damping pedal is raised.

Figure 12 depicts the leading and rear wave-fronts arising from a disturbance in the bounded domain ω.

We pass on now to the proof that the Cauchy problem is correctly formulated.

If in formula (14.13) we replace the functions φ_0 and φ_1 by φ_0^* and φ_1^* such that

$$\left|\varphi_0 - \varphi_0^*\right| < \varepsilon, \quad \left|\varphi_1 - \varphi_1^*\right| < \varepsilon,$$

$$\left|\frac{\partial \varphi_0}{\partial x} - \frac{\partial \varphi_0^*}{\partial x}\right| < \varepsilon, \quad \left|\frac{\partial \varphi_0}{\partial y} - \frac{\partial \varphi_0^*}{\partial y}\right| < \varepsilon, \quad \left|\frac{\partial \varphi_0}{\partial z} - \frac{\partial \varphi_0^*}{\partial z}\right| < \varepsilon,$$

then the new solution u^* of the Cauchy problem which is obtained from

(14.13) using the new initial conditions will differ but slightly from the solution obtained using the old conditions. For,

$$
\begin{aligned}
|u^* - u| = \frac{1}{4\pi} \Bigg| \iint_{r=at_0} &\left\{ (\varphi_0^* - \varphi_0) \frac{\partial}{\partial n} \left(\frac{1}{r}\right) - \frac{1}{r} \left(\frac{\partial \varphi_0^*}{\partial n} - \frac{\partial \varphi_0}{\partial n}\right) \right. \\
&\left. - \frac{1}{ar} \frac{\partial r}{\partial n} (\varphi_1^* - \varphi_1) \right\} dS \Bigg| \le M\varepsilon
\end{aligned}
$$

where M is a certain constant depending only on t_0.

Consequently the solution u depends on the function φ_0 continuously to order $(1, 0)$ and on the functions φ_1 continuously to order $(0, 0)$, in the sense defined in Lecture 2.

We next prove that the solution which we have derived does actually satisfy the wave equation. It is sufficient to prove this for the case when φ_0 and φ_1 are identically zero, since we shall then be able to establish the existence of the solution for any two functions φ_0 and φ_1 which continuously have second-order derivatives. For, if we prove that the solution exists for the zero initial conditions, we shall also have proved the existence of a solution of the equation

$$
\nabla^2 w - \frac{1}{a^2} \frac{\partial^2 w}{\partial t^2} = F - \nabla^2 v + \frac{1}{a^2} \frac{\partial^2 v}{\partial t^2}
$$

subject to the conditions

$$
[w]_{t=0} = 0 = \left[\frac{\partial w}{\partial t}\right]_{t=0},
$$

where v is an arbitrary function.

If v satisfies the conditions

$$
[v]_{t=0} = \varphi_0, \quad \left[\frac{\partial v}{\partial t}\right]_{t=0} = \varphi_1,
$$

and such functions obviously exist (*e.g.*, $v = \varphi_0 + t\varphi_1$), it will then follow that a function $u = w + v$ exists satisfying the Cauchy conditions and the wave equation (14.1). If the solution of this last problem does indeed exist, then it is expressed by the formula (14.13).

Putting then $\varphi_0 = \varphi_1 = 0$ in formula (14.13), we get

$$
u(x_0, y_0, z_0, t_0) = -\frac{1}{4\pi} \iiint_{r \le at_0} \frac{1}{r} F\left(x, y, z, t_0 - \frac{r}{a}\right) dx \, dy \, dz.
$$

$$(14.14)$$

We shall prove that the function $u(x_0, y_0, z_0, t_0)$ given by (14.14) does in fact satisfy the wave equation. A direct verification of this circumstance would demand an enormous amount of algebraic reduction, and we prefer to choose another way.

We shall prove an important integral identity which, in a certain sense, is equivalent to the wave equation for the function u defined by the formula (14.14).

We consider an arbitrary function $\psi(x_0, y_0, z_0, t_0)$ which vanishes everywhere except within a certain sphere c in four-dimensional space with its centre at the point (x_0, y_0, z_0, t_0) and which has everywhere continuous derivatives up to some order. This function will satisfy a certain equation

$$\nabla^2 \psi - \frac{1}{a^2} \frac{\partial^2 \psi}{\partial t^2} = \Psi \tag{14.15}$$

where Ψ is to be calculated by direct differentiation.

According to our hypothesis,

$$[\psi]_{t=T} = 0, \quad \left[\frac{\partial \psi}{\partial t} \right]_{t=T} = 0$$

for sufficiently large T.

Further, the function ψ, as the solution of Cauchy's problem for the equation (14.15), is given, for $t < T$, by the formula

$$\psi(x_0, y_0, z_0, t_0) = -\frac{1}{4\pi} \iiint_{r \leqq a(T-t_0)} \frac{1}{r} \Psi\left(x, y, z, t_0 + \frac{r}{a}\right) dx\, dy\, dz. \tag{14.16}$$

(We have not derived this equation, but it is obtained at once by, firstly, a change of variable from t to $T - t^*$ in the equation, and then, having transformed the data, by writing down the solution of the equation, and finally returning from the variable t^* to the variable t.)

In formula (14.16) we could have left out the limits of integration, since, if $r \geqq a(T - t_0)$ the function $\Psi(x, y, r, t_0 + (r/a))$ obviously vanishes, because $t_0 + (r/a) \geqq T$.

Let $F(x_0, y_0, z_0, t_0) = 0$ for $t_0 < 0$.

We multiply the function $\psi(x_0, y_0, z_0, t_0)$ by $F(x_0, y_0, z_0, t_0)$ and integrate throughout the space. Then we shall have

$$\int_{-\infty}^{+\infty} \int_{-\infty}^{+\infty} \int_{-\infty}^{+\infty} \int_{-\infty}^{+\infty} \psi(x_0, y_0, z_0, t_0) F(x_0, y_0, z_0, t_0)\, dx_0\, dy_0\, dz_0\, dt_0$$

$$= -\frac{1}{4\pi} \int_{-\infty}^{+\infty} \int_{-\infty}^{+\infty} \int_{-\infty}^{+\infty} \int_{-\infty}^{+\infty} F(x_0, y_0, z_0, t_0)$$

$$\times \left\{ \int_{-\infty}^{+\infty} \int_{-\infty}^{+\infty} \int_{-\infty}^{+\infty} \frac{1}{r} \Psi\left(x, y, z, t_0 + \frac{r}{a}\right) dx\, dy\, dz \right\} dx_0\, dy_0\, dz_0\, dt_0.$$

The last integral can be written in the form

$$-\frac{1}{4\pi} \int_{-\infty}^{+\infty} \int_{-\infty}^{+\infty} \int_{-\infty}^{+\infty} \int_{-\infty}^{+\infty} \int_{-\infty}^{+\infty} \int_{-\infty}^{+\infty} \int_{-\infty}^{+\infty} \frac{1}{r} \, \Psi\left(x, y, z, t_0 + \frac{r}{a}\right)$$

$$\times \, F(x_0, y_0, z_0, t_0) \, dx \, dy \, dz \, dx_0 \, dy_0 \, dz_0 \, dt_0$$

since the integrand is different from zero in a bounded domain of its independent variables.

If r/a or t_0 is sufficiently great, then $t_0 + (r/a)$ will also be great, since $t_0 > 0$. We change the variables, putting $t_0 + (r/a) = t$. Then we have

$$\int_{-\infty}^{+\infty} \int_{-\infty}^{+\infty} \int_{-\infty}^{+\infty} \int_{-\infty}^{+\infty} \psi(x_0, y_0, z_0, t_0) \, F(x_0, y_0, z_0, t_0) \, dx_0 \, dy_0 \, dz_0 \, dt_0$$

$$= -\frac{1}{4\pi} \int_{-\infty}^{+\infty} \int_{-\infty}^{+\infty} \int_{-\infty}^{+\infty} \int_{-\infty}^{+\infty} \Psi(x, y, z, t) \left\{ \int_{-\infty}^{+\infty} \int_{-\infty}^{+\infty} \int_{-\infty}^{+\infty} \frac{1}{r} \right.$$

$$\times \left. F\left(x_0, y_0, z_0, t - \frac{r}{a}\right) dx_0 \, dy_0 \, dz_0 \right\} dx \, dy \, dz \, dt.$$

For, the inner integral is meaningful, since the integrand for fixed x, y, z, t is different from zero only in a bounded domain. Hence

$$\int_{-\infty}^{+\infty} \int_{-\infty}^{+\infty} \int_{-\infty}^{+\infty} \int_{-\infty}^{+\infty} \psi(x, y, z, t) \, F(x, y, z, t) \, dx \, dy \, dz \, dt$$

$$= \int_{-\infty}^{+\infty} \int_{-\infty}^{+\infty} \int_{-\infty}^{+\infty} \int_{-\infty}^{+\infty} \left(\nabla^2 \psi - \frac{1}{a^2} \frac{\partial^2 \psi}{\partial t^2}\right) u(x, y, z, t) \, dx \, dy \, dz \, dt \quad (14.17)$$

where

$$u(x, y, z, t)$$

$$= -\frac{1}{4\pi} \int_{-\infty}^{+\infty} \int_{-\infty}^{+\infty} \int_{-\infty}^{+\infty} \frac{1}{r} F\left(x_0, y_0, z_0, t - \frac{r}{a}\right) dx_0 \, dy_0 \, dz_0.$$

The identity (14.17) is the fundamental integral identity which the function u satisfies.

We shall show that if u has continuous second-order derivatives, then it satisfies equation (14.1).

For, the operator

$$\nabla^2 - \frac{1}{a^2} \frac{\partial^2}{\partial t^2},$$

as it is not difficult to see, is self-adjoint (see Lecture 5, § 2). Hence the integral

$$\iiiint_{\Omega} \left[u \left(\nabla^2 \psi - \frac{1}{a^2} \frac{\partial^2 \psi}{\partial t^2} \right) - \psi \left(\nabla^2 u - \frac{1}{a^2} \frac{\partial^2 u}{\partial t^2} \right) \right] dx\, dy\, dz\, dt$$

$$= \iiiint_{\Omega} \left(\frac{\partial P_x}{\partial x} + \frac{\partial P_y}{\partial y} + \frac{\partial P_z}{\partial z} + \frac{\partial P_t}{\partial t} \right) dx\, dy\, dz\, dt$$

transforms into an integral taken over the surface S bounding the volume [see (5.15)]. If we take the volume Ω sufficiently large so that on the surface S the function ψ and all its first derivatives vanish, then the last integral is equal to zero and we get

$$\iiiint_{\Omega} u \left(\nabla^2 \psi - \frac{1}{a^2} \frac{\partial^2 \psi}{\partial t^2} \right) dx\, dy\, dz\, dt$$

$$= \iiiint_{\Omega} \psi \left(\nabla^2 u - \frac{1}{a^2} \frac{\partial^2 u}{\partial t^2} \right) dx\, dy\, dz\, dt$$

whence, by (14.17),

$$\iiiint_{\Omega} \psi(x, y, z, t) \left(\nabla^2 u - \frac{1}{a^2} \frac{\partial^2 u}{\partial t^2} - F \right) dx\, dy\, dz\, dt = 0.$$

The last integral vanishes for any function ψ; hence

$$\nabla^2 u - \frac{1}{a^2} \frac{\partial^2 u}{\partial t^2} = F,$$

and we have shown that the function u satisfies the wave equation.

It remains to show that $u = 0$ and $\partial u / \partial t_0 = 0$ when $t_0 = 0$.

That $u = 0$ when $t_0 = 0$ is an immediate consequence of formula (14.14). To show that $\partial u / \partial t_0 = 0$ when $t_0 = 0$, we change the variables in the integral on the right-hand side of (14.14), putting

$$x = x_0 + at_0 \xi, \quad y = y_0 + at_0 \eta, \quad z = z_0 + at_0 \zeta.$$

Then

$$r = \sqrt{(x - x_0)^2 + (y - y_0)^2 + (z - z_0)^2} = at_0 \varrho,$$

where

$$\varrho = \sqrt{\xi^2 + \eta^2 + \zeta^2}.$$

In the new variables

$$u(x_0, y_0, z_0, t_0)$$

$$= -\frac{a^2 t_0^2}{4\pi} \iiint_{\varrho \leq 1} \frac{1}{\varrho} F[x_0 + at_0 \xi, y + at_0 \eta, z_0 + at_0 \zeta, t_0(1 - \varrho)]\, d\xi\, d\eta\, d\zeta.$$

The right-hand member may be differentiated with respect to t_0 under the integral sign, and then, putting $t_0 = 0$, we have

$$\left[\frac{\partial u}{\partial t_0}\right]_{t_0=0} = 0.$$

For the justification of differentiating under the integral sign, assuming the boundedness of the first-order derivatives of F, see § 2 of Lecture 7.

We have purposely not gone into the question here as to how many continuous derivatives it is necessary to have in the initial data in order that our formulae shall give a solution of the problem possessing the required number of continuous derivatives. As we shall see later, this question has no real significance if we make use of the concept of the generalized solution of the wave equation, which we shall examine in a subsequent lecture.

We make one more important observation.

As we have seen, the solution of the equation for a vibrating string was just as smooth as the initial conditions, i.e., it possessed just as many continuous derivatives as did the functions entering into the initial conditions. The solution of the equation for heat conduction turned out to be smoother than the initial conditions. In this respect, the solution of the wave equation is distinguished from the other problems considered. It appears to be, in general, less smooth than the initial conditions. This is seen from the very fact that the function u is expressed in Kirchhoff's formula in terms of the integral of the normal derivative $[\partial u/\partial n]_{t=0}$. The value of a derivative of order k of a function satisfying the wave equation is thus related to the initial values of derivatives of order $(k + 1)$ of the functions in the initial conditions.

In this lecture we have analysed the Cauchy problem only for the case where the initial data are related to a surface $t = 0$. But exactly the same method enables a solution of the Cauchy problem to be constructed for the general case when the initial data are specified on a hypersurface $t = \psi(x, y, z)$, and also the solution of a problem similar to the boundary-value problem of the first kind for a hyperbolic equation in a plane. The detailed analysis of these problems we leave for the reader.

LECTURE 15

PROPERTIES OF THE POTENTIALS OF
SINGLE AND DOUBLE LAYERS

§ 1. General Remarks

In order to examine the Dirichlet problem and the Neumann problem
for domains other than a sphere or a half-space, we shall have to study in
detail the behaviour of the integrals

$$I_1 = \iint_S \frac{\partial}{\partial n}\left(\frac{1}{r}\right) f_1(S)\, dS \quad \text{and} \quad I_2 = \iint_S \frac{f_2(S)\, dS}{r} \quad (15.1)$$

which we have already met more than once. As we mentioned earlier, the
integral I_2 is called the potential of a single layer, the function $f_2(S)$ being
its density. The integral I_1 is called the potential of a double layer, and $f_1(S)$
is its density. We shall assume that the functions $f_1(S)$ and $f_2(S)$ are continuous.

We shall say that a surface S is smooth in the Lyapunov sense or simply
that it is a *Lyapunov surface* if the following conditions are satisfied:

(a) at each point of the surface S it has a tangent plane;

(b) about any point P_0 of the surface a sphere can be described with radius
h (h independent of P_0) so that the section $\sum$ of the surface S which falls
within the sphere meets lines parallel to the normal n_0 at the point P_0
no more than once;

(c) if P_1, P_2 are two points of the surface, and n_1, n_2 are unit vectors directed
along the outward normals to the surface S at these points, then the
vector $n_1 - n_2$ satisfies the inequality

$$\left| n_1 - n_2 \right| \leq A r^\delta,$$

where A and δ are constants, $0 < \delta \leq 1$, and r is the distance between
the points P_1 and P_2;

(d) the solid angle ω_σ which any part σ of the surface S subtends at an
arbitrary point P_0 is bounded:

$$\left| \omega_\sigma \right| \leq K.$$

(If straight lines emerging from P_0 met the surface S more than once, then, by taking σ to be the set of all those pieces of S subtending positive solid angles at P_0, we could have $\omega_\sigma > 4\pi$ and, in general, ω_σ could be indefinitely large. Hence the limitation (d) is essential.)

We consider first the case when S is a finite, closed, Lyapunov surface. Let Ω be the domain enclosed within S.

We shall examine in more detail the nature of the integrals occurring in the expressions for the potentials of a single and a double layer; we shall investigate their behaviour near a point P of the surface S. For convenience, we choose a coordinate system having P as the origin O, the tangent plane at P as the plane XOY, and the axis OZ directed along the inward normal. Let the local equation of the surface S be

$$z = \zeta(x, y);$$

$$\zeta(0, 0) = 0 \quad \text{and} \quad \frac{\partial \zeta(0, 0)}{\partial x} = 0 = \frac{\partial \zeta(0, 0)}{\partial y}.$$

§ 2. Properties of the Potential of a Double Layer

The potential of a double layer is given by

$$w(x_0, y_0, z_0) = \iint_S f_1(S) \frac{\partial}{\partial n}\left(\frac{1}{r}\right) dS. \tag{15.2}$$

The expression $\partial(1/r)/\partial n$ is clearly a function of two variable points: the point (x_0, y_0, z_0) occupying an arbitrary position in space, and the point (x, y, z) situated on the surface S. The derivative is taken in the direction of the inward normal to the surface S at the point (x, y, z).

We shall prove some simple propositions about this potential.

LEMMA 1. *Let φ be the angle between the direction of the inward normal at an arbitrary point of the surface S and the radius vector from this point to the point (x_0, y_0, z_0). Then the potential of the double layer can be expressed by*

$$w(x_0, y_0, z_0) = \iint_S f_1(S) \frac{\cos \varphi}{r^2} dS.$$

Proof. The cosines of the angles formed by the radius vector from (x, y, z) to (x_0, y_0, z_0) with the coordinate axes are

$$\frac{x_0 - x}{r}, \quad \frac{y_0 - y}{r}, \quad \frac{z_0 - z}{r}.$$

Hence

$$\cos \varphi = \frac{x_0 - x}{r} \cos (n, x) + \frac{y_0 - y}{r} \cos (n, y) + \frac{z_0 - z}{r} \cos (n, z)$$

$$= -\frac{\partial r}{\partial n},$$

and

$$\frac{\cos \varphi}{r^2} = -\frac{1}{r^2} \frac{\partial r}{\partial n} = \frac{\partial}{\partial n} \left(\frac{1}{r}\right).$$

Hence the lemma.

LEMMA 2. *The potential of the double layer at P_0 remains meaningful if the point P_0 coincides with a point Q_0 of the surface.*

To prove this proposition, we put $x_0 = y_0 = z_0 = 0$ in (15.3).

We select on the surface S a section S_1 containing the origin and such that on S_1 z is a single-valued function of x and y. The remaining part S_2 does not affect the convergence of the integral. Then

$$w_{S_1} = \iint_{S_1} f_1(S) \frac{\cos \varphi}{r^2} \, dS$$

$$= -\iint_{S_1} f_1(S) \left(\frac{x}{r^3} \cos (n, x) + \frac{y}{r^3} \cos (n, y) + \frac{z}{r^3} \cos (n, z)\right) dS.$$

We shall estimate the magnitude of

$$-\frac{\cos \varphi}{r^2} = \frac{x}{r^3} \cos (n, x) + \frac{y}{r^3} \cos (n, y) + \frac{z}{r^3} \cos (n, z).$$

We have

$$\cos (n, x) = \mathbf{ni}, \quad \cos (n, y) = \mathbf{nj}, \quad \cos (n, z) = \mathbf{nk},$$

where $\mathbf{i}, \mathbf{j}, \mathbf{k}$ are unit vectors directed along the coordinate axes. Further, $\mathbf{n}_0 = \mathbf{k}$ and therefore

$$\cos (n, x) = \mathbf{ni} = (\mathbf{n} - \mathbf{n}_0) \, \mathbf{i},$$

$$\cos (n, y) = \mathbf{nj} = (\mathbf{n} - \mathbf{n}_0) \, \mathbf{j},$$

$$\cos (n, z) = \mathbf{nk} = (\mathbf{n} - \mathbf{n}_0) \, \mathbf{k} + \mathbf{n}_0 \mathbf{k} = 1 + (\mathbf{n} - \mathbf{n}_0) \, \mathbf{k},$$

whence, using the conditions of Lyapunov smoothness, we have:

$$\left|\cos (n, x)\right| < Ar^\delta, \quad \left|\cos (n, y)\right| < Ar^\delta, \quad \left|\cos (n, z)\right| > 1 - Ar^\delta. \quad (15.4)$$

Next we shall estimate the magnitude of z. By the theorem on finite increments

$$z(x, y) = x \left[\frac{\partial z}{\partial x}\right]_{\xi, \eta} + y \left[\frac{\partial z}{\partial y}\right]_{\xi, \eta},$$

where (ξ, η) is a certain point on the interval joining the origin to the point (x, y) in the plane XOY. But from text-books on differential geometry we know that

$$\frac{\partial z}{\partial x} = -\frac{\cos (n, x)}{\cos (n, z)}, \quad \frac{\partial z}{\partial y} = -\frac{\cos (n, y)}{\cos (n, z)};$$

hence

$$\left| \frac{\partial z}{\partial x} \right| \leqq \frac{Ar^\delta}{1 - Ar^\delta}, \quad \left| \frac{\partial z}{\partial y} \right| \leqq \frac{Ar^\delta}{1 - Ar^\delta}.$$

For points of the surface which lie within or on the boundary of the sphere C_3 defined by the inequality

$$Ar^\delta \leqq \frac{1}{3},$$

we have

$$1 - Ar^\delta \geqq \frac{2}{3}$$

and hence

$$\left| \frac{\partial z}{\partial x} \right| \leqq \frac{3}{2} Ar^\delta, \quad \left| \frac{\partial z}{\partial y} \right| \leqq \frac{3}{2} Ar^\delta.$$

We shall show that

$$|z| \leqq 3A\varrho r^\delta \leqq \varrho = \sqrt{x^2 + y^2} \tag{15.5}$$

holds for points of the surface S which fall within the sphere C_3. Let OP denote the ray which starts at the origin and passes through the point

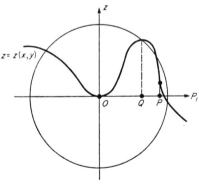

FIG. 13.

$P(x, y)$ in the plane XOY. The plane through the z-axis and the ray OP will intersect the sphere C_3 in a circle and the surface S in a certain curve (see Fig. 13). Consider the projections on to the ray OP of the points of intersection of the surface S with the surface of the sphere C_3; let Q be that particular projection which is closest to the origin. (The set of points considered is closed; hence there must be one which is closest to the origin, and it is clear that it does not coincide with the origin.)

If the point (x, y) lies on the interval OQ, then the point (ξ, η) also lies on the interval OQ, and the point $(\xi, \eta, z(\xi, \eta))$ lies within or on the boundary of the sphere C_3. Hence at this point the inequalities written above for $|\partial z/\partial x|$ and $|\partial z/\partial y|$ and also (15.5) are valid.

If the point $P(x, y)$ lies outside the interval OQ but the corresponding point of the surface lies within the sphere C_3, then the inequality (15.5) will be all the more valid:

$$\left| z(P) \right| < \left| z(Q) \right| \leqq OQ < OP = \varrho.$$

Hence (15.5) is proved.

For points of the surface lying within the sphere C_3, we have

$$\varrho \leqq r \leqq 2\varrho.$$

The left-hand inequality is obvious, and the right-hand one follows from the fact that

$$r = \sqrt{\varrho^2 + z^2} \leqq \sqrt{2\varrho^2} \leqq 2\varrho.$$

This enables the inequalities (15.4) and (15.5) to be made more precise. We have:

$$\left.\begin{aligned}
\left| z \right| &\leqq 3A\,2^{\delta}\varrho^{1+\delta} \leqq 6A\varrho^{1+\delta} \\
\left| \cos(n, x) \right| &\leqq 2^{\delta}A\varrho^{\delta} \leqq 2A\varrho^{\delta}, \quad \left| \cos(n, y) \right| \leqq 2A\varrho^{\delta}, \\
\left| \cos(n, z) \right| &> 1 - 2A\varrho^{\delta} > \tfrac{1}{3}.
\end{aligned}\right\} \quad (15.5')$$

Substituting these estimates in the expression for $(\cos \varphi)/r^2$ we get

$$\left| \frac{\cos \varphi}{r^2} \right| < 2A\varrho^{-2+\delta} + 2A\varrho^{-2+\delta} + 6A\varrho^{-2+\delta} < \frac{10A}{\varrho^{2-\delta}}. \qquad (15.6)$$

The integral over that part S' of the surface S which falls within the sphere C_3 can be written in the form

$$\iint_{S'} f_1(S) \frac{\cos \varphi}{r^2} \, dS = \iint_{\sigma'} f_1(S) \frac{\cos \varphi}{r^2} \frac{1}{\cos(n, z)} \, dx \, dy$$

where σ' is the projection of S' on to the plane XOY. By (15.5') and (15.6)

$$\left| f_1(S) \frac{\cos \varphi}{r^2} \frac{1}{\cos(n, z)} \right| \leqq \frac{30.A.M}{\varrho^{2-\delta}} \qquad (*)$$

where $M = \max\limits_{S} f_1(S)$.

Hence the first integral on the right-hand side of

$$\iint_S f_1(S) \frac{\cos \varphi}{r^2} \, \mathrm{d}S = \iint_{S'} f_1(S) \frac{\cos \varphi}{r^2} \, \mathrm{d}S + \iint_{S-S'} f_1(S) \frac{\cos \varphi}{r^2} \, \mathrm{d}S$$

exists; and since the second integral undoubtedly exists, the lemma is proved.

Remark 1. The potential of the double layer is a continuous function of the point $P_0 = Q_0$ when $Q_0(x_0, y_0, z_0)$ moves over the surface.

For the proof we pass from integration over S' to integration over σ', and notice that the inequality (*), with perhaps a bigger constant factor on its right-hand side, will hold not only for a point Q_0 coinciding with the origin but also for all points sufficiently close to the origin, in which case

$$\varrho = \sqrt{(x - x_0)^2 + (y - y_0)^2}.$$

So we can apply the criterion for uniform convergence of the integral (§ 1, Lecture 7), and then the truth of this proposition follows from Lemma 2 of Lecture 7.

As we shall show in a moment, w is a discontinuous function; it suffers a break of continuity on passing across the surface S. Let w_0 denote the value of the potential of the double layer when (x_0, y_0, z_0) in (15.3) is replaced by a point of the surface S; w_0 is a function of a point on the surface.

THEOREM 1. *The function w has limit values as the point (x_0, y_0, z_0) tends towards the surface S from the inside and from the outside, and these limits are different.*

If w_e is the limit value of w as (x_0, y_0, z_0) tends towards S from outside, and w_i is the limit value of w as (x_0, y_0, z_0) tends towards S from inside, then

$$\left. \begin{aligned} w_e &= -2\pi f_1(S) + w_0 \\ w_i &= 2\pi f_1(S) + w_0 \end{aligned} \right\} \tag{15.7}$$

Proof. We have

$$w(P) = \iint_S [f_1(S) - f_1(P_0)] \frac{\partial}{\partial n}\left(\frac{1}{r}\right) \mathrm{d}S + \iint_S f_1(P_0) \frac{\partial}{\partial n}\left(\frac{1}{r}\right) \mathrm{d}S$$

$$= w_1(P) + w_2(P), \quad \text{say,} \tag{15.8}$$

where P_0 denotes some fixed point of the surface S.

The first term, $w_1(P)$, in (15.8) is a continuous function at the point P_0. This follows because the integral is uniformly convergent at this point (see § 1, Lecture 7). For, suppose we surround the point P_0 by a domain $\sigma(\varepsilon)$ so small that

$$\left| f_1(S) - f_1(P_0) \right| < \varepsilon,$$

where S is a point belonging to the domain $\sigma(\varepsilon)$. Then for any point P lying in a certain neighbourhood $h(\varepsilon)$ of the point P_0, we shall have

$$\left| \iint_{\sigma} [f_1(S) - f_1(P_0)] \frac{\partial}{\partial n} \left(\frac{1}{r} \right) dS \right| \leq \varepsilon \iint_{\sigma} \frac{\partial}{\partial n} \left(\frac{1}{r} \right) dS$$

$$= \varepsilon \iint_{\sigma_1} d\omega - \varepsilon \iint_{\sigma_2} d\omega,$$

where r is the distance between the point P belonging to the neighbourhood $h(\varepsilon)$ of the point P_0 and the point S of the surface σ;

σ_1 is the part of σ which subtends a positive solid angle at P,

and σ_2 is the part of σ which subtends a negative solid angle at P.

For a sufficiently smooth surface S and in any case for a Lyapunov surface [by virtue of the condition (d)], both the integrals

$$\iint_{\sigma_1} d\omega \quad \text{and} \quad \iint_{\sigma_2} d\omega$$

will be bounded†. This is sufficient to establish the continuity of w_1 at the point P_0.

The second term in (15.8) can easily be calculated:

$$w_2(P) = \iint_S f_1(P_0) \frac{\partial}{\partial n} \left(\frac{1}{r} \right) dS = f_1(P_0) \iint_S d\omega = f_1(P_0) \cdot \omega_S(P),$$

where $\omega_S(P)$ is the solid angle subtended by the surface S at the point P.

Let w_2 tend to the limits w_{2e} and w_{2i} as P tends towards the surface S from outside and from inside respectively, and let $(w_2)_0$ be the value of w_2 when a point P_0 of the surface S is substituted for P. In order to evaluate these quantities, we must investigate the behaviour of $\omega_S(P)$ on crossing the surface S at the point P_0.

Let the point P_0 cross the surface S from inside to the outside. While P_0 is inside the domain bounded by the surface S, $\omega_S(P_0) = 4\pi$. As soon as P_0 crosses the surface S and falls outside it, $\omega_S(P_0) = 0$. Consequently the function ω_S suffers a break in continuity on passage across the surface.

The value of $\omega_S(P_0)$ when P_0 is a point on the surface S has still to be determined. To do this, we surround the point P_0 by a small sphere σ of radius η, and consider the curve l in which the sphere σ intersects the surface

† More detailed courses on potential theory, *e.g.*, Gyunter, "Potential Theory and its Application to the Principal Problems of Mathematical Physics", State Technical Publishing House, 1953, explain the precise conditions for the surface S in order that the integral $\iint_S \frac{\partial}{\partial n} \left(\frac{1}{r} \right) dS$ shall be bounded.

S. The curve l divides the surface of the sphere into two parts $-\sigma_1$ lying inside S and σ_2 lying outside S. The curve also divides the surface S into two parts $-S_1$ lying outside the sphere σ and S_2 lying inside σ. The exterior part S_1 together with σ_1 forms a closed surface to which the point P_0 is exterior. Consequently the solid angle subtended at P_0 by the surface S_1 is equal to the solid angle subtended at P_0 by the part σ_1 of the sphere σ.

In order to find the solid angle subtended by the surface S at P_0, it is sufficient to determine the solid angle subtended by S_1 and proceed to the limit as the radius η of the sphere σ tends to zero. From what has been said, this is equivalent to finding the limit of the angle subtended at P_0 by the part σ_1 of the sphere σ. It is not difficult to prove that this limit is equal to 2π, as follows.

We cut the sphere σ by the tangent plane to S at the point P_0. As proved earlier, the distance from points of the curve l to the tangent plane does not exceed $6A\eta^{1+\delta}$. Thus the curve l lies wholly within a zone on the sphere σ defined by $|z| \leq 6A\eta^{1+\delta}$.

We set up on σ a "geographical" system of coordinates by taking the normal to S at the point P_0 as the "polar" axis. In these coordinates, the zone within which the curve l evidently lies is bounded by a parallel of "south" latitude α_0 and a parallel of "north" latitude α_0, where

$$\alpha_0 = \sin^{-1} \frac{6A\eta^{1+\delta}}{\eta} = \sin^{-1} 6A\eta^{\delta}.$$

The angle α_0 clearly tends to zero. Consequently σ_1 tends to become a hemispherical surface, and hence the solid angle subtended at P_0 by σ_1 tends to 2π. Thus $\omega_S(P_0) = 2\pi$ if P_0 lies on the surface S.

Hence

$$(\omega_S)_e = -2\pi + (\omega_S)_0, \quad (\omega_S)_i = 2\pi + (\omega_S)_0.$$

And we have for w_2:

$$w_{2e} = -2\pi f_1(P_0) + (w_2)_0, \quad w_{2i} = 2\pi f_1(P_0) + (w_2)_0.$$

(15.7) follows at once from the formula (15.8) and the continuity of w_1. Since P_0 is an arbitrary point of the surface S, our theorem is proved.

Remark 2. From the continuity of $w_1(P)$ and the fact that $w_2(P)$ is, for a fixed P_0, a constant function both inside and outside Ω, it follows that the potential of the double layer can be continued from inside Ω on to S and also from outside Ω on to S continuously with respect to all the variables. In other words, whereas this potential is discontinuous in crossing the double layer, it does not change suddenly on crossing either of the single layers which together form the double layer; the discontinuity occurs in the double layer itself.

§ 3. Properties of the Potential of a Single Layer

We next investigate the potential of a single layer

$$v(P_0) = \iint_S \frac{1}{r} f_2(S) \, dS. \tag{15.9}$$

If the function $f_2(S)$ is continuous, then $v(P_0)$ will be continuous everywhere including the surface S.

To prove this, we surround the point P_0 by a sphere of radius η as we did in obtaining the inequality (15.6), and we denote by S' that part of the surface S which lies within this sphere. We then split the integral (15.9) into two parts, one taken over S' and the other over $S - S'$. The continuous dependence of the second integral on the parameter P_0 is obvious; we prove the continuity of the first integral.

We draw the tangent plane at any point of the surface S' and take it as the plane XOY. Then

$$\iint_{S'} \frac{1}{r} f_2(S) \, dS = \iint_{\sigma'} \frac{1}{r} f_2(S) \frac{1}{\cos(n, z)} \, dx \, dy.$$

At all points of S' we have $\cos(n, z) > \frac{1}{2}$; hence

$$\left| \frac{1}{r} f_2(S) \frac{1}{\cos(n, z)} \right| < \frac{2M}{r} \leqq \frac{2M}{\sqrt{(x - x_0)^2 + (y - y_0)^2}}.$$

Therefore, by virtue of the criterion for uniform convergence and Lemma 2 of § 1, Lecture 7, the function $v(P_0)$ must be continuous.

We notice that the first derivative of the potential of a single layer taken along any direction ν depends continuously on P_0 at any point $\tilde{P}$ which does not lie on S. For, on differentiating (15.9) formally, we get

$$\frac{\partial v(P_0)}{\partial \nu} = \iint_S \frac{\partial}{\partial \nu} \left(\frac{1}{r} \right) f_2(S) \, dS.$$

The right-hand side of this equality is continuous because the integrand is bounded and depends continuously on P_0 in the neighbourhood of the point $\tilde{P}$. For the same reason the differentiation under the integral sign is valid.

We now go on to investigate the behaviour of the derivatives of the potential of a single layer near the surface S. Let n^* denote the direction of the inward normal to S at any chosen point Q_0 of the surface. The derivative in this direction at a point P_0 not lying on the surface will be

$$\frac{\partial v(P_0)}{\partial n^*} = \iint_S \frac{\partial}{\partial n^*} \left(\frac{1}{r} \right) f_2(S) \, dS. \tag{*}$$

But

$$\frac{\partial}{\partial n^*}\left(\frac{1}{r}\right) = -\frac{1}{r^2}\frac{\partial r}{\partial n^*} = -\frac{1}{r^2}\cos\psi_0,$$

where ψ_0 is the angle between the radius vector r drawn from P to P_0 and the direction of the normal n^* at the point Q_0. For, if we introduce a co-ordinate system with origin at Q_0 and the z-axis parallel to n^*, we have

$$\frac{\partial r}{\partial n^*} = \frac{\partial r}{\partial z_0} = \frac{z_0 - z}{r} = \cos\psi_0.$$

Hence we can write

$$\frac{\partial v(P_0)}{\partial n^*} = -\iint_S f_2(S)\frac{\cos\psi_0}{r^2}\,dS. \qquad (15.10)$$

We shall prove two propositions.

LEMMA 3. *The integral*

$$\iint_S f_2(S)\frac{\cos\psi_0}{r^2}\,dS$$

occurring on the right-hand side of (15.10) *remains meaningful also when the point P_0 coincides with the point Q_0 on the surface and is then a continuous function of the point Q_0 over this surface.*

For, by virtue of (15.5'), we have for any point P of the surface situated sufficiently close to $P_0 = Q_0(0, 0, 0)$

$$\left|\frac{\cos\psi_0}{r^2}\right| = \left|\frac{z}{r^3}\right| < \frac{6A}{\varrho^{2-\delta}}.$$

This inequality is similar to (15.6) and it enables Lemma 3 to be proved in exactly the same way as Remark 1 to Lemma 2 was proved in the preceding section.

We shall denote the integral discussed in Lemma 3, when calculated for a point $P_0 \equiv Q_0$ lying on the surface S, by $\partial v/\partial n_0$:

$$\frac{\partial v}{\partial n_0} = \iint_S \frac{\partial}{\partial n^*}\left(\frac{1}{r}\right)f_2(S)\,dS.$$

We shall also denote by $\partial v(P_0)/\partial n_i$ and $\partial v(P_0)/\partial n_e$ the limiting values of the normal derivatives as the point P approaches the point P_0 along the normal from inside S and from outside S respectively.

THEOREM 2. *If* f_2 *is continuous, then*

$$\left.\begin{aligned}
\frac{\partial v}{\partial n_e} &= 2\pi f_2(P_0) + \frac{\partial v}{\partial n_0} \\
\frac{\partial v}{\partial n_i} &= -2\pi f_2(P_0) + \frac{\partial v}{\partial n_0}.
\end{aligned}\right\} \tag{15.11}$$

To prove this we compare $\partial v/\partial z_0$ with the potential w of a double layer having the density $-f_2(S)$. We form the difference

$$\frac{\partial v}{\partial z_0} - w = \iint_S f_2(S) \frac{\cos \varphi - \cos \psi_0}{r^2} \, dS \tag{15.12}$$

We shall show that this difference is a continuous function of z_0 at the point $(0, 0, 0)$. For,

$$\frac{\cos \varphi - \cos \psi_0}{r^2} = -\frac{x - x_0}{r^3} \cos(n, x) - \frac{y - y_0}{r^3} \cos(n, y)$$

$$- \frac{z - z_0}{r^3} [\cos(n, z) - 1].$$

Let the point (x_0, y_0, z_0) move along the normal to the surface, *i.e.*, put $x_0 = y_0 = 0$; then

$$\frac{\cos \varphi - \cos \psi_0}{r^2} = -\frac{x \cos(n, x)}{r^3} - \frac{y \cos(n, y)}{r^3}$$

$$- \frac{z - z_0}{r^3} [\cos(n, z) - 1].$$

Using the inequalities (15.5′) we have

$$\left| x \cos(n, x) \right| \leqq A_1 \varrho^{1+\delta}, \quad \left| y \cos(n, y) \right| \leqq A_1 \varrho^{1+\delta}, \quad \left| 1 - \cos(n, z) \right| \leqq A_1 \varrho^{\delta},$$

where A_1 is a constant. Noting further that

$$\left| z - z_0 \right| < r = \sqrt{x^2 + y^2 + (z - z_0)^2},$$

we get

$$\left| \frac{\cos \varphi - \cos \psi_0}{r^2} \right| \leqq \frac{2A_1 \varrho^{1+\delta}}{r^3} + \frac{A_1 \varrho^{\delta}}{r^2} \leqq \frac{3A_1}{\varrho^{2-\delta}}.$$

We surround the point $P_0 = Q_0$ by a sphere σ of small radius and denote by S' the part of the surface that falls within σ. We divide the integral (15.12) into the two parts, the integrals taken over S' and $S - S'$. In the first of these integrals we switch over to integration over the domain σ' in the xy-plane into which S' projects. Then by virtue of the inequality which we established above, and by the criterion for uniform convergence in § 1, Lecture 7, the integral over σ' converges uniformly with respect to the parameter z_0 and is a continuous function of z_0 at the point $x_0 = y_0 = z_0 = 0$. Hence it follows that the integral (15.12) is also a continuous function of z_0 at this point. Consequently

$$\left[\frac{\partial v}{\partial z_0} - w\right]_{z_0 = +0} = \left[\frac{\partial v}{\partial z_0} - w\right]_{z_0 = -0} = \left[\frac{\partial v}{\partial z_0} - w\right]_{z_0 = 0}.$$

Using Theorem 1 of this lecture, we obtain at once the formulae (15.11).

Remark 3. The inequality satisfied by $(\cos\varphi - \cos\psi_0)/r^2$ does not depend on the particular point of the surface considered. Consequently the derivative $\partial v/\partial z_0$ will tend to its limit value uniformly over the whole surface S.

THEOREM 3. *If the function $f_2(S)$ is bounded, then $\partial v/\partial n_0$ is continuous.*

For in the proof of Lemma 3 and also of the Remark after Lemma 2, we have in fact used only the boundedness of the function $f_2(S)$.

THEOREM 3. *Suppose that the density of a single layer satisfies the condition*

$$\left|f_2(P_1) - f_2(P_2)\right| < Kr_3^{\delta_1} \tag{15.13}$$

where K and $\delta_1 > 0$ are certain constants, P_1 and P_2 are any two points on the surface, and r_3 is the distance between P_1 and P_2. Let Q_0 be some fixed point of the surface; we draw the tangent plane and normal at this point.

Then under these conditions, at all points of the normal (with the possible exception of the point Q_0 itself) the potential will have a first-order derivative in any direction parallel to the tangent plane. This derivative will vary continuously along the normal everywhere except perhaps at the point Q_0 where it may have a removable discontinuity.

As before, we take Q_0 as the origin, the tangent plane at Q_0 as the plane XOY, and we take the x-axis in the direction in which the derivative under consideration is calculated.

We shall evaluate this derivative $\partial v/\partial x_0$ at a certain point $P_0(0, 0, z_0)$, $z_0 \neq 0$, lying on the normal. The derivative may be written in the form of the integral

$$\frac{\partial v}{\partial x_0} = \int\int_S \frac{x}{r^3} f_2(S)\, dS$$

and it is continuous at the point P_0.

To prove the theorem it is sufficient to show that this integral depending on the parameter z_0 converges uniformly at $z_0 = 0$ if its value at $z = z_0$ (when it might not exist) is chosen in a suitable manner.

We construct round the z-axis a cylinder of radius and height h such that at points on the piece S^* of the surface S which lies within this cylinder the normal to S makes an angle with the z-axis not exceeding $\pi/4$. If we also choose h to be so small that the cylinder falls within the sphere C_3 (see p. 205), we shall have $|z| \leqq \varrho$.

Let $r^* = \sqrt{x^2 + y^2 + z^2}$. We consider the integral

$$\iint_{S^*} f_2(Q_0)\, \frac{x}{r^{*3}}\, \cos (n, z)\, \mathrm{d}S.$$

This integral is clearly zero for all values of z_0 except $z_0 = 0$ for which it does not converge; for it can be rewritten in the form

$$f_2(Q_0) \iint_{x^2+y^2 \leqq h^2} \frac{x}{\sqrt{(x^2 + y^2 + z_0^2)^3}}\, \mathrm{d}x\, \mathrm{d}y,$$

where the integrand is an odd function of x.

For $z_0 = 0$ we can define the integral to be also equal to zero.

Our problem obviously reduces to showing that the integral

$$\iint_{S^*} x\, \frac{f_2(S)}{r^3}\, \mathrm{d}S$$

or

$$\iint_{S^*} x \left(\frac{f_2(S)}{r^3} - \frac{f_2(Q_0) \cos (n, z)}{r^{*3}} \right) \mathrm{d}S \qquad (15.14)$$

converges uniformly at $z_0 = 0$. This is not difficult. For, we have

$$\frac{f_2(S)}{r^3} - \frac{f_2(Q_0)}{r^{*3}} \cos (n, z) = \frac{f_2(S) - f_2(Q_0)}{r^3} + \frac{f_2(Q_0)\, [1 - \cos (n, z)]}{r^3}$$

$$+ f_2(Q_0) \cos (n, z) \left(\frac{1}{r^3} - \frac{1}{r^{*3}} \right).$$

We estimate in turn each of the terms on the right-hand side. We have

$$r = \sqrt{x^2 + y^2 + (z - z_0)^2} \geqq \varrho: \quad r^* > \varrho: \quad r_3 = \sqrt{x^2 + y^2 + z^2} \leqq \varrho \sqrt{2}:$$

$$\left| \frac{f_2(S) - f_2(Q_0)}{r^3} \right| \leqq \frac{K r_3^{\delta_1}}{r^3} \leqq \frac{K_1}{\varrho^{3-\delta_1}}.$$

From (15.4) we get

$$\left| \frac{f_2(Q_0)\, [1 - \cos (n, z)]}{r^3} \right| \leqq |f_2(Q_0)|\, A\, \frac{\varrho^\delta}{r^3} \leqq \frac{K_2}{\varrho^{3-\delta}}.$$

Further,

$$\left| \frac{1}{r^3} - \frac{1}{r^{*3}} \right| = |r - r^*| \left\{ \frac{1}{r^3 r^*} + \frac{1}{r^2 r^{*2}} + \frac{1}{r r^{*3}} \right\}.$$

The difference of the vectors r and r^* is an interval of length z. Therefore, by (15.5),

$$|r - r^*| = |z| \leqq K\varrho^{1+\delta}.$$

Hence

$$\left| \frac{1}{r^3} - \frac{1}{r^{*3}} \right| \leqq K_3 \varrho^{1+\delta} \left\{ \frac{1}{r^3 r^*} + \frac{1}{r^2 r^{*2}} + \frac{1}{rr^{*3}} \right\} \leqq K_4 \frac{1}{\varrho^{3-\delta}}.$$

Returning to the integral (15.14), we see that it may be written in the form

$$\iint_{x^2+y^2 \leqq h^2} \frac{x}{\cos (n, z)} \left(\frac{f_2(S)}{r^3} - \frac{f_2(Q_0) \cos (n, z)}{r^{*3}} \right) dx\, dy. \quad (15.15)$$

The integrand does not exceed

$$\varrho \sqrt{2} \left(\frac{K_1 + K_2 + K_4}{\varrho^{3-\delta}} \right) = \frac{K_5}{\varrho^{2-\delta}},$$

independently of z_0 (and also of the direction of x). Consequently, by Criterion 2 of Lecture 7, the integral converges uniformly relative to z_0 at the point $z_0 = 0$, and it is therefore a continuous function for $z_0 = 0$. Hence the theorem.

We next examine the gradient of the function v at a point $P_0 \neq Q_0$ lying on the normal. It can be expressed as the sum of two vectors, one directed along the normal and the other parallel to the tangent plane at the point Q_0:

$$\operatorname{grad} v(P_0) = \frac{\partial v(P_0)}{\partial n} k + (\operatorname{grad} v)_\tau$$

where k is a unit vector in the direction of the inward normal.

Remark 4. It follows from the proof of Theorem 4 that the vector $(\operatorname{grad} v)_\tau$ tends to a definite limit

$$\lim_{P_0 \to Q_0} \iint_S \left(\operatorname{grad} \frac{1}{r} \right)_\tau f_2(S)\, dS$$

when P_0 tends to Q_0 along the normal either from inside or from outside Ω, and it does so uniformly relative to the point Q_0 on the surface.

THEOREM 5. *Under the conditions of Theorem 4, the vector $\operatorname{grad} v$ has continuous limit values $(\operatorname{grad} v)_i$ and $(\operatorname{grad} v)_e$ on the surface S to which it tends uniformly when the point P tends to Q_0 from inside and from outside the domain Ω respectively.*

We consider, for definiteness, the case when the point P tends to Q_0 from inside the domain Ω.

When a point P_0 moves along the normal to a point Q_0 on the surface S, it actually follows from Theorems 3 and 4 that the vector $\operatorname{grad} v$ tends to a certain limit value, which we denote by $(\text{grad } v)_i$. And this convergence is uniform in relation to the point Q_0. Hence, by virtue of the fact that a Lyapunov surface has a continuously varying tangent plane and because of the condition (b), p. 202, it follows that the vector function $(\text{grad } v)_i$ is a continuous function of the point Q_0.

We now show that when the point P tends towards the point Q_0 along any path from inside Ω, the limit of $\text{grad } v$ will exist and will be equal to $(\text{grad } v)_i$, and also that the convergence to this limit will be uniform.

For, the value of $\text{grad } v$ at a certain point P_0 lying within a sphere of radius η with centre at Q_0 will be little different from the value of $(\text{grad } v)_i$ at the point Q_1 on the surface S which is closest to P_0 (P_0 obviously lies on the normal to S at the point Q_1). But the value of $(\text{grad } v)_i$ at Q_1 is, in its turn, little different from the value of $(\text{grad } v)_i$ at Q_0, because of the continuity of the function $(\text{grad } v)_i$ on the surface S.

The theorem is proved for the case when the point P moves towards the point Q_0 from outside the domain Ω in exactly the same way.

COROLLARY. The function v has continuous derivatives of the first order up to the boundary everywhere within, and also everywhere outside, the domain Ω.

It is sufficient to notice that the first-order partial derivative in any direction is equal to the projection of the vector $\text{grad } v$ in that direction.

THEOREM 6. *If the density $\nu(Q)$ of the potential of a single layer is a continuous function of the point Q of the surface, then the value of $\partial v/\partial n_0$ as a function of position on the surface satisfies the condition* (15.13) *of Theorem 4.*

To prove this, we consider the difference

$$\Delta(Q_1, Q_2) = \left[\frac{\partial v}{\partial n_0} \right]_{Q_1}^{Q_2} = \int\!\!\int_S \nu(Q) \left(\mathbf{n}, \frac{\mathbf{r}_1}{r_1^3} - \frac{\mathbf{r}_2}{r_2^3} \right) \mathrm{d}S,$$

where $\mathbf{r}_1$ and $\mathbf{r}_2$ are the vectors joining Q to Q_1 and Q_2,

and $\mathbf{n}$ is the unit vector along the inward normal at the point Q.

We surround the point Q_1 by a sphere C_η of radius $\eta = [r(Q_1,Q_2)]^{\frac{1}{2}}$. We suppose that the points Q_1, Q_2 are so close together that η is less than the radius h of the Lyapunov sphere (see (b), p. 202).

Let σ be the part of the surface S falling within the sphere C_η. We split the integral Δ into the two terms $\Delta(Q_1,Q_2) = I_\sigma + I_{S-\sigma}$, where

$$I_\sigma = \int\!\!\int_\sigma \nu(Q) \left(\mathbf{n}, \frac{\mathbf{r}_1}{r_1^3} - \frac{\mathbf{r}_2}{r_2^3} \right) \mathrm{d}S, \quad I_{S-\sigma} = \int\!\!\int_{S-\sigma} \nu(Q) \left(\mathbf{n}, \frac{\mathbf{r}_1}{r_1^3} - \frac{\mathbf{r}_2}{r_2^3} \right) \mathrm{d}S.$$

Let Q' be a variable point on S, and $\mathbf{r}'$ the vector joining Q and Q'. The vector $\mathbf{r}'/r'^3$ is a continuously differentiable vector in the domain $S - \sigma$. The derivatives of its components do not exceed $4/r'^3$. Hence we can apply the formula for finite increments to obtain the result

$$\left| \frac{r_1}{r_1^3} - \frac{r_2}{r_2^3} \right| \leq \frac{12C}{r_1^3} r(Q_1, Q_2).$$

Thus for the integral $I_{S-\sigma}$ we get

$$|I_{S-\sigma}| \leq C_1 r(Q_1, P_2) \iint_{r_1 \geq r(Q_1, Q_2)^{\frac{1}{2}}} \frac{dS}{r_1^3}$$

$$= C_1 r(Q_1, Q_2) \left[A + \iint_{h \geq r_1 \geq r(Q_1, Q_2)^{\frac{1}{2}}} \frac{dS}{r_1^3} \right],$$

where C, C_1 and A are constants, and h is the radius of the Lyapunov sphere. But within the Lyapunov sphere $dS/r_1^3 \leq C_2 \varrho^{-2} \, d\varrho \, d\varphi$ (see p. 206). Therefore

$$|I_{S-\sigma}| \leq C_3 r(Q_1, Q_2) \left\{ \left[\varrho^{-1} \right]_{\frac{1}{2}r(Q_1, Q_2)^{\frac{1}{2}}}^{h} + A \right\} \leq C_4 r(Q_1, Q_2)^{\frac{1}{2}}.$$

Further,

$$|I_\sigma| \leq C_3 \left\{ \left| \iint_{0 \leq r_1 \leq r(Q_1, Q_2)^{\frac{1}{2}}} \frac{(n, r_1)}{r_1^3} \, dS \right| + \left| \iint_{0 \leq r_2 \leq 2r(Q_1, Q_2)^{\frac{1}{2}}} \frac{(n, r_2)}{r_2^3} \, dS \right| \right\}.$$

By using the earlier inequalities (p. 206) we get

$$|I_\sigma| \leq K_1 \int_0^{[2r(Q_1, Q_2)]^{\frac{1}{2}}} \varrho^{-1+\delta} \, d\varrho \leq K_2 r(Q_1, Q_2)^{\frac{\delta}{2}}.$$

Hence the theorem.

§ 4. Regular Normal Derivative

We shall have to apply Green's formula later on to harmonic functions represented in the form of potentials of a single or double layer. In order to be able to do this we have to impose further restrictions on the type of surface S.

We first of all make an important remark. We obtained Green's formula (5.16) on the assumptions that the functions u and v have continuous first-order derivatives right up to the boundary and that $\nabla^2 u$ and $\nabla^2 v$ are integrable within the domain.

Let us assume that our surface S is such that the functions by means of which it is expressed in parametric form

$$x = x(\lambda, \mu), \quad y = y(\lambda, \mu), \quad z = z(\lambda, \mu) \tag{15.16}$$

have continuous derivatives up to the second order inclusive with respect to the parameters.

At each point of this surface we construct the normal n and mark off an interval of constant length ζ along the normal. The locus of the ends of these intervals is given by the equations

$$x_1 = x + \zeta \cos(n, x), \quad y_1 = y + \zeta \cos(n, y), \quad z_1 = z + \zeta \cos(n, z),$$

and describes a certain surface S_1 which can be called a surface "parallel" to S.

It is easily shown that the surface S_1 has a continuously varying tangent plane. To do this, we consider the parametric equations (15.16) for the surface S. The components of the normal vector

$$n_x(\lambda, \mu), \quad n_y(\lambda, \mu), \quad n_z(\lambda, \mu)$$

can continuously be differentiated once with respect to λ and μ. Substituting these expressions in the equations for x_1, y_1, z_1, we obtain a parametric representation for S_1 which will clearly have continuous first-order derivatives with respect to λ and μ; and this implies that the surface S_1 will have a continuously varying tangent plane.

Now let us suppose that the functions u and v are such that Δu and Δv are continuous within the domain Ω. Suppose that u and v have continuous normal derivatives on any surface S_1 "parallel" to S and lying within it.

Suppose further that the normal derivatives $[\partial u/\partial n]_{S_1}$ and $[\partial v/\partial n]_{S_1}$ defined on the surface S_1 tend uniformly to continuous limit functions $\varphi_1(S)$ and $\varphi_2(S)$ as S_1 tends to S. We shall call these functions the normal derivatives of u and v on the surface S and denote them by $[\partial u/\partial n]_S$ and $[\partial v/\partial n]_S$. We shall say in this case that the functions u and v possess *regular normal derivatives*.

THEOREM 7. *If u and v are two functions harmonic in a domain Ω and possess regular normal derivatives, then they satisfy Green's formula*

$$\iint_S \left(u \frac{\partial v}{\partial n} - v \frac{\partial u}{\partial n} \right) dS = 0.$$

To prove this, it is sufficient to apply Green's formula to a surface S_1 "parallel" to the given surface S, and then to take the limit as S_1 tends to S.

From Theorem 2, Lemma 3 and Remark 3 of this lecture it follows that, if the density $f_2(S)$ is continuous, then the potential of a single layer has regular normal derivatives.

§ 5. Normal Derivative of the Potential of a Double Layer

The condition for the existence of a regular normal derivative of the potential of a double layer is given by the following theorem.

Lyapunov's Theorem. The necessary and sufficient conditions that the solution of Laplace's equation $\nabla^2 u = 0$ in the domain Ω bounded by the surface S, which *satisfies the condition* $[u]_S = f(S)$, *(the solution of the Dirichlet problem), should have a regular normal derivative are that the potential W of the double layer, expressed by using the function* $f(S)$,

$$W = \frac{1}{4\pi} \iint_S f(S) \frac{\partial}{\partial n} \left(\frac{1}{r}\right) dS,$$

should have a regular normal derivative both from outside and from inside and that the values of these regular normal derivatives should coincide.

To prove this theorem, we consider the function

$$w(P_0) = \begin{cases} u - W & \text{if } P_0 \text{ is inside } \Omega, \\ -W & \text{if } P_0 \text{ is outside } \Omega, \end{cases}$$

$w(P_0)$ will be harmonic both inside and outside the domain Ω. Further, it will be continuous throughout space except on the surface S, where it has a removable discontinuity. This follows from Remark 1 of this lecture and the fact that the jump of the function $w(P_0)$ on crossing the surface S is, as may readily be seen, equal to $[u(P_0)]_S - f(S) = 0$.

If the function $u(P_0)$ has a regular normal derivative $\partial u/\partial n$, then the function $w(P_0)$ will coincide with the potential of a single layer

$$-\frac{1}{4\pi} \iint_S \frac{1}{r} \frac{\partial u}{\partial n} dS.$$

For in this case, by Green's formulae (9.4) and (9.6) we have

$$\frac{1}{4\pi} \iint_S f \frac{\partial}{\partial n} \left(\frac{1}{r}\right) dS - \frac{1}{4\pi} \iint_S \frac{1}{r} \frac{\partial u}{\partial n} dS = \begin{cases} u(P_0) & \text{if } P_0 \text{ is inside } \Omega, \\ 0 & \text{if } P_0 \text{ is outside } \Omega, \end{cases}$$

whence

$$-\frac{1}{4\pi} \iint_S f \frac{\partial}{\partial n} \left(\frac{1}{r}\right) dS$$

$$= \begin{cases} u(P_0) + \dfrac{1}{4\pi} \displaystyle\iint_S \frac{1}{r} \frac{\partial u}{\partial n} dS & \text{if } P_0 \text{ is inside } \Omega, \\[3mm] -\dfrac{1}{4\pi} \displaystyle\iint_S \frac{1}{r} \frac{\partial u}{\partial n} dS & \text{if } P_0 \text{ is outside } \Omega. \end{cases}$$

Thus the potential, W, of the double layer is expressible as the sum of two functions having a regular normal derivative from both sides of the surface S, and consequently it too has a regular normal derivative from both sides of S. The jump of this derivative, by Theorem 2 of this lecture, is zero.

Consequently, the limit values of the normal derivative outside and inside the surface coincide.

We have proved that the conditions of the theorem are necessary. We now show that they are sufficient.

We note first of all that if a continuous, regular normal derivative of the function W exists, then the function w may be regarded outside Ω as the solution of the Neumann problem with the specified continuous values of the external normal derivative

$$\left[\frac{\partial w}{\partial n}\right]_s = \left[\frac{\partial W}{\partial n}\right]_s.$$

But, as will be proved in Lecture 19 independently of the present proposition, the solution of such a problem is presentable as the potential V^* of a single layer with a continuous density† $v(S)$. This potential will coincide everywhere with the function $w(P_0)$, since their difference $V^*(P_0) - w(P_0)$ is a function which is harmonic in Ω, continuous throughout space, and identically zero outside Ω. This implies the existence of a normal derivative $\partial w/\partial n_0$ regular from inside Ω. But inside Ω we have

$$\frac{\partial u}{\partial n} = \frac{\partial W}{\partial n} + \frac{\partial w}{\partial n}.$$

Hence, since w and W have regular normal derivatives, u must also have a regular normal derivative, as was to be shown.

§ 6. Behaviour of the Potentials at Infinity

The last important property of the potentials of a single and a double layer which we shall analyse now is their behaviour at infinity. We prove that if the surface S is bounded, the potential of a single layer will decrease at least as rapidly as $1/R_0$, and the potential of a double layer will decrease at least as rapidly as $1/R_0^2$, when $R_0 = \sqrt{x_0^2 + y_0^2 + z_0^2}$ tends to infinity.

For, when R_0 is sufficiently great, we clearly have

$$r = \sqrt{(x - x_0)^2 + (y - y_0)^2 + (z - z_0)^2}$$

$$= \sqrt{(x_0^2 + y_0^2 + z_0^2) - 2(xx_0 + yy_0 + zz_0) + (x^2 + y^2 + z^2)}$$

$$= \sqrt{x_0^2 + y_0^2 + z_0^2} \, \sqrt{1 - \frac{2(xx_0 + yy_0 + zz_0) - (x^2 + y^2 + z^2)}{(x_0^2 + y_0^2 + z_0^2)}}$$

$$\geqq R_0 \sqrt{1 - \frac{2RR_0 - R^2}{R_0^2}} \geqq \frac{R_0}{2}.$$

† See page 261.

Hence

$$|v| = \left| \iint_S \frac{1}{r} f_2(S) \, dS \right| \leq \frac{2}{R_0} \iint_S |f_2(S)| \, dS$$

$$|w| = \left| \iint_S \frac{\cos \varphi}{r^2} f_1(S) \, dS \right| \leq \frac{4}{R_0^2} \iint_S |f_1(S)| \, dS,$$

as was to be proved.

REDUCTION OF THE DIRICHLET PROBLEM AND THE NEUMANN PROBLEM TO INTEGRAL EQUATIONS

§ 1. Formulation of the Problems and the Uniqueness of their Solutions

Let S be a closed and sufficiently smooth surface. Let Ω_1 be the volume enclosed by S, and Ω_2 the infinite domain external to S which is also bounded by S.

We consider four problems:

1. *The Internal Dirichlet Problem.*

To find a function u harmonic in Ω_1 and satisfying the condition

$$[u]_S = f_1(S).$$

2. *The External Dirichlet Problem.*

To find a function u harmonic in Ω_2 and satisfying the conditions

(a) $$[u]_S = f_1(S)$$

(b) $$\lim_{R \to \infty} u = 0.$$

3. *The Internal Neumann Problem.*

To find a function u harmonic in Ω_1 and satisfying the condition

$$\left[\frac{\partial u}{\partial n} \right]_S = f_2(S).$$

4. *The External Neumann Problem.*

To find a function u harmonic in Ω_2 and satisfying the conditions

(a) $$\left[\frac{\partial u}{\partial n} \right]_S = f_2(S)$$

(b) $$\lim_{R \to \infty} u = 0.$$

222

Before showing how actually to find the solutions of these problems we investigate some of their properties.

THEOREM 1. *The solution of the Dirichlet problem, internal or external, is unique.*

The proof is quite simple. It follows from the principle of the maximum. The difference of two distinct solutions, if they existed, would in the case of the internal problem be a harmonic function equal to zero on S; in the case of the external problem, the difference of the two solutions would be harmonic, and zero on S and at infinity. Consequently the difference of solutions cannot take within the domain either positive or negative values, since otherwise it would attain its positive maximum or its negative minimum, and this is impossible. Hence in both cases the difference of the two solutions is always zero.

THEOREM 2. *A solution of the external Neumann problem which has continuous first-order derivatives right up to the boundary is unique: and a solution of the external problem is determined to within an arbitrary additive constant.*

As regards the surface S, it is sufficient to assume that it satisfies the Lyapunov conditions. We deal with the internal problem first. Using Green's formula, which is applicable under the conditions we have here adopted, we can transform an integral $\iint_S v \dfrac{\partial v}{\partial n} \, dS$ as follows:

$$
\begin{aligned}
\iint_S & v \frac{\partial v}{\partial n} \, dS \\
&= -\iiint_{\Omega_1} \left\{ \frac{\partial}{\partial x}\left(v \frac{\partial v}{\partial x}\right) + \frac{\partial}{\partial y}\left(v \frac{\partial v}{\partial y}\right) + \frac{\partial}{\partial z}\left(v \frac{\partial v}{\partial z}\right) \right\} dx \, dy \, dz \\
&= -\iiint_{\Omega_1} \left\{ v \nabla^2 v + \left(\frac{\partial v}{\partial x}\right)^2 + \left(\frac{\partial v}{\partial y}\right)^2 + \left(\frac{\partial v}{\partial z}\right)^2 \right\} dx \, dy \, dz.
\end{aligned}
$$

If now v is a harmonic function and if $\partial v/\partial n$ vanishes on the boundary, then

$$
\iiint_{\Omega_1} \left\{ \left(\frac{\partial v}{\partial x}\right)^2 + \left(\frac{\partial v}{\partial y}\right)^2 + \left(\frac{\partial v}{\partial z}\right)^2 \right\} dx \, dy \, dz = 0,
$$

and this implies

$$
\frac{\partial v}{\partial x} = \frac{\partial v}{\partial y} = \frac{\partial v}{\partial z} = 0.
$$

Hence it follows that v is a constant.

Now let u_1, u_2 be two solutions of the internal Neumann problem. Then their difference v is a harmonic function whose normal derivative vanishes on the boundary of the domain. As we have just proved, such a function is a constant. Hence the theorem in the case of the internal problem.

The internal Neumann problem is not always soluble. A necessary and sufficient condition for its solubility is that

$$\int\int_S f_2(S)\, dS = 0.$$

The necessity of this condition follows in an obvious way from Green's formula (5.16) on putting $v \equiv 1$. We shall prove its sufficiency in Lecture 19.

For the external Neumann problem we take a sphere Σ of radius A, where A is a sufficiently large number, and let Ω_3 be the volume enclosed between Σ and S. By the foregoing,

$$\int\int_\Sigma v\frac{\partial v}{\partial n}\, dS + \int\int_S v\frac{\partial v}{\partial n}\, dS$$

$$= -\int\int\int_{\Omega_3}\left\{v\,\nabla^2 v + \left(\frac{\partial v}{\partial x}\right)^2 + \left(\frac{\partial v}{\partial y}\right)^2 + \left(\frac{\partial v}{\partial z}\right)^2\right\} dx\, dy\, dz.$$

If v is a harmonic function which tends to zero at infinity and which is such that $[\partial v/\partial n]_S = 0$, then the left-hand side of the above equation is as small as we please by choice of A. For, by Theorem 1, Lecture 12, on Σ we have

$$|v| \leqq \frac{M}{A}, \qquad \left|\frac{\partial v}{\partial n}\right| \leqq \frac{M}{A^2}$$

and

$$\left|\int\int_\Sigma v\frac{\partial v}{\partial n}\, dS\right| \leqq \frac{M^2}{A^3}\int\int_\Sigma dS = 4\pi\,\frac{M^2}{A}.$$

Hence

$$\int\int\int_{\Omega_3}\left[\left(\frac{\partial v}{\partial x}\right)^2 + \left(\frac{\partial v}{\partial y}\right)^2 + \left(\frac{\partial v}{\partial z}\right)^2\right] dx\, dy\, dz < \varepsilon$$

for any $\varepsilon > 0$, and this is possible only if

$$\frac{\partial v}{\partial x} = \frac{\partial v}{\partial y} = \frac{\partial v}{\partial z} = 0.$$

Hence v is a constant. But this constant can only be zero, for otherwise v would not tend to zero at infinity. The difference of two solutions of the external Neumann problem is therefore zero, and hence the solution of this problem is unique.

Remark. The assertion in Theorem 2 remains true if the limitation on the first derivative of the function v is weakened while that on the surface S is made more rigorous, namely, by supposing that the surface satisfies the conditions of § 4, Lecture 15, and that the solution has a regular normal derivative. For, in proving Theorem 2 we used only Green's formula and this remains applicable under the conditions stipulated in this remark. It would also be possible to prove the uniqueness of the solution of the external Neumann problem under even weaker conditions.

§ 2. The Integral Equations for the Formulated Problems

The properties of potentials which we established in the last lecture enable us to solve the Dirichlet and Neumann problems for any domains which are bounded by sufficiently smooth surfaces by reducing the problems to the form of integral equations.

For, suppose we wish to find, for example, the solution of the internal Dirichlet problem. We assume that the required function u is the potential of a double layer with an as yet unknown density $\mu(S)$:

$$u = w = \int\int_S \frac{\mu(S_1) \cos \varphi}{r^2} \, dS_1.$$

As we already know, the potential of a double layer is a harmonic function. We shall impose on w the condition that its limit value from within the domain shall be equal to $f_1(S)$:

$$w_i = f_1(S).$$

From equation (14.7) we have

$$w_i = 2\pi\mu(S) + w_0 = 2\pi\mu(S) + \int\int_S \frac{\mu(S_1) \cos \varphi}{r^2} \, dS_1$$

where r is the distance between the points S and S_1 of our surface. Thus we have for $\mu(S)$ the equation

$$\mu(S) = \frac{f_1(S)}{2\pi} - \frac{1}{2\pi} \int\int_S \frac{\mu(S_1) \cos \varphi}{r^2} \, dS_1. \tag{16.1}$$

If, for brevity, we denote $(1/2\pi)f_1(S)$ by $F_1(S)$ and $[(1/2\pi)\cos\varphi]/r^2$ by $K(S, S_1)$ (this last expression being obviously a function of the two points S, S_1 on the surface), then we arrive at the equation

$$\mu(S) = F_1(S) - \int\int_S K(S, S_1) \, \mu(S_1) \, dS. \tag{16.2}$$

Integral equations of this form for the unknown function $\mu(S)$ are known as Fredholm integral equations of the second kind. We shall shortly begin to study such equations.

In exactly the same way, the Dirichlet problem for an external domain bounded by the surface S, *i.e.*, for an infinite domain bounded by S, can be reduced to a Fredholm equation of the second kind. For, again seeking a solution in the form of the potential of a double layer with the condition $w_e = f_1(S)$, we similarly obtain for the unknown density $\mu(S)$

$$w_e = -2\pi\mu(S) + w_0,$$

whence

$$\mu(S) = -\frac{f_1(S)}{2\pi} + \frac{1}{2\pi}\iint_S \frac{\mu(S_1)\cos\varphi}{r^2}\,dS.$$

Denoting $-\dfrac{f_1(S)}{2\pi}$ by $\Phi_1(S)$ we obtain

$$\mu(S) = \Phi_1(S) + \iint_S K(S, S_1)\,\mu(S_1)\,dS. \qquad (16.3)$$

This is an equation of the same type as (16.2).

The internal and external Neumann problems can also be reduced to the solution of integral equations.

We shall seek a solution of the *internal Neumann problem* in the form of the potential of a single layer

$$u = v = \iint_S \frac{v(S_1)\,dS_1}{r}.$$

As in the preceding case, we have

$$\frac{\partial v}{\partial n_i} = -2\pi v(S) + \frac{\partial v}{\partial n_0} = f_2(S),$$

whence

$$v(S) = -\frac{f_2(S)}{2\pi} + \frac{1}{2\pi}\iint_S \frac{v(S_1)\cos\psi_0}{r^2}\,dS_1. \qquad (16.4)$$

The angle ψ_0 is obtained from the angle φ by replacing the point S by S_1.

Hence, putting

$$-\frac{f_2(S)}{2\pi} = F_2(S)$$

we get for $v(S)$ the equation

$$v(S) = F_2(S) + \iint_S K(S_1, S)\,v(S_1)\,dS_1. \qquad (16.5)$$

Finally, if we seek a solution of the *external Neumann problem* in the form of the potential v of a single layer, we get

$$\frac{\partial v}{\partial n_0} = 2\pi v(S) + \iint_S v(S_1)\frac{\cos\psi_0}{r^2}\,dS_1 = f_2(S)$$

or

$$v(S) = \frac{f_2(S)}{2\pi} - \frac{1}{2\pi} \int\int_S \frac{v(S_1) \cos \psi_0 \, dS_1}{r^2}.$$

Putting

$$\frac{f_2(S)}{2\pi} = \Phi_2(S),$$

we get for $v(S)$ the equation

$$v(S) = \Phi_2(S) - \int\int_S K(S_1, S) \, v(S_1) \, dS_1. \qquad (16.6)$$

If we succeed in finding functions μ and v to satisfy the equations (16.2), (16.3), (16.5), or (16.6), then the corresponding problems of mathematical physics will be solved. The theory of such integral equations will be studied in Lectures 18 and 19.

LAPLACE'S EQUATION AND POISSON'S EQUATION IN A PLANE

§ 1. The Principal Solution

We have already discussed Laplace's equation and Poisson's equation in space in sufficient detail. But it frequently happens in practice that the function u does not depend at all on one of the variables, say z, and then

$$\frac{\partial u}{\partial z} = \frac{\partial^2 u}{\partial z^2} = 0.$$

In this case these equations become equations in two independent variables. The problems which we previously posed for space can now be posed in the plane XOY for the equation

$$\nabla^2 u = \frac{\partial^2 u}{\partial x^2} + \frac{\partial^2 u}{\partial y^2} = \varrho(x, y),$$

where $\varrho(x, y)$ may sometimes be identically zero.

We shall consider certain properties of such two-dimensional problems which distinguish them from the three-dimensional case.

Exactly the same as in space, it is easy to prove that a function which is harmonic in a certain domain D of the plane XOY attains its maximum and minimum values on the boundary of this domain. Hence it follows, by the former arguments, that the solution of the Dirichlet problem for any bounded domain is unique. However, as we shall see later, the Dirichlet problem for an unbounded domain, as it was previously formulated, no longer has any meaning. To pose the problem of finding a harmonic function which shall be equal to zero at infinity is meaningless for the two-dimensional case. The fact is that a solution which will vanish at infinity does not, in general, exist, and any question about the uniqueness of such a solution is pointless.

We have here two other lemmas similar to those proved in Lecture 9.

LEMMA 1. *The function* $\log_e 1/r = -\log r$, *where* $r = \sqrt{(x - x_0)^2 + (y - y)^2}$, *is a harmonic function of the variables x and y.*

For,

$$-\frac{\partial^2}{\partial x^2} \log_e r = -\frac{1}{r^2} + \frac{2(x - x_0)^2}{r^4},$$

$$-\frac{\partial^2}{\partial y^2} \log_e r = -\frac{1}{r^2} + \frac{2(y - y_0)^2}{r^4},$$

whence, by addition,

$$\nabla^2 \log_e \frac{1}{r} = 0.$$

LEMMA 2. *If a function u is continuous and has continuous second-order derivatives within a domain D, boundary s, containing the point (x_0, y_0), then*

$$u(x_0, y_0) = -\frac{1}{2\pi} \iint_D \log_e \left(\frac{1}{r}\right) \nabla^2 u \, dx \, dy$$

$$+ \frac{1}{2\pi} \int_s \left\{ u \frac{\partial \left(\log_e \frac{1}{r}\right)}{\partial n} - \log_e \left(\frac{1}{r}\right) \frac{\partial u}{\partial n} \right\} ds \quad (17.1)$$

$\partial/\partial n$ here denotes the derivative along the inward normal to the boundary curve s.

If the point (x_0, y_0) lies outside the domain D, then

$$-\frac{1}{2\pi} \iint_D \log_e \left(\frac{1}{r}\right) \nabla^2 u \, dx \, dy$$

$$+ \frac{1}{2\pi} \int_s \left\{ u \frac{\partial}{\partial n} \left(\log_e \frac{1}{r}\right) - \log_e \left(\frac{1}{r}\right) \frac{\partial u}{\partial n} \right\} ds = 0. \quad (17.2)$$

The proof of this lemma is exactly analogous to that of Lemma 3 in Lecture 9, and we shall not go through it.

It can also be proved, as before, that if the function u has a singularity at the origin, is harmonic in the neighbourhood of the origin, and satisfies the inequality

$$|u| \leqq \frac{A}{R^n} \quad \text{where} \quad R = \sqrt{x^2 + y^2},$$

then u can be expressed in the form

$$u = \sum_{m=0}^{n} \sum_{i+j=m} a_{ij} \frac{\partial^m}{\partial x^i \partial y^j} \left(\log_e \frac{1}{R}\right) + u^*, \quad (17.3)$$

where u^* is a function which is harmonic throughout the domain including the origin.

The following theorems are valid.

THEOREM 1. *If $u(R, \theta)$ is a harmonic function, then so is $v(R, \theta) = u(1/R, \theta)$* ($R$ and θ here are polar coordinates in the plane.)

THEOREM 2. *If $u(R, \theta)$ is harmonic for large values of R and satisfies the condition*

$$|u| \leq AR^n,$$

then it can be expressed in the form

$$u = \sum_{m=1}^{n} \sum_{i+j=m} a_{ij} R^{2m} \frac{\partial^m}{\partial x^i \, \partial y^j} \left(\log_e \frac{1}{R} \right) + u^*,$$

where u^* is bounded (and harmonic).

§ 2. The Basic Problems

The problem of finding a solution over the whole plane for Poisson's equation in two independent variables

$$\nabla^2 u = \varrho(x, y)$$

which will vanish at infinity is, in general, insoluble, and we shall not deal it. We may remark that the integral

$$\int_{-\infty}^{+\infty} \int_{-\infty}^{+\infty} \varrho \, \log_e \frac{1}{r} \, dx \, dy,$$

extending over the whole plane (if ϱ is different from zero only in a finite domain, then the domain of the integral will effectively be finite) is nevertheless a particular solution of Poisson's equation, but it will in general increase without limit at infinity. This integral is known as the *logarithmic potential of a distributed mass*.

The Dirichlet problem for a half-plane, with certain limitations on the boundary function, has a solution among the class of functions which vanish at infinity. Let the function $f_1(x)$ satisfy the inequality

$$|f_1(x)| < \frac{A}{x^a} \quad \text{where } a > 0.$$

The solution of the equation

$$\nabla^2 u = 0$$

subject to the condition

$$[u]_{y=0} = f_1(x),$$

and vanishing at infinity, will have the form

$$u(x_0, y_0) = \frac{1}{2\pi} \int_{-\infty}^{+\infty} \frac{\partial}{\partial n}\left(\log_e \frac{1}{r}\right) f_1(x) \, dx = \frac{1}{\pi} \int_{-\infty}^{+\infty} \frac{\cos \varphi}{r} f_1(x) \, dx$$

$$= \frac{1}{\pi} \int_{-\infty}^{+\infty} f_1(x) \, d\omega(x, x_0, y_0).$$

where φ is the angle between the radius vector drawn from the point $(x, 0)$ to the point (x_0, y_0) and the direction of the inward normal to the boundary of the domain at the point $(x, 0)$.

The reader will easily be able to derive the proof for himself.

The Neumann problem for the half-plane not only has no solutions which vanish at infinity; it has not even any bounded solutions. If we write the solution of this problem formally as an integral which we may regard as the potential of a single layer in the plane† [compare (13.9)]

$$-\frac{1}{\pi} \int_{-\infty}^{+\infty} \log_e \frac{1}{r} f_2(x) \, dx,$$

then this integral will increase without bound as the point (x_0, y_0) moves away to infinity. We shall not deal with this problem.

The Dirichlet problem for a circle is solved by a method similar to that used in solving the same problem for a sphere. If (x_1, y_1) is the point inverse to (x_0, y_0) with respect to a unit circle round the origin, *i.e.*, if

$$x_1 = \frac{x_0}{x_0^2 + y_0^2}, \quad y_1 = \frac{y_0}{x_0^2 + y_0^2},$$

then for a point on this circle, as before, $r = R_0 r_1$,

$$r = \sqrt{(x - x_0)^2 + (y - y_0)^2}, \quad r_1 = \sqrt{(x - x_1)^2 + (y - y_1)^2},$$

and R_0 is the radius vector of the point (x_0, y_0).

Let (x_0, y_0) be an internal point of the circle $R \leq 1$. Applying Green's formula to the solution of the equation $\nabla^2 u = \varrho$ we get

$$u(x_0, y_0) = -\frac{1}{2\pi} \iint_{R \leq 1} \varrho \log_e \frac{1}{r} \, dx \, dy$$

$$+ \frac{1}{2\pi} \int_{R=1} \left\{ u \frac{\partial}{\partial n}\left(\log_e \frac{1}{r}\right) - \frac{\partial u}{\partial n} \log_e \frac{1}{r} \right\} dS. \tag{17.4}$$

Since (x_1, y_1) lies outside the circle, the function $\log_e 1/R_0 r_1$ is a harmonic

---

† We deal with this, the so-called logarithmic potential, in § 3 of this lecture.

function everywhere inside it, and by virtue of the remark made about the formula (17.2) we obtain

$$-\frac{1}{2\pi} \iint_{R \leq 1} \varrho \log_e \frac{1}{R_0 r_1} \, dx \, dy$$

$$+ \frac{1}{2\pi} \int_{R=1} \left\{ u \frac{\partial}{\partial n} \left(\log_e \frac{1}{R_0 r_1} \right) - \frac{\partial u}{\partial n} \log_e \frac{1}{R_0 r_1} \right\} dS = 0. \qquad (17.5)$$

Subtracting (17.5) from (17.4) and writing

$$G(x, y, x_0, y_0) = \frac{1}{2\pi} \log_e \frac{1}{R_0 r_1} - \frac{1}{2\pi} \log_e \frac{1}{r},$$

gives

$$u(x_0, y_0) = \iint_{R \leq 1} \varrho G(x, y, x_0, y_0) \, dx \, dy - \frac{1}{2\pi} \int_{R=1} u \frac{\partial G}{\partial n} \, dS. \qquad (17.6)$$

Green's function G is a symmetrical function of the points (x, y) and (x_0, y_0); consequently it is a harmonic function of (x_0, y_0). Clearly, it is identically zero for all x, y if (x_0, y_0) lies on the boundary.

If the problem of finding a solution of Poisson's equation satisfying the condition

$$[u]_S = f_1(s), \qquad (17.7)$$

is soluble, then the solution will have the form (17.6). In the particular case when $[u]_s = 0$, the solution is given by the formula

$$u(x_0, y_0) = \iint_{R \leq 1} \varrho G \, dx \, dy. \qquad (17.8)$$

It is easily verified that the function u given by (17.8) does in fact solve the problem. For,

$$\iint_{R \leq 1} \varrho G \, dx \, dy = -\frac{1}{2\pi} \iint_{R \leq 1} \varrho \log_e \frac{1}{r} \, dx \, dy$$

$$+ \frac{1}{2\pi} \iint_{R \leq 1} \varrho \log_e \frac{1}{R_0 r_1} \, dx \, dy.$$

It can be shown that the first term is a logarithmic potential of a distributed mass, satisfying the equation $\nabla^2 v = \varrho$, and the second is a harmonic function. Consequently, (17.8) gives a solution of Poisson's equation, and it is immediately obvious that this solution satisfies the condition $[u]_s = 0$.

We transform formula (17.6) by substituting in it an explicit expression for G. In polar coordinates, replacing x, y by R and θ, and x_0, y_0 by R_0 and

θ_0, we get for the coordinates of the point x_1, y_1 the expression $1/R_0, \theta_0$. Also

$$\log_e \frac{1}{r} = \frac{1}{2} \log_e [R^2 + R_0^2 - 2RR_0 \cos (\theta - \theta_0)],$$

$$\log_e \frac{1}{R_0 r_1} = -\frac{1}{2} \log_e [R^2 R_0^2 - 2RR_0 \cos (\theta - \theta_0) + 1].$$

Green's function may then be written in the form

$$G = \frac{1}{4\pi} \{\log_e [R^2 + R_0^2 - 2RR_0 \cos (\theta - \theta_0)]$$

$$- \log_e [R^2 R_0^2 - 2RR_0 \cos (\theta - \theta_0) + 1]\}$$

and

$$\left[\frac{\partial G}{\partial n}\right]_{R=1} = \frac{1}{2\pi} \left[\frac{-R + R_0 \cos (\theta - \theta_0)}{R^2 + R_0^2 - 2RR_0 \cos (\theta - \theta_0)} \right.$$

$$\left. - \frac{-RR_0^2 + R_0 \cos (\theta - \theta_0)}{R^2 R_0^2 - 2RR_0 \cos (\theta - \theta_0) + 1} \right]_{R=1}$$

$$= \frac{1}{2\pi} \frac{R_0^2 - 1}{R_0^2 - 2R_0 \cos (\theta - \theta_0) + 1}.$$

The solution of the Dirichlet problem for the circle then becomes

$$u(R_0, \theta_0) = \frac{1}{2\pi} \int_{-\pi}^{+\pi} \frac{1 - R_0^2}{R_0^2 - 2R_0 \cos (\theta - \theta_0) + 1} f_1(\theta) \, d\theta,$$

a result known as *Poisson's formula*. It can be verified that this formula does really give the solution of the problem in the same way as was used for the three-dimensional case.

The external Dirichlet problem for a circle is solved in a similar way. This is the problem of finding a function u which is harmonic outside the circle $R = 1$, satisfies the condition $[u]_s = f_1(s)$ on the boundary, and which is *bounded* at infinity. This last condition makes the two-dimensional external Dirichlet problem essentially different from the three-dimensional one. We shall not stop to go into this problem. The formula giving its solution will have the form

$$u(R_0, \theta_0) = \frac{1}{2\pi} \int_{-\pi}^{+\pi} \frac{R_0^2 - 1}{R_0^2 - 2R_0 \cos (\theta - \theta_0) + 1} f_1(\theta) \, d\theta.$$

From Poisson's formula for harmonic functions inside and outside a circle, all the results which were obtained in the three-dimensional case follow, with hardly any change in the method of proof.

§ 3. The Logarithmic Potential

The concept of potentials can also be carried over to functions of two variables.

An integral of the form

$$v = \int_s \log_e \left(\frac{1}{r}\right) \mu(s) \, ds$$

is called the *logarithmic potential of a single layer*. It is a harmonic function outside and within the domain D bounded by the curve s. The function is continuous on crossing s, but its normal derivative suffers a break in continuity. If we form the integral

$$\int_s \frac{\partial}{\partial n} \left(\log_e \frac{1}{r}\right) \mu(s) \, ds$$

where the derivative is taken for varying x_0 and y_0, then it turns out to be meaningful if the point (x_0, y_0) lies on the boundary. If we denote its value by $[\partial v/\partial n]_0$, we get

$$\left[\frac{\partial v}{\partial n}\right]_e = \pi\mu + \left[\frac{\partial v}{\partial n}\right]_0$$

$$\left[\frac{\partial v}{\partial n}\right]_i = -\pi\mu + \left[\frac{\partial v}{\partial n}\right]_0.$$

The quantity $[\partial v/\partial n]_0$ can be expressed in the form

$$\int_s \mu \frac{\cos \psi_0}{r} \, ds$$

where ψ_0 is the angle between the radius vector from the point (x, y) to the point (x_0, y_0) and the normal at the latter point. The function $(\cos \psi_0)/r$ is a bounded function provided only that the curve s is sufficiently smooth.

In general, the logarithmic potential of a single layer is not bounded at infinity.

An integral of the form

$$w = \int_s \frac{\partial}{\partial n} \left(\log_e \frac{1}{r}\right) \mu(s) \, ds = \int_s \frac{\mu(s) \cos \varphi}{r} \, ds$$

where φ is the angle between the radius vector from the point (x, y) to the point (x_0, y_0) and the normal at (x, y), is called the *logarithmic potential of a double layer*. It is a harmonic function both inside and outside the domain D bounded by the curve s. On s the function suffers a break in continuity. If (x_0, y_0) lies on the boundary curve, which we assume to be sufficiently smooth, then $(\cos\varphi)/r$ is a bounded function and the integral w is meaningful. Denoting its value in this case by w_0, we shall have

$$w_e = -\pi\mu + w_0,$$

$$w_i = \pi\mu + w_0,$$

The logarithmic potential of a double layer vanishes at infinity.

And as with the three-dimensional case, the Dirichlet and Neumann problems can also be formulated in the plane. There are, however, certain special features in the external problems. In the external Dirichlet problem, instead of requiring that the function u shall vanish at infinity, we have to impose the condition that it shall be bounded in the neighbourhood of an infinitely distant point. The Dirichlet problem then has a determinate and unique solution.

In the external Neumann problem we have, as before, to seek a solution which vanishes at infinity, but, in contrast to the earlier case, this problem will, in general, have no solution. The necessary and sufficient condition for a solution to exist is that

$$\int_s f_2(s)\, \mathrm{d}s = 0$$

where $f_2(s)$ is the value of the normal derivative on the boundary.

In this respect the internal and external two-dimensional problems are more similar to one another than they are to the three-dimensional analogues.

The reader would find it beneficial to prove the foregoing assertions for himself.

The Dirichlet and Neumann problems can, as before, be reduced to integral equations. We leave this reduction too for the reader to carry out.

For the integration of the two-dimensional Laplace equation there is another extremely powerful method, based on an application of the theory of complex variables. Here we shall only indicate the essence of this method, without going into details.

We consider any analytic function $w(z) = u + iv$ of the complex variable $z = x + iy$. Taking x and y as the independent variables and applying the Laplace operator to w, we get

$$\nabla^2 w = \frac{\partial^2 w}{\partial x^2} + \frac{\partial^2 w}{\partial y^2} = w''(z) + i^2 w''(z) = 0.$$

It follows that the function $w(z)$ is a harmonic function of the variables x and y; consequently its real and imaginary parts, $u(x, y)$ and $v(x, y)$, will also be harmonic within the domain in which w is analytic.

We now introduce a new independent complex variable $\zeta = \xi + i\eta$ and put $z = \psi(\zeta)$, where ψ is any analytic function. Then

$$x = x(\xi, \eta), \quad y = y(\xi, \eta),$$

and the analytic function $w(z)$ will go over into an analytic function of the variable ζ:

$$w^*(\zeta) \equiv w(\psi(\zeta)).$$

Hence the functions

$$\left. \begin{aligned} u^*(\xi, \eta) &= u(x(\xi, \eta), y(\xi, \eta)), \\ v^*(\xi, \eta) &= v(x(\xi, \eta), y(\xi, \eta)) \end{aligned} \right\} \tag{17.9}$$

will again be harmonic functions of the variables ξ and η.

In the theory of functions of a complex variable it is shown that the formulae (17.9) define a conformal mapping of the x, y plane on to the ξ, η plane, and that any conformal mapping can be obtained in this way. It follows from this argument that a harmonic function of the variables x, y within a certain domain remains harmonic when this domain undergoes a conformal transformation.

For any simply connected domain Ω of the x, y plane we can obtain a solution of the Dirichlet problem in the following way. We find a conformal mapping

$$x = x(\xi, \eta), \quad y = y(\xi, \eta)$$

which carries the domain Ω over into a circle. It is proved in the theory of functions of a complex variable that such a mapping always exists. The function $u^*(\xi, \eta)$ must be harmonic within the circle and take specified values on its circumferences; it can be constructed by using Poisson's formula. Then going back to the variables x, y, we obtain the required solution of our problem.

THE THEORY OF INTEGRAL EQUATIONS

§ 1. General Remarks

We have already seen in earlier lectures that the solution of certain problems of mathematical physics can be made to depend on the solution of equations of the form

$$\varphi(P) = \int_D K(P, P_1) \, \varphi(P_1) \, dP_1 + f(P), \qquad (18.1)$$

where D is a certain domain over which the points P and P_1 vary. The function $K(P, P_1)$ of two variable points of this domain is called the *kernel*, and $\varphi(P)$ is the unknown function.

Suppose that the domain D lies in n-dimensional space, and let the co-ordinates of the point P and P_1 be $(x_1, x_2, \ldots, x_n)$ and $(x_1', x_2', \ldots, x_n')$ respectively; then the kernel $K(P, P_1)$ is a function of the $2n$ variables $(x_1, x_2, \ldots, x_n, x_1', x_2', \ldots, x_n')$, and the functions $\varphi(P)$ and $f(P)$ are functions of n variables, $\varphi(x_1, x_2, \ldots, x_n), f(x_1, x_2, \ldots, x_n)$; and these variables may vary only in such a way that the points $P(x_1, x_2, \ldots, x_n), P_1(x_1', x_2', \ldots, x_n')$ do not go outside the domain D.

If, in particular, the domain D is one-dimensional and connected, then the position of the point P is defined by a single coordinate x, and the integral equation takes the form

$$\varphi(x) = \int_a^b K(x, x') \, \varphi(x') \, dx' + f(x).$$

We are about to begin a systematic investigation of equations of the form (18.1), which are generally known as *Fredholm integral equations of the second kind*.

We shall assume the functions K and f to be real. We shall introduce other restrictions such as boundedness, continuity, and so on, as the need arises.

Later we shall encounter equations which can have not one, but many solutions. But in all such cases we shall not regard functions as being different if they are *equivalent*, *i.e.*, if they take different values only on a set of measure zero.

For Fredholm integral equations, as for all linear equations, the following theorem holds:

THEOREM 1. *The general solution of the equation* (18.1) *has the form*

$$\varphi(P) = \varphi_0(P) + \varphi^*(P)$$

where $\varphi_0(P)$ *is some particular solution of the equation* (18.1)
and $\varphi^*(P)$ *is the general solution of the equation*

$$\varphi(P) = \int_D K(P, P_1)\, \varphi(P_1)\, \mathrm{d}P_1. \tag{18.2}$$

We shall call (18.2) the *homogeneous equation corresponding* to equation (18.1).

It is clear from this that if the corresponding homogeneous equation has no solution other than the trivial one $\varphi^*(P) = 0$, then the equation (18.1) cannot have more than one solution.

§ 2. The Method of Successive Approximations

We shall examine first, in §§ 2, 3 and 4, a few special cases; this will enable us to pass on to a more general treatment of the problem.

Instead of equation (18.1) we shall consider a more general equation of the form

$$\varphi(P) = \lambda \int_D K(P, P_1)\, \varphi(P_1)\, \mathrm{d}P_1 + f(P), \tag{18.3}$$

the so-called *equation with a parameter*.

Let us suppose that the domain D of variation of the point P is bounded; we shall denote the volume of this domain by the same letter D. Let us further suppose that the kernel $K(P, P_1)$ is a summable function of the pair of variable points P, P_1 in the space of $2n$ variables, and that

$$\int_D \left| K(P, P_1) \right| \mathrm{d}P_1 \leq M$$

where the constant M does not depend on the position of the point P.

In the applications of the theory with which we shall be concerned, the function $K(P, P_1)$ will have only a finite number of manifolds of a smaller number of variables (*i.e.*, of points, curves, surfaces, *etc.*) on which it suffers a break in continuity. If we exclude these manifolds together with a volume, as small as we please, surrounding them, then the function $K(P, P_1)$ will be continuous in the remaining part of the domain.

We shall assume that $f(P)$ is a bounded, measurable function:

$$\sup \left| f(P) \right| = L < \infty.$$

This condition will be fulfilled if, for example, f is bounded in D and is bounded everywhere except, perhaps, on a finite number of manifolds of lesser dimensionality.

We shall also regard the function $\varphi(P)$ as being bounded and measurable.

If the value of λ is small, the idea naturally occurs of seeking a solution of the equation (18.3) in the form of a power series in λ

$$\varphi(P) = \varphi_0(P) + \sum_{k=1}^{\infty} \lambda^k \varphi_k(P). \tag{18.4}$$

Substituting this expression in (18.3), we get

$$\varphi_0(P) + \sum_{k=1}^{\infty} \lambda^k \varphi_k(P) = \lambda \int_D K(P, P_1) \left[\varphi_0(P_1) + \sum_{k=1}^{\infty} \lambda^k \varphi_k(P_1) \right] dP_1 + f(P),$$

from which we get by comparing coefficients of powers of λ on the two sides,

$$\left.\begin{aligned}
\varphi_0(P) &= f(P), \\[1ex]
\varphi_1(P) &= \int_D K(P, P_1)\, \varphi_0(P_1)\, dP_1 \\
\quad \cdot &\quad \cdot \quad \cdot \quad \cdot \quad \cdot \\
\varphi_k(P) &= \int_D K(P, P_1)\, \varphi_{k-1}(P_1)\, dP_1 \\
\quad \cdot &\quad \cdot \quad \cdot \quad \cdot \quad \cdot
\end{aligned}\right\} \tag{18.5}$$

The relations (18.5) allow all the functions $\varphi_k(P)$ to be calculated step by step. For, all the integrals on the right-hand sides are meaningful, since the integrands are the products of bounded functions by summable functions. Estimating the right-hand sides of these equations in turn, we get

$$\left| \varphi_0(P) \right| \leq L, \quad \left| \varphi_1(P) \right| \leq \int_D \left| K(P, P_1) \right| L\, dP_1 \leq LM,$$

and, in general,

$$\left| \varphi_k(P) \right| \leq LM^k.$$

Hence by the Lebesgue–Fubini theorem (Lecture 6), the functions $\varphi_k(P)$ are measurable.

For brevity, we denote the function

$$\psi(P) = \int_D K(P, P_1)\, f(P_1)\, dP_1$$

by the symbol Af. And as we have seen, $|Af| \leqq M \sup |f|$. We shall call A an operator on the function f, and we shall consider powers of the operator A:

$$A(Af) = A^2f, \quad A(A^2f) = A^3f, \ldots, A(A^{m-1}f) = A^mf.$$

Clearly,

$$A^{m+n}f = A^m(A^nf).$$

In this notation, (18.5) takes the form

$$\varphi_k(P) = A^kf. \tag{18.6}$$

Also, as we have seen

$$|A^kf| \leqq M^k \sup |f|. \tag{18.7}$$

This operator A has an important property:

THEOREM 2. *If a sequence of bounded, measurable functions f_k tends uniformly to a limit function f_0:*

$$f_0 = \lim_{k \to \infty} f_k,$$

then

$$Af_0 = \lim_{k \to \infty} Af_k$$

and the sequence Af_k also converges uniformly.

For,

$$|Af_k - Af_0| = |A(f_k - f_0)| \leqq M \sup |f_k - f_0|.$$

COROLLARY. If the series

$$\sum_{k=1}^{\infty} u_k(P) = u(P)$$

converges uniformly, then

$$Au(P) = \sum_{k=1}^{\infty} Au_k(P).$$

In our new notation, (18.3) takes the form

$$\varphi - \lambda A\varphi = f,$$

or, writing $E\varphi \equiv \varphi$, where E is the identical or unit operator, we have

$$(E - \lambda A) \varphi = f.$$

If we substitute in (18.4) the expression for φ_k from (18.6), we get, and so far merely formally, the equation

$$\varphi(P) = f + \lambda Af + \lambda^2 A^2f + \cdots + \lambda^k A^kf + \cdots. \tag{18.8}$$

The absolute value of the terms in the series on the right-hand side will, by (18.7) and if $|\lambda M| < 1$, be less than the terms of the convergent, positive numerical series

$$\sum_{k=0}^{\infty} |\lambda|^k M^k \sup |f| = \sup |f| \frac{1}{1 - |\lambda| M}.$$

Consequently, for a fixed λ satisfying the condition $|\lambda| < 1/M$, the series in (18.8) converges uniformly, and (18.8) does indeed define a certain measurable function $\varphi(P)$. Also,

$$|\varphi| \leq \sum_{k=0}^{\infty} |\lambda|^k M^k \sup |f| = \sup |f| \frac{1}{1 - |\lambda| M}.$$

We now verify that the function $\varphi(P)$ which we have obtained does satisfy equation (18.3). Substituting the expression for φ from (18.8) in the left-hand side of (18.3), we get

$$(E - \lambda A) \left[f + \sum_{k=1}^{\infty} \lambda^k A^k f \right].$$

If we show that this expression is identically equal to f, our assertion will be proved. The series in the square brackets converges uniformly. By Theorem 2 we have

$$(E - \lambda A) \left[f + \sum_{k=1}^{\infty} \lambda^k A^k f \right] = f + \sum_{k=1}^{\infty} \lambda^k A^k f - \lambda A f - \sum_{k=1}^{\infty} \lambda^{k+1} A^{k+1} f = f.$$

$$(18.9)$$

Thus we have proved that the function $\varphi(P)$ given by (18.8) is a solution of the integral equation (18.3).

The uniqueness of the bounded solution subject to the condition $|\lambda| M < 1$ is easily proved. For, assuming that the integral equation (18.3) is satisfied by φ, we apply to both sides the operator

$$E + \sum_{k=1}^{\infty} \lambda^k A^k,$$

which is taken to have the meaning that

$$\left(E + \sum_{k=1}^{\infty} \lambda^k A^k \right) f \equiv f + \sum_{k=1}^{\infty} \lambda^k A^k f.$$

We then obtain

$$\left(E + \sum_{k=1}^{\infty} \lambda^k A^k \right) [(E - \lambda A) \varphi] = \left(E + \sum_{k=1}^{\infty} \lambda^k A^k \right) f.$$

Removing the brackets on the left-hand side, we find

$$\left(E + \sum_{k=1}^{\infty} \lambda^k A^k \right) (E - \lambda A) \varphi = \varphi + \sum_{k=1}^{\infty} \varphi^k A^k \varphi - \lambda A \varphi - \sum_{k=1}^{\infty} \lambda^{k+1} A^{k+1} \varphi = \varphi.$$

$$(18.10)$$

Hence

$$\varphi = \left(E + \sum_{k=1}^{\infty} \lambda^k A^k \right) f.$$

Consequently the solution must be expressed by formula (18.8) and is unique.

If we introduce the notation

$$B = E - \lambda A,$$

then the operator

$$E + \sum_{k=1}^{\infty} \lambda^k A^k$$

will naturally be denoted by B^{-1}. The equations (18.9), (18.10) then take the form

$$BB^{-1}f = f \qquad \qquad (18.11)$$

$$B^{-1}Bf = f \qquad \qquad (18.12)$$

where f is any measurable bounded function. These formulae serve to justify our notation.

§ 3. Volterra Equations

In certain particular cases it may happen that the series (18.8) converges over the whole plane of the complex variable λ. We examine a case of this sort.

Let the domain D of a single variable be the ray $x > 0$,

let

$$\varphi(x) = \lambda \int_0^{\infty} K(x, y)\, \varphi(y)\, \mathrm{d}y + f(x),$$

and suppose that the kernel $K(x, y)$ has the property that

$$K(x, y) \equiv 0 \quad \text{if} \quad y > x,$$

and that it is bounded,

$$\left| K(x, y) \right| \leq M.$$

Then the equation will take the form

$$\varphi(x) = \lambda \int_0^x K(x, y)\, \varphi(y)\, \mathrm{d}y + f(x).$$

Equations of this type are known as *Volterra Equations*.

Assuming as before that the function f is bounded (less than the constant L) and is measurable, we can estimate the magnitude of $A^k f$. We have:

$$\left| Af \right| = \left| \int_0^x K(x, y) f(y) \, dy \right| \leqq \int_0^x LM \, dy = L \, \frac{Mx}{1}$$

$$\left| A^2 f \right| = \left| \int_0^x K(x, y) \, (Af) \, dy \right| \leqq \int_0^x ML \, \frac{My}{1} \, dy = L \, \frac{M^2 x^2}{1.2}$$

$$\left| A^3 f \right| = \left| \int_0^x K(x, y) \, (A^2 f) \, dy \right| \leqq \int_0^x ML \, \frac{M^2 y^2}{1.2} \, dy = L \, \frac{M^3 x^3}{1.2.3}$$

$$\cdot \qquad \cdot \qquad \cdot \qquad \cdot \qquad \cdot \qquad \cdot \qquad \cdot \qquad \cdot \qquad \cdot$$

$$\left| A^k f \right| \leqq L \, \frac{M^k x^k}{k!}.$$

Consequently, any term of the series (18.8)

$$f + \lambda Af + \lambda^2 A^2 f + \cdots + \lambda^k A^k f + \cdots$$

does not exceed in absolute magnitude the corresponding term of the series

$$L + L \, \frac{\lambda Mx}{1!} + L \, \frac{\lambda^2 M^2 x^2}{2!} + \cdots + L \, \frac{\lambda^k M^k x^k}{k!},$$

which converges over the whole λ-plane. It follows that the series (18.8) for any fixed x converges uniformly in any finite circle of the λ-plane and is a completely analytic function of λ. For a fixed λ it converges uniformly on any finite interval of values of x, and this implies, as before, that it is the unique solution of the integral equation considered.

§ 4. Equations with Degenerate Kernel

We consider one more particular class of integral equations, for which the theory is easily constructed. The results obtained turn out to be valid also in more general cases. We shall examine the integral equation (18.3) when its kernel has the special form

$$K(P, P_1) = \sum_{l=1}^{N} \varphi_l(P) \, \psi_l(P_1). \tag{18.13}$$

A kernel of this form, (18.13), is said to be *degenerate*.

Let us say that the system of functions $\varphi_1(P), \varphi_2(P), \ldots, \varphi_N(P)$ is *linearly independent* if there are no constants $\alpha_1, \alpha_2, \ldots, \alpha_N$ which are not simultaneously zero such that the linear combination

$$\alpha_1 \varphi_1(P) + \alpha_2 \varphi_2(P) + \cdots + \alpha_N \varphi_N(P)$$

is equivalent to zero (see p. 237 for the meaning of "equivalent"). We can regard the system of functions $\varphi_1(P)$, $\varphi_2(P)$, ..., $\varphi_N(P)$ and also the system $\psi_1(P), \psi_2(P), ..., \psi_N(P)$ as being linearly independent in the domain D, for otherwise one or more of these functions could be expressed as a linear combination of the remainder and we should be led to a kernel of the same form but with a smaller number of terms.

Unless some special proviso is made, we shall in future regard the functions $\varphi_1, ..., \varphi_N, \psi_1, ..., \psi_N$ as being bounded and having only isolated discontinuities on a finite number of smooth surfaces.

The equation (18.3) for a degenerate kernel takes the form

$$\varphi(P) = \lambda \sum_{l=1}^{N} \varphi_l(P) \int_D \psi_l(P_1)\, \varphi(P_1)\, dP_1 + f(P) \qquad (18.14)$$

and consequently the difference $\varphi(P) - f(P)$ must be a linear combination of the functions $\varphi_l(P)$, $l = 1, 2, ..., N$, with constant coefficients. Putting $\varphi(P) - f(P) = \lambda \sum_{k=1}^{N} \alpha_k \varphi_k(P)$ and substituting this expression in (18.14), we get

$$\sum_{k=1}^{N} \alpha_k \varphi_k(P) = \lambda \sum_{l=1}^{N} \varphi_l(P) \int_D \left\{ \sum_{k=1}^{N} \alpha_k \varphi_k(P_1) \right\} \psi_l(P_1)\, dP_1$$

$$+ \sum_{l=1}^{N} \varphi_l(P) \int_D \psi_l(P_1)\, f(P_1)\, dP_1.$$

Since the functions $\varphi_l(P)$ are linearly independent, the coefficients of the same functions must be the same on both sides of this equation, and so we obtain the system of equations

$$\alpha_l - \lambda \sum_{k=1}^{N} \alpha_k m_{kl} = f_l, \quad l = 1, 2, ..., N, \qquad (18.15)$$

where

$$m_{kl} = \int_D \varphi_k(P_1)\, \psi_l(P_1)\, dP_1, \qquad (18.16)$$

$$f_l = \int_D f(P_1)\, \psi_l(P_1)\, dP_1. \qquad (18.17)$$

The system (18.15) is equipollent with the integral equation (18.14). We denote the matrix of the system (18.15) by $M(\lambda)$:

$$M(\lambda) = \left\| \begin{matrix} 1 - m_{11}\lambda & - m_{12}\lambda & \cdots & - m_{1N}\lambda \\ - m_{21}\lambda & 1 - m_{22}\lambda & \cdots & - m_{2N}\lambda \\ \cdot & \cdot & \cdots & \cdot \\ - m_{N1}\lambda & - m_{N2}\lambda & \cdots & 1 - m_{NN}\lambda \end{matrix} \right\|.$$

The solubility of our system depends on the determinant det $M = |M(\lambda)|$ of this matrix. Clearly there are two cases:

I. det $M \neq 0$,

II. det $M = 0$.

In case I we have the following theorem:

THEOREM 3. *The system* (18.15) *for values of* λ *for which* det $M \neq 0$ *is uniquely soluble for any* f_l, *and the equation* (18.14) *is soluble for any function* f.

In particular, the equation

$$\varphi(P) = \lambda \int_D K(P, P_1)\, \varphi(P_1)\, \mathrm{d}P_1, \tag{18.18}$$

which we called the homogeneous equation corresponding to (18.3), will in this case have only a trivial solution.

The first assertion of this theorem is proved in the algebra text-books; and the second assertion follows immediately from the first.

In case II, for values of λ for which det $M = 0$, the system (18.15) is not soluble for all f_l, and consequently the equation (18.3) is not soluble for all functions f.

In this case the system of homogeneous equations

$$\alpha_l - \lambda \sum_{k=1}^{N} m_{kl}\alpha_k = 0 \quad l = 1, 2, \ldots, N \tag{18.19}$$

has $N - q$ linearly independent solutions, where q is the rank of the matrix $M(\lambda)$. Let these solutions be $\alpha_1^{(s)}, \alpha_2^{(s)}, \ldots, \alpha_N^{(s)}, s = 1, 2, \ldots, N - q$. The equation (18.18) will obviously also have exactly $N - q$ linearly independent solutions.

As is well-known, when their determinant is zero, a system of inhomogeneous equations may have no solution. We recall some of the necessary and sufficient conditions for the solubility of the system (18.15).

By virtue of the condition det $M = 0$ the left-hand members of (18.15) are not independent; a linear combination of them can be formed which will vanish identically. For, if we multiply the equations of (18.15) by β_l and add, we get

$$\sum_{l=1}^{N} \alpha_l\beta_l - \lambda \sum_{l=1}^{N}\sum_{k=1}^{N} m_{kl}\alpha_k\beta_l = \sum_{k=1}^{N} \alpha_k\beta_k - \lambda \sum_{k=1}^{N}\sum_{l=1}^{N} m_{kl}\alpha_k\beta_l$$
$$= \sum_{k=1}^{N} \alpha_k \left[\beta_k - \lambda \sum_{l=1}^{N} m_{kl}\beta_l \right] = \sum_{l=1}^{N} f_l\beta_l.$$

The β_l can be chosen so that

$$\beta_k - \lambda \sum_{l=1}^{N} m_{kl}\beta_l = 0, \quad k = 1, 2, \ldots, N. \tag{18.20}$$

The determinant of the system of equations (18.20) is equal to det $M = 0$. It is proved in the theory of algebraic equations that the number of linearly independent solutions of (18.20) is again equal to $N - q$. Let these solutions be $\beta_1^{(s)}, \beta_2^{(s)}, \ldots, \beta_N^{(s)}, s = 1, 2, \ldots, (N-q)$.

A necessary condition for the equations (18.15) to be soluble is that

$$\sum_{l=1}^{N} f_l \beta_l^{(s)} = 0, \qquad s = 1, 2, \ldots, (N - q), \tag{18.21}$$

and it is proved in the theory of algebraic equations that this condition is also sufficient.

Just as the system of equations (18.15) corresponds to the equation (18.3), and the system (18.19) to the equation (18.18), so a correspondence can be established between the system (18.20) and the equation

$$\psi(P_1) = \lambda \int_D K(P, P_1)\, \psi(P)\, dP \tag{18.22}$$

which we shall call the homogeneous equation allied to the equation (18.18). Substituting the expression for the kernel $K(P, P_1)$ and repeating exactly the previous argument, we can show that the solution of the equation (18.22) must have the form

$$\psi^{(s)}(P_1) = \sum_{l=1}^{N} \beta_l^{(s)} \psi_l(P_1). \tag{18.23}$$

where the $\beta_l^{(s)}$ are numbers satisfying (18.20). Hence the following theorem.

THEOREM 4. *The homogeneous equation* (18.18) *and the equation* (18.22) *allied with it have the same number of linearly independent solutions. This number* $r = N - q$, *where q is the rank of the matrix* $M(\lambda)$, *and N is the number of terms in the degenerate kernel* (18.13).

We emphasize that the homogeneous equations (18.18) and (18.22) have a non-trivial solution only for those values of λ for which the determinant of the matrix $M(\lambda)$ vanishes. These values of λ are called *eigenvalues* or *characteristic values* for the equation. The number of linearly independent solutions $r = N - q$ corresponding to a given characteristic value is called its *rank*.

If we substitute the expression for f_l in the equation (18.21) we get

$$\int_D f(P) \sum_{l=1}^{N} \beta_l^{(s)} \psi_l(P)\, dP = 0 \tag{18.24}$$

or

$$\int_D f(P)\, \psi^{(s)}(P)\, dP = 0, \quad s = 1, 2, \ldots, (N - q). \tag{18.25}$$

It is clear that (18.21) and (18.25) are equipollent.

We shall say that two functions are *orthogonal if the integral of their product over the domain D is equal to zero.*

We have thus proved the following theorem:

THEOREM 5. *The necessary and sufficient condition for the equations* (18.14) *to be soluble in case II, i.e., when* det $M = 0$, *is that its free member f should be orthogonal to all solutions of the allied homogeneous equation.*

It is clear that in this case the general solution of the equation (18.14) takes the form

$$\varphi(P) = \varphi^{(o)}(P) + \sum_{s=1}^{N-q} C_s \varphi^{(s)}(P), \tag{18.26}$$

where $\varphi^{(0)}(P)$ is some particular solution,

and $\varphi^{(s)}(P)$, $s = 1, 2, ..., (N - q)$, are particular solutions of the homogeneous equation (18.18).

The conditions (18.25) are independent of one another, for we can show that if a_s is a set of $(N - q)$ arbitrary numbers, then a function $f(P)$ can be found such that

$$\int_D f(P)\, \psi^{(s)}(P)\, \mathrm{d}P = a_s, \quad s = 1, 2, ..., (N - q). \tag{18.25'}$$

We first prove that for any given set of numbers f_l a function $f(P)$ can be found for which the equations (18.17) are satisfied.

We shall seek the function $f(P)$ in the form

$$f(P) = \sum_{i=1}^{N} \gamma_i \psi_i(P).$$

We multiply this equation by ψ_l and integrate over the domain D. We obtain a system of equations for the γ_i:

$$f_l = \sum_{i=1}^{N} \gamma_i \int \psi_i(P)\, \psi_l(P)\, \mathrm{d}P, \quad l = 1, 2, ..., N,$$

which are soluble since their determinant is the Gram determinant† for the

---

† *Translator's note.* For *N* functions $\psi_1, \psi_2, ..., \psi_N$, the determinant with $\int_\Omega \psi_i \psi_j\, \mathrm{d}\Omega$ as the element of its *i*th row and *j*th column is known as the Gramian or Gram determinant. This determinant is zero if and only if the functions ψ_i are linearly independent in Ω, provided suitable restrictions are imposed on the functions ψ_i:

e.g., if Ω is a bounded, closed set, that

        (*a*) each ψ_i be continuous, and

        (*b*) each ψ_i be measurable, and $|\psi_i|$ be summable.

See p. 243 for the meaning of 'linearly dependent'.

system of linearly independent functions ψ_i and is therefore positive. Thus the solubility of the system (18.25′) will be proved if we show that the system

$$\sum_{l=1}^{N} f_l \cdot \beta_l^{(s)} = a_s, \quad s = 1, 2, \ldots, (N - q) \tag{18.21′}$$

is soluble. The latter system is soluble, since, due to the linear independence of the system of numbers $\beta_1^{(s)}, \beta_2^{(s)}, \ldots, \beta_N^{(s)}$ $(s = 1, 2, \ldots, N - q)$, the rank of the matrix $\|\beta_l^{(s)}\|$ of the system of equations (18.21′) is equal to $(N - q)$, i.e., to the number of equations, and consequently the rank of the principal matrix of the system is equal to the rank of the augmented matrix.

Remark. Theorem 3 follows essentially from Theorems 4 and 5. For, if the number of linearly independent solutions of the associated equation and of the corresponding homogeneous equation is zero (the number of solutions of these two equations is always the same), then the conditions for orthogonality drop out, and the inhomogeneous equation will be uniquely soluble.

Up to now, in considering equations with degenerate kernels, we have assumed that the functions $\varphi_l(P)$ and $\psi_l(P)$ do not contain the parameter λ. Let us now suppose that the functions $\varphi_l(P)$ and $\psi_l(P)$ do depend on the complex parameter λ and that they are analytic over the domain of variation of this parameter. Then the determinant of M will also be an analytic function of λ in this domain Ω. *Then inside Ω, the second case, i.e. $\det M = 0$, can occur only at isolated points*, since $\det M$, as an analytic function, can have only isolated zeros. This result can be expressed as a theorem.

THEOREM 6. *If in an equation with a degenerate kernel the functions of which the kernel is composed are analytic functions of the parameter λ in a certain domain of the λ-plane, and if the equation is soluble uniquely with any right-hand side (even if only for one λ), then the second case ($\det M = 0$) can occur only at isolated points of domain Ω.*

As we shall see, Theorems 3, 4, 5 and 6 hold good not only for equations with degenerate kernels but also for more general equations. In the more general case they are known respectively as Fredholm's first, second, third and fourth theorems.

§ 5. A Kernel of Special Type. Fredholm's Theorems

Having discussed the solution of integral equations within a circle $|\lambda| < 1/M$ and also the solution of equations with degenerate kernels, we now pass on to the analysis of a more general case. Suppose that the kernel of equation (18.3) has the form

$$K(P, P_1) = \sum_{l=1}^{N} \chi_l(P) \xi_l(P_1) + K_1(P, P_1) \tag{18.27}$$

where

$$\int_D |K_1(P, P_1)|\, dP_1 \leqq M \tag{18.28}$$

and

$$\int_D |K_1(P, P_1)|\, dP \leqq M. \tag{18.29}$$

Then equation (18.3) takes the form

$$\varphi(P) - \lambda \int_D K_1(P, P_1)\, \varphi(P_1)\, dP_1$$

$$= \lambda \sum_{l=1}^N \chi_l(P) \int_D \xi_l(P_1)\, \varphi(P_1)\, dP_1 + f(P) \tag{18.30}$$

We shall consider equation (18.27) for values of λ satisfying the condition $|\lambda| < 1/M$ (M now has a meaning different from that in § 2). We again introduce a symbolic notation

$$\varphi(P) - \lambda \int_D K_1(P, P_1)\, \varphi(P_1)\, dP_1 \equiv (E - \lambda A_1)\varphi \equiv B_1\varphi,$$

$$\chi(P) + \sum_{k=1}^\infty \lambda^k A_1^k \chi \equiv B_1^{-1}\chi.$$

We also use the corresponding notation for the associated equation

$$\psi(P_1) - \lambda \int_D K_1(P, P_1)\, \psi(P)\, dP \equiv (E - \lambda A_1^*)\psi = B_1^*\psi.$$

It is easily verified that, if we put

$$B_1^{*-1}\xi(P_1) \equiv \xi(P_1) + \sum_{k=1}^\infty \lambda^k A_1^{*k}\xi(P_1).$$

then, similar to (18.11) and (18.12),

$$B_1^* B_1^{*-1}\xi(P_1) \equiv \xi(P_1), \quad B_1^{*-1}B_1^*\psi(P_1) = \psi(P_1).$$

We shall also use the formulae

$$\int_D [B_1\varphi(P)]\, \psi(P)\, dP = \int_D \varphi(P)\, [B_1^*\psi(P)]\, dP \tag{18.31}$$

$$\int_D [B_1^{-1}\chi(P)]\xi(P)\, dP = \int_D \chi(P)\, [B_1^{*-1}\xi(P)]\, dP. \tag{18.32}$$

These are easily proved. We have

$$\int_D \psi(P)[A_1\varphi(P)]\,dP = \int_D \psi(P)\left\{\int_D K_1(P,\,P_1)\,\varphi(P_1)\,dP_1\right\}dP$$

$$= \int_D \varphi(P_1)\left\{\int_D K_1(P,\,P_1)\,\psi(P)\,dP\right\}dP_1$$

$$= \int_D \varphi(P_1)\,[A_1^*\psi(P_1)]\,dP_1.$$

Hence we immediately obtain

$$\int_D \psi(P)\,\{[E - \lambda A_1]\varphi\}\,dP = \int_D \varphi(P)\,\{[E - \lambda A_1^*]\psi\}\,dP.$$

The formula (18.32) is proved in the same way.

Using this notation (18.30) may be written in the form

$$B_1\varphi = \lambda \sum \chi_i(P)\int_D \xi_i(P_1)\,\varphi(P_1)\,dP_1 + f(P). \qquad (18.33)$$

We introduce a new unknown function by putting $B_1\varphi(P) = \chi(P)$, whence $\varphi(P) = B_1^{-1}\chi(P)$. Substituting this expression in the equation (18.30) and using (18.32), we get

$$\chi(P) = \lambda \sum \chi_i(P)\int_D \xi_i(P_1)\,[B_1^{-1}\chi(P_1)]\,dP_1 + f(P)$$

$$= \lambda \sum \chi_i(P)\int_D [B_1^{*-1}\xi_i(P_1)]\,\chi(P_1)\,dP_1 + f(P). \qquad (18.34)$$

Thus the equation (18.30) has gone over into an equation with degenerate kernel (18.34) for the new unknown function $\chi(P)$ and having the same free term, and the solution $\varphi(P)$ of equation (18.30) has gone over into the solution $\chi(P)$ of equation (18.34). Conversely, if $\chi(P)$ is a solution of (18.34), then the function $\varphi(P) = B_1^{-1}\chi(P)$ will be a solution of equation (18.30). To prove this, it is sufficient to substitute for $\chi(P)$ in (18.34) its expression $\chi(P) = B_1\varphi(P)$.

We shall show that Theorems 3, 4, 5, 6, which, by what has already been proved, hold good for the new integral equation as regards $\chi(P)$, will also hold for the original equation (18.30) if $|\lambda| < 1/M$.

We set up the associated homogeneous equation for (18.34):

$$\psi(P_1) - \lambda \sum B_1^{*-1}\xi_i(P_1)\int_D \chi_i(P)\,\psi(P)\,dP = 0 \qquad (18.35)$$

and denote its left-hand side by $\omega(P_1)$. The function $\omega(P_1)$ vanishes if and only if

$$B_1^* \omega(P_1) = 0. \tag{18.36}$$

This implies that the equation (18.35) has just the same solutions as the equation (18.36) has. But

$$B_1^* \omega(P_1) = B_1^* \psi(P_1) - \lambda \sum B_1^* B_1^{*-1} \xi_i(P_1) \int_D \chi_i(P) \, \psi(P) \, \mathrm{d}P$$

$$= B_1^* \psi(P_1) - \lambda \sum \xi_i(P_1) \int_D \chi_i(P) \, \psi(P) \, \mathrm{d}P,$$

and consequently (18.36) is the associated homogeneous equation for (18.30). Thus the associated homogeneous equation for (18.34) and the associated homogeneous equation for (18.30) are equipollent.

Using the theory which we have developed we can establish all the Fredholm theorems for the original equation.

Fredholm's Second Theorem. The number q of linearly independent solutions of the homogeneous equation corresponding to equation (18.30) is the same as the number of linearly independent solutions of the equation (18.36) associated with it.

It is sufficient to show that the number of linearly independent solutions is the same for the homogeneous equation corresponding to (18.34) and for (18.36), which are equivalent to the equations under consideration. But the number of such solutions for the homogeneous equation corresponding to (18.34) and for (18.35) is the same by virtue of Theorem 4, which we have proved for an equation with a degenerate kernel. Hence the theorem.

Fredholm's Third Theorem. The necessary and sufficient condition for the equation (18.30) to be soluble is that its free member should be orthogonal to all solutions of the associated homogeneous equation (18.36):

$$\int_D f(P) \, \psi_i(P) \, \mathrm{d}P = 0, \quad i = 1, 2, \ldots, q. \tag{18.37}$$

We remark that the solutions $\psi_i(P)$ of the equation (18.36) are solutions of equation (18.35), and the free members of (18.30) and (18.34) are the same. Hence, by Theorem 5 which has been established for an equation with a degenerate kernel, the conditions (18.37) are necessary and sufficient conditions for the solubility of (18.34). But equation (18.30) is soluble if (18.34) is soluble, and *vice versa*. Hence the conditions (18.37) are also necessary and sufficient conditions for the solubility of (18.30). Hence the theorem.

Fredholm's First Theorem. If the equation (18.30) is soluble for any function $f(P)$ on its right-hand side, then the solution is unique, and this implies that the corresponding homogeneous equation has only a trivial solution. Conversely, *if the homogeneous equation has only a trivial solution, then the equation is uniquely soluble for any function f(P).*

As mentioned earlier, this theorem is a corollary of Fredholm's second and third theorems.

Our previous Theorem 6 can now be formulated extremely simply in the present case:

Fredholm's Fourth Theorem. The characteristic values of λ have no limit-points inside the circle $|\lambda| < 1/M$.

The proof follows from the fact that $\lambda = 0$ will not be a characteristic value for (18.30), since for this value of λ there will be no other solutions than $\chi(P) = f(P)$. By Theorem 6 for equations with degenerate kernels, the characteristic values for equation (18.34), and hence also for equation (18.30), cannot have limit-points inside the circle $|\lambda| < 1/M$, where the functions which compose the kernel are regular. Hence the theorem.

We have proved Fredholm's four theorems for integral equations whose kernel can be expressed in the form (18.27). The kernels encountered in applications of the theory can often be approximated to by a sum of the form

$$\sum_{l=1}^{N} \chi_l(P)\, \xi_l(P_1)$$

with any degree of accuracy, in the sense that the number M, the upper bound of the integrals

$$\int_D |K_1(P,\, P_1)|\, \mathrm{d}P_1 \quad \text{and} \quad \int_D |K_1(P,\, P_1)|\, \mathrm{d}P$$

can be as small as we please.

For kernels of this type all the Fredholm theorems are valid in a circle of arbitrarily large radius, and the unique limit-point for the characteristic values of λ can be taken to be at infinity.

We shall prove that this will be so if, for example, the following conditions are fulfilled (as is often the case in applications of the theory):

(1) The domain D is a certain n-dimensional manifold in a certain k-dimensional cube Ω_1:

$$0 \leqq x_1 \leqq 1, \quad 0 \leqq x_2 \leqq 1, \,...,\quad 0 \leqq x_k \leqq 1$$

(or a continuous 1–1 mapping between D and such a manifold can be set up);

(2) the kernel $K(P,\, P_1)$ is a function which is continuous in a $2n$-dimensional manifold and which can be continuously continued in the $2k$-dimensional

cube Ω_2:

$$0 \leq x_i \leq 1, \quad i = 1, 2, \ldots, k$$

$$0 \leq y_i \leq 1 \quad i = 1, 2, \ldots, k,$$

where x_i denotes the coordinates of the point P in the cube Ω_1 and y_i the coordinates of the point P_1 in the same cube.

It is clear that both these conditions will be satisfied if, for example, the domain D is a surface in three-dimensional space and the function $K(P, P_1)$ is continuous for variation of P and P_1 over this surface.

We now pass on to the proof of our assertion.

By a well-known theorem due to Weierstrass on the approximation of continuous functions by means of polynomials, it is always possible to represent a function $K(P, P_1)$ which is continuous in the closed $2k$-dimensional cube by means of polynomials in $2k$ variables to any specified degree of approximation. But polynomials in $x_i, i = 1, 2, \ldots, k$ and $y_i, i = 1, 2, \ldots, k$, can always be put into the form of a sum of products of functions which depend only on $x_1, x_2, \ldots, x_k$ by functions which depend only on $y_1, y_2, \ldots y_k$, as was to be proved.

It is clear from the very method of constructing the solution of the Fredholm equation in our case that this solution will be an analytic function of the parameter λ over the whole plane, with the exception of the singular points, which are the characteristic values of λ and which are isolated points.

§ 6. Generalization of the Results

In order to obtain the results set forth in § 5 of this lecture, it is not necessary to require that the kernel $K_1(P, P_1)$ shall satisfy both the conditions (18.28) and (18.29). It is sufficient if just one of them, say the first, is satisfied. We shall suppose that (18.28) is satisfied, but we make no stipulation about the integral

$$\int_D |K_1(P, P_1)| \, dP.$$

We put

$$\chi(P_1) = A^*\psi \equiv \int_D K_1(P, P_1) \, \psi(P) \, dP.$$

We shall prove the following fundamental property of the operator A^*: if the function $\psi(P)$ is summable, then the function $\chi(P_1)$ is also summable and

$$\int_D |\chi(P_1)| \, dP_1 \leq M \int_D |\psi(P)| \, dP. \tag{18.38}$$

To prove this we consider the function of $2n$ variables $K_1(P, P_1) \psi(P)$. This function is measurable, since it is the product of two measurable functions. Moreover, it is summable as a function of $2n$ variables. For, the inequality

$$\int_D |\psi(P)| \left\{ \int_D |K_1(P, P_1)| \, dP_1 \right\} dP \leqq M \int_D |\psi(P)| \, dP$$

clearly holds. The repeated integral on the left-hand side evidently exists. Hence, by the Remark made after the Lebesgue–Fubini theorem (Lecture 6, § 8, p. 123), the double integral

$$\iint_{DD} |\psi(P) K(P, P_1)| \, dP \, dP_1$$

exists, and consequently so does the double integral

$$\iint_{DD} \psi(P) K_1(P, P_1) \, dP \, dP_1$$

and moreover the order of integration may be changed. This implies that both sides of the inequality

$$\int_D \left| \left\{ \int_D K_1(P, P_1) \psi(P) \, dP \right\} \right| dP_1 \leqq \int_D \left\{ \int_D |K_1(P, P_1) \psi(P)| \, dP \right\} dP_1,$$

are meaningful and the inequality is valid. But

$$\int_D \left\{ \int_D |K_1(P, P_1) \psi(P)| \, dP \right\} dP_1 = \int_D |\psi(P)| \left\{ \int_D |K_1(P, P_1)| \, dP_1 \right\} dP,$$

which, together with (18.28) immediately gives (18.38).

THEOREM 6. *If the sequence ψ_k converges in the mean to a function ψ_0, i.e., if the integral $\int_D |\psi_0 - \psi_k| \, dP$ converges to zero, then the sequence $A^* \psi_k$ converges in the mean to $A^* \psi_0$.*

For,

$$\int_D |A^* \psi_0 - A^* \psi_k| \, dP = \int_D |A^*(\psi_0 - \psi_k)| \, dP \leqq M \cdot \int_D |\psi_0 - \psi_k| \, dP,$$

from which our assertion follows.

In particular, if the series $v = \sum_{k=1}^{\infty} u_k$ is convergent in the mean, then $A^* v = \sum_{k=1}^{\infty} A^* u_k$ and the series on the right converges in the mean.

Consider the integral equation

$$B^* \psi \equiv (E - \lambda A^*) \psi = \chi \tag{18.39}$$

associated with equation (18.3). Just as we found a solution of (18.3), we can obtain a solution of this equation in the form of a series

$$\psi \equiv B^{*-1}\chi = \chi + \lambda A^*\chi + \lambda^2 A^{*2}\chi + \cdots + \lambda^k A^{*k}\chi + \cdots. \quad (18.40)$$

We shall prove that the formula (18.40) really does give the solution of (18.39) within the circle $|\lambda| < 1/M$.

We show first that the series on the right-hand side of (18.40) converges. Applying the inequality (18.38) successively gives

$$\int_D |A^*\chi|\, dP \leqq M \int_D |\chi|\, dP, \int_D |A|^{*2}\chi\, dP \leqq M^2 \int_D |\chi|\, dP, \ldots,$$

$$\int_D |A^{*k}\chi|\, dP \leqq M^k \int_D |\chi|\, dP, \ldots.$$

Hence

$$\int_D \left| \sum_{k=m}^{m+p} \lambda^k A^{*k}\chi \right| dP \leqq \int_D \left\{ \sum_{k=m}^{m+p} |\lambda^k A^{*k}\chi| \right\} dP = \sum_{k=m}^{m+p} |\lambda|^k \int_D |A^{*k}\chi|\, dP$$

$$\leqq \left(\sum_{k=m}^{m+p} |\lambda|^k M^k \right) \int_D |\chi|\, dP.$$

In the circle $|\lambda| < 1/M$ we have

$$\int_D \left| \sum_{k=m}^{m+p} \lambda^k A^{*k}\chi \right| dP < \frac{|\lambda|^m M^m}{1 - |\lambda| M} \int_D |\chi|\, dP.$$

For a sufficiently large m the right-hand side of this inequality can be made less than any previously specified number. Consequently, by the theorem on convergence in the mean (Lecture 6, Theorem 23), the series (18.40) converges in the mean, and its sum, denoted by $B^{*-1}\chi$, is a summable function of the variable point P_1.

We next prove the identities

$$B^* B^{*-1}\xi = \xi, \quad (18.41)$$

$$B^{*-1} B^* \psi = \psi. \quad (18.42)$$

We have

$$B^{*-1} B^* \psi = (\psi - \lambda A^*\psi) + \lambda A^*(\psi - \lambda A^*\psi) + \lambda^2 A^{*2}(\psi - \lambda A^*\psi) + \cdots$$

$$+ \lambda^k A^{*k}(\psi - \lambda A^*\psi) + \cdots = \psi.$$

Further, using Theorem 6 and the convergence in the mean of the series, we get

$$\xi + \lambda A^*\xi + \lambda^2 A^{*2}\xi + \cdots + \lambda^k A^{*k}\xi + \cdots$$

we find

$$(E - \lambda A^*) (\xi + \lambda A^* \xi + \lambda^2 A^{*2} \xi + \cdots + \lambda^k A^{*k} \xi + \cdots) = \xi,$$

whence

$$B^* B^{*-1} \xi \equiv \xi.$$

The last identity expresses the fact that the function $B^{*-1} \chi$ is a solution of equation (18.39). The identity (18.42) shows that $B^{*-1} \chi$ is the unique solution, for from (18.39), on applying to both sides the operator B^{*-1}, we get $\psi = B^{*-1} \chi$.

The remaining considerations for equations whose kernel satisfies only the one condition (18.28) are the same as those adduced in § 5.

§ 7. Equations with Unbounded Kernels of a Special Form

We consider a kernel $K(P, P_1)$ where the points P, P_1 belong to a bounded domain D, and let r be the distance between the points P and P_1. We suppose that the kernel $K(P, P_1)$ is continuous everywhere over the aggregate of points P, P_1 for which $r \neq 0$, and that when $r \to 0$ the kernel may tend to infinity but will satisfy the inequality

$$\left| K(P, P_1) \right| < \frac{A}{r^\alpha}$$

over the whole domain D, where $0 < \alpha < n$, n being the dimensionality of the space.

We shall prove that for a kernel of this type all our conditions are satisfied and that in this case the kernel $K(P, P_1)$ can be put into the form (18.27), the constant M in the inequalities (18.28) and (18.29) being as small as we please.

We consider the function

$$K^*(P, P_1) = \min \left[K(P, P_1), \frac{A}{\delta^\alpha} \right]$$

where δ is a given positive number.

We shall estimate the integral

$$\int_D \left| K(P, P_1) - K^*(P, P_1) \right| dP_1.$$

The difference $K(P, P_1) - K^*(P, P_1)$ can be different from zero only in the domain $r < \delta$, for at all other points $K(P, P_1) \leq A/\delta^\alpha$ and this implies that $K^*(P, P_1) = K(P, P_1)$. The function $K(P, P_1) - K^*(P, P_1)$ is everywhere non-negative and for $r \leqq \delta$ it satisfies the inequality

$$0 \leqq K(P, P_1) - K^*(P, P_1) \leqq A\left(\frac{1}{r^\alpha} - \frac{1}{\delta^\alpha}\right).$$

Hence we get

$$\int_D \left| K(P, P_1) - K^*(P, P_1) \right| dP_1 = \int_D \left[K(P, P_1) - K^*(P, P_1) \right] dP_1$$

$$\leqq A \int_{r \leqq \delta} \frac{1}{r^\alpha} \, dP_1 \leqq \frac{\varepsilon}{2}$$

for any given ε, provided only that δ is sufficiently small.

The function $K^*(P, P_1)$, being the minimum of two continuous functions, is continuous for $r \neq 0$. In the neighbourhood of the surface $r = 0$ it is a constant, A/δ^α, and consequently it is also continuous there. Thus, from § 5, this function can be put into the form

$$K^*(P, P_1) = \sum_{i=1}^N \varphi_1(P) \, \psi_1(P_1) + K_2(P, P_1),$$

where

$$\int_D \left| K_2(P, P_1) \right| dP_1 \leqq \frac{\varepsilon}{2}.$$

Consequently,

$$K(P, P_1) = \sum_{i=1}^N \varphi_i(P) \, \psi_i(P_1) + K_2(P, P_1) + [K(P, P_1) - K^*(P, P_1)].$$

Putting
$$K_1(P, P_1) = K_2(P, P_1) + [K(P, P_1) - K^*(P, P_1)],$$

we shall have

$$\int_D \left| K_1(P, P_1) \right| dP_1 \leqq \int_D \left| K_2(P, P_1) \right| dP_1 + \int_D \left| K(P, P_1) - K^*(P, P_1) \right| dP_1$$

$$\leqq \frac{\varepsilon}{2} + \frac{\varepsilon}{2} = \varepsilon.$$

Thus for kernels of the type considered, Fredholm's theorems are valid over the whole plane of the complex parameter λ, just as they were for continuous kernels: and this is what we had to show.

Remark. It follows from the inequalities that have been established that the operators of the form considered in this section carry any bounded, summable function over into some continuous function.

APPLICATION OF THE THEORY
OF FREDHOLM EQUATIONS
TO THE SOLUTION OF THE DIRICHLET
AND NEUMANN PROBLEMS

§ 1. Derivation of the Properties of Integral Equations

The theory which we have developed for Fredholm equations enables us to proceed to an immediate solution of the Dirichlet and Neumann problems, which we earlier reduced to the problem of finding solutions of certain integral equations. We begin with a few auxiliary propositions.

We shall always suppose in future that any surface S is either smooth in the Lyapunov sense or satisfies the conditions of § 4 of Lecture 15.

LEMMA 1. *Let the surface S satisfy the conditions of § 4 of Lecture 15, let $v(S)$ be a continuous function, and let the potential*

$$v = \int_S \frac{v(S)\, dS}{r} \tag{19.1}$$

of the single layer have an external normal derivative $[\partial v/\partial n]_e$ which is identically zero on S.

Then the density $v(S)$ is identically zero.

Proof. The potential of the single layer is a harmonic function outside the surface S and tends to zero at infinity like $1/r$. Its normal derivative from outside S is zero on S. In accordance with the definition in § 4 of Lecture 15 and by Remark 3 in § 3 of that lecture, this normal derivative will, under the conditions of this lemma, be regular.

By virtue of the Remark appended to the theorem on uniqueness of solution of the Neumann problem (Lecture 16, § 1, Theorem 2), the potential under consideration coincides outside the domain bounded by S with the trivial solution of the Neumann problem, *i.e.*, it is identically zero.

The potential of a single layer is a function which is continuous throughout space. Consequently its limit value from within the domain is also zero. We now apply the uniqueness theorem for the internal Dirichlet problem. The potential under consideration is a harmonic function inside S. Its limit value on the surface is zero. Consequently it is identically zero. Hence the limit value of its normal derivative from inside S is also zero.

Using the fact that

$$\left[\frac{\partial v}{\partial n}\right]_e - \left[\frac{\partial v}{\partial n}\right]_i = 4\pi v(S) \tag{19.2}$$

and noting that the normal derivative has zero as its limit value both from inside and from outside S, we see that the density $v(S)$ is zero, as was to be proved.

If the limit value of $[\partial v/\partial n]_i$ on S is zero, then by the theorem on the uniqueness of solution of the Neumann problem, the function v is a constant inside S.

LEMMA 2. *If $v(S)$ satisfies the same conditions as in Lemma 1, and if*

$$\left[\frac{\partial v}{\partial n}\right]_i = 0 \quad \text{and if} \quad v_i = 0,$$

then the density is identically zero.

The proof is almost the same as for Lemma 1. The potential v is a continuous function, and this implies $v_e = 0$. From the uniqueness of solution of the external Dirichlet problem it follows that v is identically zero outside S too. But then formula (19.2) gives $v(S) = 0$, as was to be shown.

Remark. In proving Lemma 1 and 2 we have used the Remark appended to the uniqueness theorem (Lecture 16, § 1, Theorem 2). It would also be possible to weaken the conditions regarding S and treat it as a Lyapunov surface; we should then use Theorem 6 of Lecture 15, the identity $f_2 \equiv 0$, equation (16.4), the Corollary to Theorem 5 of Lecture 15 and Theorem 2 of Lecture 16.

Lemmas 1 and 2 enable us to proceed immediately to the investigation of the Fredholm equations for the Dirichlet and Neumann problems, which we derived in Lecture 16. We obtained the equations:

for the internal Dirichlet problem:

$$\mu(S_0) + \iint_S K(S_0, S)\, \mu(S)\, \mathrm{d}S = \varphi(S_0); \tag{19.3}$$

for the external Neumann problem:

$$v(S_0) + \iint_S K(S, S_0)\, v(S)\, \mathrm{d}S = \psi(S_0); \tag{19.4}$$

for the external Dirichlet problem:

$$\mu(S_0) - \iint_S K(S_0, S)\, \mu(S)\, \mathrm{d}S = \varphi(S_0); \tag{19.5}$$

for the internal Neumann problem:

$$\nu(S_0) - \iint_S K(S, S_0)\, \nu(S)\, dS = \psi(S). \tag{19.6}$$

The kernel $K(S_0, S)$ has the form

$$K(S_0, S) = \frac{1}{2\pi} \cdot \frac{\cos(n, r)}{r^2}. \tag{19.7}$$

By (15.6) we have

$$|K(S_0, S)| \leq \frac{A}{r^{2-\delta}}.$$

Consequently the kernel $K(S_0, S)$ is one of the bounded kernels of the special type considered in Lecture 18, since here we have, using the notation of the previous lecture (see § 7, Lecture 18),

$$n = 2, \quad \alpha = 2 - \delta < n.$$

Thus for these equations all the Fredholm theorems are valid.

The results obtained in Lecture 18 were for integral equations in which the domain of integration D was a plane. Nevertheless the results there obtained are applicable in the present case, because a Lyapunov surface may be divided into a finite number of pieces and we can then pass from integration over any piece of the surface to integration over that plane domain into which this piece projects, on a tangent plane to the piece at some point Q. In this process, by the third inequality of (15.5′), we can suppose that the function under the integral sign is multiplied by a quantity

$$0 < \frac{1}{\cos(n_Q, n)} < 3,$$

where n_Q is the inward normal at the point Q,
and n is the direction of the normal at the current point of integration.

§ 2. Investigation of the Equations

THEOREM 1. *Equations (19.3) and (19.4) always have unique solutions.*
Proof. Suppose that the integral equation

$$\nu(S_0) + \iint_S K(S, S_0)\, \nu(S)\, dS = 0 \tag{19.8}$$

has a bounded solution $\nu(S)$. By the property of the integral operator with kernel $K(S, S_0)$ mentioned at the end of § 7, Lecture 18, the second term in (19.8) is continuous, and consequently the first term is also continuous.

The potential of a single layer

$$v = \int_{S} \frac{\nu(S)}{r} \, dS$$

with a continuous density $\nu(S)$ has a regular normal derivative, as we saw in § 4 of Lecture 15. The limit values of the normal derivative of this potential are expressed (as we saw in Theorem 2 of Lecture 15) by a density of the form

$$2\pi \left[\nu(S_0) + \iint_{S} K(S, S_0) \, \nu(S) \, dS \right]$$

and by equation (19.8) this is zero. By Lemma 1 from the previous section, $\nu(S) \equiv 0$, and so the homogeneous equation corresponding to (19.4) does not have a non-trivial solution.

By the first Fredholm theorem it then follows that the associated homogeneous equation, obtained by putting $\varphi = 0$ in (19.3), also has no nontrivial solution. Further, by the same Fredholm theorem, it follows that each of the equations (19.3) and (19.4) will have a determinate, unique, and bounded solution when any bounded functions stand on their right-hand side. Hence we can obtain the solutions of these equations and thus the solutions of *the internal Dirichlet problem* and the *external Neumann problem*. (See p. 220.)

Passing on to the solution of the *external Dirichlet problem and the internal Neumann problem,* we prove the following theorem.

THEOREM 2. *Each of the equations*

$$\left. \begin{aligned} \mu(S_0) - \iint_{S} K(S_0, S) \, \mu(S) \, dS &= 0 \\ \nu(S_0) - \iint_{S} K(S, S_0) \, \nu(S) \, dS &= 0 \end{aligned} \right\} \tag{19.9}$$

has one and only one principal function.

We first note that the two equations (19.9) are associated, and consequently they have the same number of linearly independent solutions.

For the first of the equations (19.9) we can take unity as such a principal solution. For, the potential of a double layer

$$w = \iint_{S} \frac{\partial}{\partial n} \left(\frac{1}{r} \right) \mu(S) \, dS$$

with $\mu = 1$ gives the size of the solid angle subtended by the surface S, and consequently its value from the outside is zero.

But the left-hand side of the first of the equations (19.9) is precisely the limit value of the potential of a double layer from the outside, and therefore $\mu = 1$ must make the left-hand side vanish. Consequently the number of principal functions for each of the equations (19.9) is not less than one.

We now show that neither of the equations (19.9) can have two linearly independent solutions.

Repeating exactly the same argument as was used to prove the continuity of the function $v(S)$ in Theorem 1, and using Theorem 6 of Lecture 15, we can show that all solutions of the second of the equations (19.9) satisfy the condition (15.13).

If there were two principal functions $v_1(S)$ and $v_2(S)$, then both the potentials

$$v_1 = \iint_S \frac{v_1(S)}{r}\, dS \quad \text{and} \quad v_2 = \iint_S \frac{v_2(S)}{r}\, dS$$

would be harmonic functions within the domain bounded by S and would have within this domain first-order derivatives continuous right up to the boundary. Moreover, the limit values $[\partial v_1/\partial n_1]_i$ and $[\partial v_2/\partial n]_i$ for both functions would be zero; and by the theorem on uniqueness of solution of the Neumann problem both these functions could only be constants. Replacing v_i, $i = 1, 2$ by $\alpha_i\, v_i$ where α_i, $i = 1, 2$, are constant multipliers, we can make both potentials v_1 and v_2 equal to unity within the domain. Let us suppose that this has been done. Then the potential

$$|v = v_1 - v_2 = \iint_S \frac{1}{r}\, [v_1(S) - v_2(S)]\, dS$$

with the density $v_1(S) - v_2(S)$ would be equal to zero inside S. The limit value v_i would then also be zero. And by Lemma 2, the density $v_1(S) - v_2(S)$ must also be zero, and this proves our theorem.

We have proved the existence of an *eigenfunction* $v_0(S)$ for the second equation of (19.9). We may suppose that the value of the potential

$$v_0 = \iint_S \frac{v_0(S)}{r}\, dS.$$

within the domain Ω bounded by the surface S is equal to unity. The function $v_0(S)$ gives the corresponding distribution of electric charge on the surface of a conductor filling the domain Ω bounded by the surface S. The problem of finding the potential of a charged conductor is known as *Robin's problem*, and the potential v_0 is called *Robin's potential*.

The internal Neumann problem consisted in finding a harmonic function such that

$$\left[\frac{\partial u}{\partial n}\right]_i = f_2(S). \tag{19.10}$$

THEOREM 3. *The necessary and sufficient condition that the internal Neumann problem should have a solution is that the right-hand side of* (19.10), *i.e., the function $f_2(S)$, should satisfy the equation*

$$\iint_S f_2(S) \, dS = 0. \tag{19.11}$$

That the condition is necessary follows from the fact that, if we apply Green's formula (5.16) to u and 1 (both these functions are harmonic), then we get

$$0 = \iint_S \frac{\partial u}{\partial n} \, dS = \iint_S f_2(S) \, dS.$$

Its sufficiency follows because unity is the unique solution of the first of the equations (19.9) which is associated with (19.6). Hence, if the free member $\psi(S)$ of the equation (19.6) is orthogonal to unity, then by Fredholm's theorem this equation is soluble. But $\psi(S)$ differs from $f_2(S)$ only by a factor; consequently the condition (19.11) is necessary and sufficient for the solution of the equation (19.6), and, with it, for the solution of the internal Neumann problem.

We now pass on to the solution of the *external Dirichlet problem*.

As we have seen, the integral equation (19.5) may have no solution, because the corresponding homogeneous equation has a trivial solution. This might have been expected from the very beginning. The required harmonic function u which satisfies the condition

$$[u]_e = f(S) \tag{19.12}$$

is unique, as we have seen. It will tend to zero at infinity, generally speaking, like A/r (see Theorem 1, Lecture 12), for example, and like the harmonic function $1/r$ which is equal to unity on the unit sphere.

But we are trying to represent the function in the form of the potential of a double layer, and this decreases like A/r^2. Evidently such a representation may not be possible. So we now try to find a solution of the external Dirichlet problem in the form

$$u = \frac{a}{r_0} + u_1$$

where a is an indeterminate constant,

r_0 is the distance of the point (x, y, z) from the origin (chosen inside S), and u_1 is the potential of a double layer

$$u_1 = \iint_S \frac{\partial}{\partial n} \left(\frac{1}{r} \right) \mu(S) \, dS.$$

The value of u_1 on the surface S will be

$$[u_1]_S = f(S) - \left[\frac{a}{r_0} \right]_S.$$

The equation (19.5) therefore becomes

$$\mu(S_0) - \iint_S K(S_0, S) \mu(S) \, dS = \varphi(S_0) + \frac{a}{2\pi} \left[\frac{1}{r_0} \right]_S. \quad (19.13)$$

We require that the free member of this equation shall be orthogonal to $v_0(S)$, the eigenfunction of the second of the equations (19.9). We get:

$$\iint_S \varphi(S_0) \, v_0(S_0) \, dS_0 + \frac{a}{2\pi} \iint_S \frac{v_0(S_0)}{r_0} \, dS_0 = 0.$$

The integral in the second term is equal to unity by hypothesis, and we get

$$a = -2\pi \iint_S \varphi(S_0) \, v_0(S_0) \, dS_0.$$

Having determined the constant a in this way, we get from (19.13) a soluble equation. By solving it, we get at the same time the solution of the external Dirichlet problem.

GREEN'S FUNCTION

1. The Differential Operator with One Independent Variable

In many of the problems of mathematical physics which we have encountered, the unknown function was determined not merely by the requirement that it should satisfy a certain differential equation but, in addition, by the conditions on the boundaries of the domain in which the unknown function was defined.

Most of such problems were concerned with the solution of equations of elliptical type. But there are also boundary-value problems for equations of other types; and it is useful to be able to find a solution of such an equation which will satisfy prescribed conditions on the boundary of the domain. In order to investigate in detail the most important properties of these boundary-value problems, we set ourselves the task of presenting these solutions in explicit form.

We begin with a very simple case. We shall seek a solution of an ordinary differential equation of the second order

$$Ly \equiv p(x)\, y'' + q(x)\, y' + r(x)\, y = f(x) \qquad (20.1)$$

in the interval $0 \leq x \leq 1$, the solution satisfying certain conditions on the boundaries $x = 0$ and $x = 1$.

We assume that the function $p(x)$ is continuous, has second-order derivatives, and does not vanish, in the interval $0 \leq x \leq 1$. Subsequently, in certain cases, we shall remove the last condition on $p(x)$. We assume that the function $q(x)$ is continuous and has first-order derivatives, and that the function $r(x)$ is continuous, in the same interval.

Although, formally, this problem belongs to the theory of ordinary differential equations, we shall nevertheless examine it in detail. But we must first make one or two important preliminary observations.

We set up the operator M adjoint with the operator L (see § 2, Lecture 5). This adjoint operator will have the form such that

$$Mz \equiv [p(x)\, z]'' - [q(x)\, z]' + r(x)\, z. \qquad (20.2)$$

The operators M and L are connected by the relation

$$zLy - yMz = \frac{d}{dx}\left(pz\frac{dy}{dx} - y\frac{d(pz)}{dx} + qyz\right).$$

Green's formula in the interval $[0, 1]$ for these two operators will have the form

$$\{p_0y_0'z_0 - p_0y_0z_0' + (q_0 - p_0')y_0z_0\} - \{p_1y_1'z_1 - p_1y_1z_1' + (q_1 - p_1')y_1z_1\}$$
$$+ \int_0^1 (zLy - yMz)\,dx = 0 \qquad (20.3)$$

where the suffixes 0 and 1 indicate that the values of the functions are to be taken for $x = 0$ and for $x = 1$ respectively; for example,

$$p_0 = p(0), \quad y_0 = y(0), \quad z_1' = z'(1).$$

If the functions y and z are chosen so that the expressions in the curly brackets vanish, then (20.3) simplifies to

$$\int_0^1 (zLy - yMz)\,dx = 0. \qquad (20.4)$$

It may happen that the formula (20.4) holds good for any pair of functions y and z belonging to two families of functions $\{y\}$ and $\{z\}$. We say that two such families are *adjoint families*.

We shall examine some examples of adjoint families of functions for the operator L of (20.1).

We consider families of functions y satisfying on the boundaries one of the two linear conditions:

(I) $\qquad\qquad\qquad y_0 = 0,$

(II) $\qquad\qquad\qquad p_0y_0' + \beta_0y_0 = 0$ $\qquad\qquad$ (20.5)

at the end $x = 0$, and similarly

(I) $\qquad\qquad\qquad y_1 = 0,$

(II) $\qquad\qquad\qquad p_1y_1' + \beta_1y_1 = 0$ $\qquad\qquad$ (20.6)

at the end $x = 1$, where β_0 and β_1 are given numbers.

The first curly bracket in (20.3) may be written in the form

$$-y_0[p_0z_0' + (p_0' - q_0 + \beta_0)z_0] + z_0[p_0y_0' + \beta_0y_0],$$

and the second similarly. Thus (20.3) becomes

$$-y_0[(pz)' + (\beta_0 - q) z]_{x=0} + z_0[py' + \beta_0 y]_{x=0} + \int_0^1 (zLy - yMz) \, \mathrm{d}x$$

$$+ y_1[(pz)' + (\beta_1 - q) z]_{x=1} - z_1[py' + \beta_1 y]_{x=1} = 0. \quad (20.7)$$

A number of simple results may be deduced from this expression.

If the family $\{y\}$ satisfies the first condition at the end $x = 0$, then in order that only the term outside the integral with $x = 0$ should vanish, we may take as the family $\{z\}$ any family of functions which satisfy the condition

(Ia) $$z_0 = 0.$$

If the family $\{y\}$ satisfies the second condition at the end $x = 0$, then it is sufficient to impose on the family $\{z\}$ the condition

(IIa) $$p_0 z_0' + (p_0' - q_0 + \beta_0) z_0 = 0.$$

If the family $\{y\}$ satisfies both conditions at $x = 0$, $i.e.$, if $y_0' = y_0 = 0$, then no condition need be imposed on the family $\{z\}$ at $x = 0$. Conversely, if no restrictions are put on the family $\{y\}$, then the functions $\{z\}$ must satisfy the conditions $z_0 = z_0' = 0$.

The end $x = 1$ can be examined in exactly the same way, and the conditions which we have analysed will take the form:

(I) $$y_1 = 0,$$

(II) $$p_1 y_1' + \beta_1 y_1 = 0,$$

(Ia) $$z_1 = 0,$$

(IIa) $$p_1 z_1' + (p_1' - q_1 + \beta_1) z_1 = 0.$$

The conditions (I a) or (II a) imposed at both ends of the interval are sufficient to ensure that the family $\{z\}$ shall be adjoint to the family $\{y\}$ if the latter satisfy the conditions (I) or (II). These conditions will also be necessary if we take as the family $\{y\}$ the set of all functions which satisfy the conditions (I) or (II) at both ends of the interval and which have continuous derivatives up to the second order inclusive. We shall in what follows define the families $\{y\}$ and $\{z\}$ in precisely this way.

The conditions (20.5) may for brevity be written in the form of the single condition

$$[\alpha_0 py' + \beta_0 y]_{x=0} = 0 \quad (20.5')$$

and the conditions (20.6) as

$$[\alpha_1 py' + \beta_1 y]_{x=1} = 0, \quad (20.6')$$

where $\alpha_0, \alpha_1, \beta_0, \beta_1$ are certain numbers; and without loss of generality we may suppose that α_0 and α_1 are only equal either to 0 or 1.

The adjoint conditions (Ia) and (IIa) at the end $x = 0$ take the form

$$\alpha_0 p_0 z_0' + (\alpha_0 p_0' - \alpha_0 q_0 + \beta_0) z_0 = 0 \qquad (20.5'')$$

and at the end $x = 1$ they become

$$\alpha_1 p_1 z_1' + (\alpha_1 p_1' - \alpha_1 q_1 + \beta_1) z_1 = 0. \qquad (20.6'')$$

We might also consider conditions of a more general type to be imposed on the family $\{y\}$. These would connect the values of y and y' at both ends of the interval. We shall not, however, bother to enumerate and classify such conditions, but merely indicate how they may be obtained.

The bilinear form

$$\Phi(y_0, y_0', y_1, y_1'; \quad z_0, z_0', z_1, z_1') \equiv p_0 y_0' z_0 - p_0 y_0 z_0' + (q_0 - p_0') y_0 z_0$$
$$- p_1 y_1' z_1 + p_1 q_1 z_1' - (q_1 - p_1') y_1 z_1,$$

consisting of two sets of four variables, vanishes for arbitrary values of the variables of the one set if and only if all the variables of the other set are zero. If there are linear relations between the four variables of one set, say, y_0, y_0', y_1, y_1', then on expressing some of the variables in terms of the remainder, we can transform the given form into another form which is linear relative to y but which depends on a smaller number of variables. The condition for the annihilation of this new form is that the coefficients of these remaining variables shall vanish. Thus, the sum of the number of conditions which are imposed on $\{y\}$ and $\{z\}$ is equal to four.

Consider, for example, the equation

$$Ly \equiv y'' + k^2 y = 0.$$

Here, $p = 1, q = 0, r = k^2$. Let the conditions imposed on the family $\{y\}$ be:

$$y_0 = y_1, \quad y_0' = y_1'.$$

In our theory Φ will have the form

$$\Phi \equiv y_0(z_0' - z_1') + y_0'(z_0 - z_1) = 0.$$

Consequently, the adjoint conditions will be:

$$z_0 = z_1, \quad z_0' = z_1'.$$

§ 2. Adjoint Operators and Adjoint Families

The concept of adjoint families or of adjoint conditions relates not merely to ordinary differential equations of the second order. It can easily be carried over to the case when L is a linear differential operator with partial derivatives of any order. We have defined above the concept of the operator M adjoint to a given differential operator L. The integral formula (20.4), which we have already shown to be true for two adjoint operators, in our illustrative example, will form the basis of the general definition of adjoint operators.

DEFINITION. *The operator M is the adjoint of L, and the two families of functions $\{u\}$ and $\{v\}$, defined in a domain Ω, are adjoint relative to L and M in this domain, if*

$$\int_\Omega v\,Lu\,d\Omega = \int_\Omega u\,Mv\,d\Omega \tag{20.8}$$

for any functions u and v from the respective families.

The conditions, by means of which the two families which are adjoint relative to the operators L and M are defined, are called the *adjoint conditions*.

DEFINITION. *An operator L is said to be self-adjoint if it coincides with its own adjoint, i.e.,*

$$Mu \equiv Lu.$$

Similarly, for any self-adjoint operator, a family of functions which coincide with their own adjoints is said to be *self-adjoint*.

We shall examine a few more very simple examples.

EXAMPLE 1.

$$Ly = p_0 y^{(n)} + p_1 y^{(n-1)} + \cdots + p_n y.$$

We choose the family of functions $\{y\}$ in the interval $0 \leq x \leq 1$ so that

$$y_0 = y_0' = \cdots = y_0^{(n-1)} = 0. \tag{20.9}$$

It is easily seen that the adjoint operator will be

$$Mz = (-1)^n \frac{d^n(p_0 z)}{dx^n} + (-1)^{n-1} \frac{d^{n-1}(p_1 z)}{dx^{n-1}} + \cdots + p_n z.$$

Noting that

$$\varphi \frac{d^m\psi}{dx^m} + (-1)^m \psi \frac{d^m\varphi}{dx^m} = \frac{d}{dx}\left[\varphi \frac{d^{m-1}\psi}{dx^{m-1}} - \frac{d\varphi}{dx}\frac{d^{m-2}\psi}{dx^{m-2}} + \frac{d^2\varphi}{dx^2}\frac{d^{m-3}\psi}{dx^{m-3}} - \cdots \right],$$

the indefinite integral

$$\int (zLy - xMz)\,dx$$

can easily be obtained in closed form. Without working it out in full, we see that it will be expressed as a bilinear form in the derivatives

$$y, y', \ldots, y^{(n-1)}, \quad z, z'\ldots, , z^{(n-1)},$$

the coefficients being functions of the variable x.

By the conditions (20.9) this bilinear form will vanish for $x = 0$. To make it vanish also at $x = 1$, it is sufficient to put

$$z_1 = z_1' = \cdots = z_1^{(n-1)} = 0. \tag{20.10}$$

(These conditions are also necessary if no restriction is laid on the values of the function y and its derivatives at $x = 1$.)

Under these conditions

$$\int_0^1 (zLy - yMz)\, \mathrm{d}x = 0.$$

Thus the conditions (20.10) are adjoint with (20.9).

EXAMPLE 2. We consider the operator

$$Lu \equiv \nabla^2 u \tag{20.11}$$

and investigate a family of functions $\{u\}$ which satisfy the condition

$$\left[\frac{\partial u}{\partial n} + \alpha u + \beta \frac{\partial u}{\partial s} \right]_s = 0, \tag{20.12}$$

in a domain Ω of the two variables x, y which is bounded by the curve, s, where $\partial u/\partial n$ is the derivative along the inward normal at a point of the boundary s, and $\partial u/\partial s$ is the derivative along the tangent taken in the positive sense of description of the curve s. Green's formula gives:

$$\iint_\Omega (u\nabla_v^2 - v\nabla_u^2)\, \mathrm{d}\Omega = \int_s \left(v\frac{\partial u}{\partial n} - u\frac{\partial v}{\partial n} \right) \mathrm{d}s$$

$$= \int_s \left\{ v\left(\frac{\partial u}{\partial n} + \alpha u + \beta\frac{\partial u}{\partial s} \right) - u\frac{\partial v}{\partial n} - v\left(\alpha u + \beta\frac{\partial u}{\partial s} \right) \right\} \mathrm{d}s.$$

Integrating by parts the term

$$\int_s \beta v\, \frac{\partial u}{\partial s}\, \mathrm{d}s,$$

and noting that the integrated terms drop out because of the periodicity of the functions β, u, v on the boundary, we obtain

$$\iint_\Omega (u\nabla^2 v - v\nabla^2 u)\, \mathrm{d}\Omega$$

$$= \int_s \left\{ v\left[\frac{\partial u}{\partial n} + \alpha u + \beta\frac{\partial u}{\partial s} \right] - u\left[\frac{\partial v}{\partial n} + \alpha v - \frac{\partial(\beta v)}{\partial s} \right] \right\} \mathrm{d}s.$$

It is natural to say that the expression

$$\left[\frac{\partial u}{\partial n} + \left(\alpha - \frac{\partial \beta}{\partial s} \right) v - \beta \frac{\partial v}{\partial s} \right]_s$$

is *adjoint to* (20.12) *relative to the operator* ∇^2.

The families adjoint relative to the operator ∇^2 will be: the family $\{u\}$ satisfying the condition

$$\left[\frac{\partial u}{\partial n} + \alpha u + \beta \frac{\partial u}{\partial s} \right]_s = 0,$$

and the family $\{v\}$ satisfying the condition

$$\left[\frac{\partial u}{\partial n} + \left(\alpha - \frac{\partial \beta}{\partial s} \right) v - \beta \frac{\partial v}{\partial s} \right]_s = 0.$$

For two such families we have:

$$\iint_\Omega v \, \nabla^2 u \, d\Omega = \iint_\Omega u \, \nabla^2 v \, d\Omega.$$

EXAMPLE 3. If $\{u\}$ is a family of functions not restricted by any conditions, then the adjoint family in the domain Ω relative to the Laplace operator will satisfy the two conditions

$$[v]_s = 0, \qquad \left[\frac{\partial v}{\partial n} \right]_s = 0.$$

This follows from the fact that the condition for adjointness, *i.e.*,

$$\iint_\Omega (u\nabla^2 v - v\nabla^2 u) \, d\Omega = 0$$

will be satisfied for an arbitrary choice of the function u only if the stated conditions are satisfied.

These examples should have made the concepts of adjoint families and adjoint operators sufficiently clear. We pass on now to a more detailed examination of the problem.

§ 3. The Fundamental Lemma on the Integrals of Adjoint Equations

We consider the following problem:

To find the solution of the equation (20.1) satisfying the conditions, either

$$\left. \begin{array}{l} [\alpha_0 py' + \beta_0 y]_{x=0} = a_0, \\[2mm] [\alpha_1 py' + \beta_1 y]_{x=1} = a_1, \end{array} \right\} \tag{20.13}$$

or

$$[\alpha_0 py' + \beta_0 y]_{x=0} = 0, \\
[\alpha_1 py' + \beta_1 y]_{x=1} = 0. \quad\Big\} \tag{20.13'}$$

Problems of a similar type are encountered, for example, if we have to find the equilibrium form for a string of variable density and variable tension, at the ends of which certain relations between the displacement and the component of the tension along the y-axis are to be satisfied. The variable tension could arise from the string being inclined to the horizontal, so that the weight of each element would affect the tension above this element.

LEMMA 1. *Let*

$$Ly \equiv py'' + qy' + ry \quad \text{and} \quad Mz \equiv (pz)'' - (qz)' + rz$$

be two adjoint operators, and suppose that the coefficients $p(x)$, $q(x)$, $r(x)$ are continuous in the interval $0 \leq x \leq 1$ and that $p(x)$ has continuous derivatives of the first and second order and that $q(x)$ has a continuous first-order derivative in the same interval. Suppose, further, that $p(x)$ does not vanish anywhere in this interval.
Then, if $y_1(x)$ satisfies the equation

$$Ly = 0 \tag{20.14}$$

and the condition

$$[\alpha_0 py' + \beta_0 y]_{x=0} = 0, \tag{20.15}$$

then the function

$$z_1(x) = \frac{y_1(x)}{p} \exp \int_c^x \frac{q}{p} \, dx \tag{20.16}$$

is a solution of the adjoint equation

$$Mz = 0 \tag{20.17}$$

satisfying the condition

$$[(\alpha_0 pz)' + (\beta_0 - \alpha_0 q) z]_{x=0} = 0 \tag{20.18}$$

adjoint to the condition (20.15).
[Compare (20.5') and (20.5'')].
    The lemma may be proved by a straightforward method. We have

$$(pz_1)' = y_1' \exp \int_c^x \frac{q}{p} \, dx + y_1 \frac{q}{p} \exp \int_c^x \frac{q}{p} \, dx,$$

$$Mz_1 = [(pz_1)' - qz_1]' + rz_1 = \frac{1}{p} [py_1'' + qy_1' + ry_1] \exp \int_c^x \frac{q}{p} \, dx = 0.$$

Thus the function (20.16) does indeed satisfy equation (20.17). We next verify that condition (20.18) is satisfied:

$$[(\alpha_0 p z_1)' + (\beta_0 - \alpha_0 q) z_1]_{x=0}$$

$$= \left[\alpha_0 y_1' + \alpha_0 y_1 \frac{p}{q} + \left(\frac{\beta_0}{p} - \frac{\alpha_0 q}{p} \right) y_1 \right]_{x=0} \exp \int_c^0 \frac{q}{p} \, dx$$

$$= \left[\frac{1}{p} (\alpha_0 p y_1' + \beta_0 y_1) \right]_{x=0} \exp \int_c^0 \frac{q}{p} \, dx = 0.$$

Hence the lemma.

Similarly, if y_2 is the solution of the equation (20.14), satisfying the condition

$$[\alpha_1 p y_2' + \beta_1 y_2]_{x=1} = 0, \tag{20.19}$$

then

$$z_2 = \frac{y_2}{p} \exp \int_c^x \frac{q}{p} \, dx \tag{20.20}$$

will be the solution of equation (20.17) satisfying the condition

$$[(\alpha_1 p z)' + (\beta_1 - \alpha_1 q) z]_{x=1} = 0, \tag{20.21}$$

which is adjoint to (20.19).

If y_1 and y_2 are linearly independent, then the formulae (20.16) and (20.20) can be given a rather different form. Using the familiar fact that the Wronskian, $y_1' y_2 - y_2' y_1$, satisfies the equation

$$C(y_1' y_2 - y_2' y_1) = \exp \left[-\int_c^x \frac{q}{p} \, dx \right] \tag{20.22}$$

we shall have, by a suitable choice of the arbitrary constant factor,

$$z_1 = \frac{y_1}{p[y_1' y_2 - y_2' y_1]}, \quad z_2 = \frac{y_2}{p[y_1' y_2 - y_2' y_1]}. \tag{20.23}$$

Remark. Lemma 1 is evidently the dual of this result, and it may be formulated thus:

If z_1 and z_2 are solutions of the equation $Mz = 0$ satisfying the conditions (20.18) and (20.21) respectively, then

$$y_1 = \frac{z_1}{p} \exp \int_c^x \frac{2p' - q}{p} \, dx = Cp z_1 \exp \left(-\int_c^x \frac{q}{p} \, dx \right)$$

$$y_2 = \frac{z_2}{p} \exp \int_c^x \frac{2p' - q}{p} \, dx = Cp z_2 \exp \left(-\int_c^x \frac{q}{p} \, dx \right)$$

where C is a certain constant, are solutions of the equation $Ly = 0$ satisfying the conditions (20.15) and (20.19) respectively.

It follows from (20.23) that

$$\frac{d}{dx}\left(\log_e \frac{z_1}{z_2}\right) = \frac{d}{dx}\left(\log_e \frac{y_1}{y_2}\right).$$

i.e.,

$$\frac{z_1'}{z_1} - \frac{z_2'}{z_2} = \frac{y_1'}{y_1} - \frac{y_2'}{y_2},$$

or

$$\frac{z_1'z_2 - z_2'z_1}{z_1 z_2} = \frac{y_1'y_2 - y_2'y_1}{y_1 y_2} = \frac{1}{py_1 z_2} = \frac{1}{py_2 z_1}$$

from which we get:

$$y_1 = \frac{z_1}{p(z_1'z_2 - z_2'z_1)}, \quad y_2 = \frac{z_2}{p(z_1'z_2 - z_2'z_1)}. \tag{20.23'}$$

It may also be noted that

$$y_1 z_2 = z_1 y_2,$$

$$y_1' z_2 - y_2' z_1 = z_1' y_2 - y_1 z_2' = \frac{1}{p};$$

these results may be immediately verified.

COROLLARY. *If the equation (20.14) has a non-trivial solution, satisfying the conditions (20.15) and (20.19), then the equation (20.17) has a non-trivial solution satisfying the conditions (20.18) and (20.21): and conversely. In this case both equations have the same number of linearly independent solutions.*

It is important in our further discussion that we should distinguish the two cases when:

1. Equation (20.14) has no non-trivial solution satisfying the conditions (20.15) and (20.19).

2. Equation (20.14) has such a solution.

We shall prove later that in the first case equation (20.1) always has a definite unique solution satisfying the conditions (20.13). In the second case this will not be so.

We shall call the problem of finding a solution of (20.14) subject to the conditions (20.15) and (20.19) the *homogeneous problem.*

Let y_1 be a solution of the homogeneous problem. It is not difficult to see that such a solution can be determined only to within a constant factor. For, if y_1 and y_2 are two linearly independent solutions of equation (20.14), then their Wronskian is not zero; consequently, the ratios y_1'/y_1 and y_2'/y_2

are not equal. Hence neither y_2 nor any linear combination $C_1 y_1 + C_2 y_2$ with $C_2 \neq 0$ will satisfy the boundary conditions.

By the Corollary to Lemma 1,

$$z_1 = \frac{y_1}{p} \exp \int_c^x \frac{q}{p} \, dx$$

is the unique solution of the adjoint problem.

We now prove for the second case under consideration a theorem which is formulated under the supposition that $\alpha_0 = \alpha_1 = 1$.

THEOREM 1. *A necessary condition for the equation* (20.1) *to have a solution satisfying the conditions* (20.13) *is that*

$$[a_0 z_1]_{x=0} + \int_0^1 f(x) z_1 \, dx - [a_1 z_1]_{x=1} = 0. \tag{20.24}$$

Let the function $y_1(x)$ be a non-trivial solution of the homogeneous problem, *i.e.*, it satisfies equation (20.14) and the conditions (20.15) and (20.19); then the function $z_1(x)$ defined by (20.16) will satisfy the adjoint equation and the adjoint boundary conditions (20.18) and (20.21).

Applying formula (20.7) to the function $y(x)$, which is the solution of (20.1) and satisfies the conditions (20.13), and to $z_1(x)$, we obtain immediately (20.24).

We leave the reader to formulate and prove the similar theorems for the cases when α_0, or α_1, or both α_0 and α_1, are zero.

§ 4. The Influence Function

We now examine in more detail the first of the cases mentioned on p. 274.

Let x_0 be an arbitrary point of the interval $0 \leq x \leq 1$, and let the function z_ε satisfy

$$M z_\varepsilon = \begin{cases} 0 & \text{if} \quad |x - x_0| > \varepsilon \\ \dfrac{1}{2\varepsilon} & \text{if} \quad |x - x_0| \leq \varepsilon; \end{cases}$$

then

$$\int_0^1 y \, M z_\varepsilon \, dx = \frac{1}{2\varepsilon} \int_{x_0-\varepsilon}^{x_0+\varepsilon} y \, dx.$$

Using the integral mean-value theorem, and in the resulting equality passing to the limit as $\varepsilon \to 0$, we get

$$\lim_{\varepsilon \to 0} \int_0^1 y M z_\varepsilon \, dx = y(x_0).$$

If z_ε and y belong to adjoint families, *i.e.*, if they satisfy the conditions (20.15), (20.19), (20.18), and (20.21), then the formula (20.7) gives

$$y(x_0) = \lim_{\varepsilon \to 0} \int_0^1 y M z_\varepsilon \, dx = \lim_{\varepsilon \to 0} \int_0^1 z_\varepsilon L y \, dx = \lim_{\varepsilon \to 0} \int_0^1 z_\varepsilon f(x) \, dx.$$

For the sake of definiteness we shall in future suppose that the numbers α_0 and α_1 occurring in the conditions (20.13) are both equal to 1 (we shall leave the reader to analyse the three remaining possible cases: $\alpha_0 = 0, \alpha_1 = 1$; $\alpha_0 = 1, \alpha_1 = 0$; $\alpha_0 = \alpha_1 = 0$).

Then for the same z_ε and for an arbitrary function y, we get from (20.7):

$$y(x_0) = \lim_{\varepsilon \to 0} \left[z_\varepsilon(0) \, (py' + \beta_0 y)_{x=0} + \int_0^1 z_\varepsilon L y \, dx - z_\varepsilon(1) \, (py' + \beta_1 y)_{x=1} \right].$$

$$(20.25)$$

If it now happens that as $\varepsilon \to 0$, the function z_ε converges uniformly to a limit function

$$z_{+0}(x, x_0) = \lim_{\varepsilon \to 0} z_\varepsilon,$$

which depends, of course, on the parameter x_0, then it will be possible to pass to the limit in the formulae for $y(x_0)$. The solution of the equation (20.1) satisfying the conditions (20.13) will be expressible as

$$y(x_0) = z_{+0}(0, x_0) \, a_0 + \int_0^1 z_{+0}(x, x_0) f(x) \, dx - z_{+0}(1, x_0) \, a_1. \quad (20.26)$$

We may remark that the equation involving z_ε with which we began this section may be considered as the equation for the equilibrium of a string which satisfies end-conditions (fixing conditions) adjoint to the conditions (20.13′) and which is acted on by a transverse force of magnitude 1 distributed over an interval of the string of length 2ε with its mid-point at the point x_0. In this case we can interpret $z_\varepsilon(x, x_0)$ as being the deviation of the string at the point x under the influence of a unit force acting in the ε-neighbourhod of the point x_0. The function $z_{+0}(x, x_0)$ is thus equal to the deviation (or *influence*) produced by unit force concentrated at the point x_0. It will become clear later that if x is regarded as the parameter and x_0 as the argument, the function $z_{+0}(x, x_0)$ can still be regarded as an *influence function* for a string with a different density and tension, whose deviation $y(x_0)$ under the action of a force $f(x)$ satisfies equation (20.1). This enables the formula (20.26) to be interpreted physically: the integral in it denotes the total influence at the point x_0 due to the distributed force $f(x)$.

Without calculating z_ε, we shall at once make clear what properties the function $z_{+0}(x, x_0)$ must have.

Property 1. Knowing the properties of z_ε, we shall naturally expect that the function $z_{+0}(x, x_0)$ regarded as a function of x will satisfy the equation $Mz = 0$ everywhere except at the point x_0 and will also satisfy boundary conditions adjoint to (20.13′).

We explain one more property of z_{+0}. By integrating Mz_ε over the interval $(x_0 - \delta, x_0 + \delta)$, where $\delta > \varepsilon$, we get

$$\int_{x_0-\delta}^{x_0+\delta} Mz_\varepsilon \, dx = 1.$$

On the other hand,

$$\int_{x_0-\delta}^{x_0+\delta} Mz_\varepsilon \, dx = \left[(pz_\varepsilon)'\right]_{x=x_0-\delta}^{x=x_0+\delta} - \int_{x_0-\delta}^{x_0+\delta} [(qz_\varepsilon)' - rz_\varepsilon] \, dx,$$

whence

$$1 = \left[(pz_\varepsilon)'\right]_{x=x_0-\delta}^{x=x_0+\delta} - \int_{x_0-\delta}^{x_0+\delta} [(qz_\varepsilon)' - rz_\varepsilon] \, dx.$$

Assuming that z_ε and z_ε' are uniformly bounded, and passing to the limit as $\varepsilon \to 0$, we get:

$$\left[(pz_{+0})'\right]_{x=x_0-0}^{x=x_0+0} = 1.$$

But the function $p(x)$ and its derivative are continuous, and therefore, regarding z_{+0} as continuous, we get:

$$\left[(pz_{+0})'\right]_{x=x_0-0}^{x=x_0+0} = p(x_0)\left[z_{+0}'\right]_{x=x_0-0}^{x=x_0+0}.$$

Hence follows:
Property 2.

$$\left[z_{+0}'\right]_{x=x_0-0}^{x=x_0+0} = \frac{1}{p(x_0)}.$$

The heuristic considerations which we have adopted make the following a natural supposition:

If it is possible to construct a function $G(x, x_0)$ having Properties 1 and 2 which the function z_{+0} must have, then an equation similar to (20.25) must hold for the function $G(x, x_0)$ and any function $y(x)$:

$$y(x_0) = G(0, x_0)\left[py' + \beta_0 y\right]_{x=0} + \int_0^1 G(x, x_0) \, Ly \, dx$$

$$- G(1, x_0)\left[py' + \beta_1 y\right]_{x=1} \qquad (20.25')$$

and consequently, if the problem under consideration has a solution, then this solution may be written in the form

$$y(x_0) = G(0, x_0)a_0 + \int_0^1 G(x, x_0) f(x) \, dx - G(1, x_0)a_1. \qquad (20.26')$$

§ 5. Definition and Construction of Green's Function

We pass on to the strict definition and the construction of Green's function $G(x, x_0)$ and to the proof of the formulae (20.25') and (20.26'). In defining and constructing the function $G(x, x_0)$ we shall have in view the general case for the boundary conditions; the formulae (20.25') and (20.26') themselves are valid only for the particular case when $\alpha_0 = \alpha_1 = 1$.

DEFINITION. A function $G(x, x_0)$ will be called *Green's function for the operator Ly and the boundary conditions* (20.13) if it satisfies the following conditions:

1. As a function of the variable x, for any x_0, it satisfies the conditions (20.18) and (20.21) adjoint with the conditions (20.13') for the operator Ly, i.e.,

$$[(\alpha_0 pG)'_x + (\beta_0 - \alpha_0 q) G]_{x=0} = 0$$

$$[(\alpha_1 pG)'_x + (\beta_1 - \alpha_1 q) G]_{x=1} = 0.$$

2. The function $G(x, x_0)$ and its derivatives up to the second order are continuous in the intervals $0 \leq x < x_0$, $x_0 < x \leq 1$, and G satisfies an equation adjoint to $Ly = 0$, i.e.,

$$Mz = \frac{d^2[p(x) z]}{dx^2} - \frac{d[q(x) z]}{dx} + r(x) z = 0.$$

3. At the point $x = x_0$ the function $G(x, x_0)$ as a function of x is itself continuous, but its first derivative has a discontinuity, the jump being

$$G'_x(x_0 + 0, x_0) - G'_x(x_0 - 0, x_0) = \frac{1}{p(x_0)}. \qquad (20.27)$$

The actual construction of Green's function as we defined it is not difficult. It must clearly have the form

$$G(x, x_0) = a(x_0) z_1(x), \quad 0 \leq x \leq x_0$$

$$G(x, x_0) = b(x_0) z_2(x), \quad x_0 \leq x \leq 1.$$

A function G satisfying these two equations will have the first two properties stipulated in the definition, for *any* values of a and b. And we can now choose a and b to satisfy the third requirement, i.e., to satisfy the equations

$$a(x_0) z_1(x_0) - b(x_0) z_2(x_0) = 0,$$

$$a(x_0) z'_1(x_0) - b(x_0) z'_2(x_0) = -\frac{1}{p(x_0)}.$$

Hence we obtain

$$a(x_0) = \frac{-z_2(x_0)}{p(x_0)\,[z_2(x_0)\,z_1'(x_0) - z_2'(x_0)\,z_1(x_0)]}$$

$$b(x_0) = \frac{-z_1(x_0)}{p(x_0)\,[z_2(x_0)\,z_1'(x_0) - z_2'(x_0)\,z_1(x_0)]}.$$

The Wronskian which appears in the denominator of these fractions is not zero since $z_1(x)$ and $z_2(x)$ are linearly independent solutions.

Comparing these expressions with (20.23'), we see that

$$a(x_0) = y_2(x_0), \quad b(x_0) = y_1(x_0),$$

where $y_1(x)$ and $y_2(x)$ are solutions of the equation $Ly = 0$ which satisfy the conditions (20.15) and (20.19).

We have thus proved that Green's function exists and have obtained for it the explicit from

$$G(x, x_0) = \begin{cases} y_2(x_0)\,z_1(x), & 0 \leqq x \leqq x_0, \\ y_1(x_0)\,z_2(x), & x_0 \leqq x \leqq 1. \end{cases} \tag{20.28}$$

From (20.28) we have the following:

THEOREM. *Green's function* $G(x, x_0)$ *with* x_0 *taken as the argument is the Green's function for the operator Mz and the conditions* (20.18) *and* (20.21). *In other words, when the problem is replaced by the adjoint problem, the arguments of Green's function merely change places.*

Other properties of Green's function can be derived from (20.28).

Property 1. Green's function $G(x, x_0)$ is a continuous function relative to the aggregate of variables in the square $0 \leqq x \leqq 1$, $0 \leqq x_0 \leqq 1$ in the plane xOx_0.

Property 2. The first derivatives of Green's function are continuous inside, and on the sides containing the right angle, both triangles into which the diagonal $x = x_0$ divides the square $0 \leqq x \leqq 1$, $0 \leqq x_0 \leqq 1$. Further, the first derivatives can be continued without discontinuity on to this diagonal $x = x_0$. Hence, in particular,

(a) $\qquad\qquad G_{x_0}(x_0 + 0, x_0) \quad \text{and} \quad G_{x_0}(x_0 - 0, x_0)$

are continuous functions of x_0;

(b) $\qquad\qquad G_{x_0}(x_0 - 0, x_0) = G_{x_0}(x_0, x_0 + 0),$

$$G_{x_0}(x_0 + 0, x_0) = G_{x_0}(x_0, x_0 - 0).$$

Property 3. The derivative G'_{x_0} on the diagonal $x = x_0$ satisfies the condition

$$G'_{x_0}(x_0, x_0 + 0) - G'_{x_0}(x_0, x_0 - 0) = \frac{1}{p(x_0)}.$$

The first two properties are obvious. To prove the third, we recall that the function $G(x, x_0)$ as a function of x_0 is Green's function for the operator $M_0 z$, and the coefficient of $z''(x_0)$ in the expression for this operator is $p(x_0)$. Hence the equation to be proved holds by definition of Green's function [see (20.27)].

We now prove that the equation (20.25′) is satisfied by any function $y(x)$ which has continuous derivatives up to the second order inclusive. We write down the equation (20.7), substituting for $z(x)$ the function $G(x, x_0)$, separately for the intervals $0 \leqq x \leqq x_0 - 0$ and $x_0 + 0 \leqq x \leqq 1$. The formula is valid on each of these intervals, since $G(x, x_0)$ has continuous second-order derivatives within each interval and has continuous first-order derivatives everywhere including the ends. Taking into account the first two points of the definition of Green's function, we have for the two intervals respectively the formulae:

$$G(0, x_0)\,[py' + \beta_0 y]_{x=0} + \int_0^{x_0} G(x, x_0)\,Ly\,\mathrm{d}x + y(x_0)\,p(x_0)\,G'_x(x_0 - 0, x_0)$$

$$+ y(x_0)\,p'(x_0)\,G(x_0, x_0) - y(x_0)\,q(x_0)\,G(x_0, x_0)$$

$$- G(x_0, x_0)\,p(x_0)\,y'(x_0) = 0$$

and

$$-y(x_0)\,p(x_0)\,G'_x(x_0 + 0, x_0) - y(x_0)\,p'(x_0)\,G(x_0, x_0)$$

$$+ G(x_0, x_0)\,p(x_0)\,y'(x_0) + y(x_0)\,q(x_0)\,G(x_0, x_0)$$

$$+ \int_0^1 G(x, x_0)\,Ly\,\mathrm{d}x - G(1, x_0)\,[py' + \beta_1 y]_{x=1} = 0.$$

Adding both equations and taking (20.27) into account we obtain the required equation (20.25′).

Since equation (20.25′) holds good, it immediately follows that, if the solution exists, then it is given by the formula (20.26′) and, consequently, it is unique. We shall prove that the solution exists. Since the problem can always be reduced to the problem with homogeneous boundary conditions (20.13′), it is sufficient to show that the function

$$y(x_0) = \int_0^1 G(x, x_0)\,f(x)\,\mathrm{d}x \tag{20.29}$$

satisfies equation (20.1) and the conditions (20.13′).

We differentiate the expression with respect to x_0. We have:

$$y(x_0) = \int_0^{x_0} G(x, x_0) f(x)\, dx + \int_{x_0}^1 G(x, x_0) f(x)\, dx,$$

$$y'(x_0) = \int_0^{x_0} \frac{\partial G}{\partial x_0} f(x)\, dx + \int_{x_0}^1 \frac{\partial G}{\partial x_0} f(x)\, dx$$
$$+ G(x_0, x_0) f(x_0) - G(x_0, x_0) f(x_0),$$

i. e. $$y'(x_0) = \int_0^{x_0} \frac{\partial G}{\partial x_0} f(x)\, dx + \int_0^1 \frac{\partial G}{\partial x_0} f(x)\, dx. \qquad (20.30)$$

Further, using Property 2, (a) and (b), we can write

$$y''(x_0) = \int_0^{x_0} \frac{\partial^2 G}{\partial x_0^2} f(x)\, dx + \int_0^1 \frac{\partial^2 G}{\partial x_0^2} f(x)\, dx + f(x_0) \left[\frac{\partial G}{\partial x_0} \right]_{x = x_0 - 0}$$

$$- f(x_0) \left[\frac{\partial G}{\partial x_0} \right]_{x = x_0 + 0}$$

$$= \int_0^1 \frac{\partial^2 G}{\partial x_0^2} f(x)\, dx + f(x_0)\, \frac{1}{p(x_0)}. \qquad (20.31)$$

Using these expressions, we find:

$$p(x_0)\, y''(x_0) + q(x_0)\, y'(x_0) + r(x_0)\, y(x_0) = f(x_0) + \int_0^1 L_0 G(x, x_0) f(x)\, dx.$$

But $G(x, x_0)$ as a function of its second argument x_0 satisfies the equation $L_0 G = 0$; whence

$$p(x_0)\, y''(x_0) + q(x_0)\, y'(x_0) + r(x_0)\, y(x_0) = f(x_0).$$

That the conditions (20.13') are satisfied follows from the first point of the definition of Green's function.

We have examined the problem in the first of the cases distinguished on p. 274. We have established that in this case there is always a definite solution of equation (20.1) satisfying the conditions (20.13) or (20.13') and it may be expressed in the form (20.26'), where $G(x, x_0)$ is Green's function as we have defined it.

§ 6. The Generalized Green's Function for a Linear Second-Order Equation

We shall now analyse in more detail the *second* case.

Green's function in this case does not exist, for $z(x)$ is the *only* function — apart from its possible multiplication by a constant factor — which satisfies the first two requirements in the definition of Green's function and the continuity condition, but it does *not* satisfy the condition (20.27).

LEMMA 2. *If $y_0(x)$ and $z_0(x)$ are solutions of the equations $Ly = O$ and $Mz = O$ satisfying the homogeneous conditions (20.15), (20.19) and (20.18) (20.21) respectively,*
then

$$\int_0^1 y_0(x) z_0(x) \, dx \neq 0. \tag{20.32}$$

This follows from the fact that

$$\int_0^1 y_0(x) z_0(x) \, dx = c \int_0^1 \frac{y_0^2}{p} \left[\exp \left(\int_c^x \frac{q}{p} \, dx \right) \right] dx,$$

and since the integrand is constant in sign, the integral cannot vanish.

LEMMA 3. *The equation*

$$Mz = z_0$$

cannot have a solution satisfying the homogeneous conditions (20.18) and (20.21), and the equation

$$Ly = y_0$$

cannot have a solution satisfying the homogeneous conditions (20.15) and (20.19).

For, otherwise, by applying formula (20.7) and substituting in it for y the solution y_0, we should obtain

$$\int_0^1 y_0 Mz \, dx = \int_0^1 z_0 y_0 \, dx = 0,$$

and this would contradict Lemma 2. The second part is proved in the same way.

DEFINITION. *The generalized Green's function for the operator Ly and the boundary conditions (20.13) is a function $G_1(x, x_0)$ satisfying the following conditions:*

1. As a function of the variable x, for any x_0, the function $G_1(x, x_0)$ satisfies the conditions adjoint to (20.13')

$$\left.\begin{array}{l} [(\alpha_0 p(x)G_1)' + (\beta_0 - \alpha_0 q)G_1]_{x=0} = 0, \\ [(\alpha_1 p(x)G_1)' + (\beta_1 - \alpha_1 q)G_1]_{x=0} = 0. \end{array}\right\} \tag{20.33}$$

2. In the intervals $0 \leq x < x_0$, $x_0 < x \leq 1$ the function $G_1(x, x_0)$ and its second-order derivatives are continuous, and G_1 satisfies the equation

$$Mz = a(x_0) z_0(x); \tag{20.34}$$

we define the function $a(x_0)$ later.

3. At the point $x = x_0$ the function $G_1(x, x_0)$ itself as a function of x is continuous, but its derivative is discontinuous, with

$$\frac{\partial G_1(x_0 + 0, x_0)}{\partial x} - \frac{\partial G_1(x_0 - 0, x_0)}{\partial x} = \frac{1}{p(x_0)}. \qquad (20.35)$$

4.
$$\int_0^1 G_1(x, x_0)\, y_0(x)\, dx = 0. \qquad (20.36)$$

The condition (20.35) enables the multiplier $a(x_0)$ to be determined.

We pass on to the construction of the function G_1 in explicit form. Let $y^{(1)}$ and $y^{(2)}$ be two particular solutions of the equation

$$Ly = y_0 \qquad (20.37)$$

which satisfy the conditions:

$$[y^{(1)}]_{x=0} = [y^{(1)\prime}]_{x=0} = 0,$$

$$[y^{(2)}]_{x=1} = [y^{(2)\prime}]_{x=1} = 0$$

and similarly let $z^{(1)}$ and $z^{(2)}$ be two particular solutions of the equation

$$Mz = z_0 \qquad (20.38)$$

which satisfy the conditions:

$$[z^{(1)}]_{x=0} = [z^{(1)\prime}]_{x=0} = 0,$$

$$[z^{(2)}]_{x=1} = [z^{(2)\prime}]_{x=1} = 0.$$

It is clear that the difference $y^{(2)} - y^{(1)} = y_1$ is a particular solution of equation (20.14) and is linearly independent of y_0. For, if it were possible that $y^{(2)} - y^{(1)} = \alpha y_0$, then both functions $y^{(1)}$ and $y^{(2)}$ would have to satisfy the conditions (20.15) and (20.19) at the ends, and this is impossible by Lemma 3.

The specification of y_0 determines the functions $y^{(2)}$, $y^{(1)}$, and y_1. We put

$$z_0 = \frac{y_0}{p(y_0' y_1 - y_1' y_0)}, \quad z_1 = \frac{y_1}{p(y_0' y_1 - y_1' y_0)}.$$

It may be immediately verified that the functions $y^{(1)}$ and $y^{(2)}$ can be expressed in terms of y_0 and y_1 in the form

$$y^{(1)}(x_0) = y_0(x_0) \int_0^{x_0} y_0(x)\, z_1(x)\, dx - y_1(x_0) \int_0^{x_0} y_0(x)\, z_0(x)\, dx,$$

$$y^{(2)}(x_0) = y_0(x_0) \int_1^{x_0} y_0(x)\, z_1(x)\, dx - y_1(x_0) \int_1^{x_0} y_0(x)\, z_0(x)\, dx.$$

For, it is clear firstly that $y^{(1)}(0) = y^{(2)}(1) = 0$.

Further,
$$y^{(1)\prime\prime}(0) = [y_0(0)]^2\, z_1(0) - y_1(0)\, y_0(0)\, z_0(0) = 0,$$

$$y^{(2)\prime\prime}(1) = [y_0(1)]^2\, z_1(1) - y_1(1)\, y_0(1)\, z_0(1) = 0.$$

Moreover,
$$y^{(1)\prime} = y_0'(x_0) \int_0^{x_0} y_0(x)\, z_1(x)\, dx - y_1'(x_0) \int_0^{x_0} y_0(x)\, z_0(x)\, dx,$$

$$y^{(1)\prime\prime} = y''(x_0) \int_0^{x_0} y_0(x)\, z_1(x)\, dx - y_1''(x_0) \int_0^{x_0} y_0(x)\, z_0(x)\, dx$$
$$+ [y_0'(x_0)\, z_1(x_0) - y_1'(x_0)\, z_0(x_0)]\, y_0(x_0),$$

from which
$$Ly^{(1)} = y_0(x),$$

as was to be shown.

The formula for $y^{(2)}$ is proved in the same way.

If we now form the difference

$$y_1(x_0) = y^{(2)}(x_0) - y^{(1)}(x_0)$$
$$= -y_0(x_0) \int_0^1 y_0(x)\, z_1(x)\, dx + y_1(x_0) \int_0^1 y_0(x)\, z_0(x)\, dx,$$

we see that
$$\int_0^1 y_0(x)\, z_1(x)\, dx = 0,$$

$$\int_0^1 y_0(x)\, z_0(x)\, dx = 1,$$

i.e.,
$$\int_0^1 y_0(x)\, z^{(1)}(x)\, dx = \int_0^1 y_0(x)\, z^{(2)}(x)\, dx = C.$$

We shall prove that the function

$$G_1 = \begin{cases} y_0(x)\, z^{(1)}(x) + y^{(2)}(x_0)\, z_0(x) - Cy_0(x_0)\, z_0(x), & x \leqq x_0, \\ y_0(x)\, z^{(2)}(x) + y^{(1)}(x_0)\, z_0(x) - Cy_0(x_0)\, z_0(x), & x \geqq x_0 \end{cases}$$

satisfies all the four conditions enumerated above.

(1) The conditions (20.33) are obviously satisfied since $z_0(x)$ satisfies them and so do $z^{(1)}$ at $x = 0$ and $z^{(2)}$ at $x = 1$.

(2) The equation (20.34) is satisfied by virtue of (20.38). It is clear that

$$MG_1 = y_0(x_0)\, z_0(x).$$

(3) The continuity of G_1 is obvious. For the jump of its derivative we have

$$G'_{1x}(x_0 + 0, x_0) - G'_{1x}(x_0 - 0, x_0)$$

$$= y_0(x_0) [z^{(2)\prime\prime}(x_0) - z^{(2)\prime\prime}(x_0)] + [y^{(1)\prime}(x_0) - y^{(2)\prime}(x_0)]z'_0(x_0)$$

$$= y_0(x_0) z'_1(x_0) - y_1(x_0) z'_0(x_0) = \frac{1}{p(x_0)}.$$

Finally,

$$(4) \int_0^1 G_1(x, x_0) y_0(x) \, dx$$

$$= -Cy_0(x_0) + y_0(x_0) \int_0^1 y_0(x) z^{(2)}(x) \, dx$$

$$+ y_0(x_0) \int_0^{x_0} [z^{(1)}(x) - z^{(2)}(x)] y_0(x) \, dx$$

$$+ [y^{(2)}(x_0) - y^{(1)}(x)] \int_0^{x_0} z_0(x) y_0(x) \, dx + y^{(1)}(x_0) \int_0^1 y_0(x) z_0(x) \, dx$$

$$= -Cy_0(x_0) + Cy_0(x_0) + y^{(1)}(x_0) - y^{(1)}(x_0) = 0.$$

It is clear that the generalized Green's function $G_1(x, x_0)$ as a function of the variable x_0 will serve also as the generalized Green's function for the adjoint operator Mz and the boundary conditions (20.18) and (20.21). By means of it the solution of equation (20.1) which satisfies the conditions (20.13′) can be expressed in the form

$$y(x_0) = \int_0^1 G_1(x, x_0) f(x) \, dx + C_1 y_0(x_0), \qquad (20.39)$$

provided only that $f(x)$ satisfies the condition

$$\int_0^1 f(x) z_0(x) \, dx = 0. \qquad (20.40)$$

It follows from the proof, incidentally, that the condition (20.40) is not only necessary but is sufficient for the solubility of equation (20.1) subject to the conditions (20.13′). We shall not give the proof of all these assertions, since it is exactly the same as the one we gave for Green's function itself.

For an equation of the second order under the conditions investigated, the homogeneous problem can have only one non-trivial solution. For equations of higher order, cases can occur where the corresponding homogeneous problem has several linearly independent solutions, $y_i(x)$, $i = 1, ..., n$.

The generalized Green's function must then satisfy the equation

$$MG = \sum_{i=1}^{n} y_i(x_0)\, z_i(x),$$

and instead of the single condition (20.36) we shall have several similar conditions. We shall not analyse these cases in detail, but pass on now to consider a few examples.

§ 7. Examples

EXAMPLE 1. We seek a solution of the equation

$$Ly \equiv y'' = f(x) \tag{20.41}$$

subject to the conditions

$$[y]_{x=0} = a_0, \quad [y]_{x=1} = a_1. \tag{20.42}$$

The only solution of the equation $y'' = 0$ subject to the conditions $[y]_{x=0} = [y]_{x=1} = 0$ is zero.
By the general theory, Green's function exists:

$$G = \begin{cases} x(x_0 - 1), & x \leqq x_0 \\ x_0(x - 1), & x \geqq x_0. \end{cases} \tag{20.43}$$

The solution of the problem will be:

$$y(x_0) = a_0(1 - x_0) + a_1 x_0 + (x_0 - 1)\int_0^{x_0} f(x)x\,dx + x_0\int_{x_0}^1 f(x)(x - 1)\,dx. \tag{20.44}$$

EXAMPLE 2. We seek a solution of equation (20.41) satisfying the conditions

$$[y']_{x=0} = a_0, \quad [y']_{x=1} = a_1. \tag{20.45}$$

In this case the equation $y'' = 0$ has a non-trivial solution satisfying the conditions

$$[y']_{x=0} = [y']_{x=1} = 0, \tag{20.46}$$

namely,

$$y = 1.$$

Consequently in this case, Green's function does not exist, and we shall have to construct a generalized Green's function. We have:

$$y_0 = 1, \quad z_0 = 1 \quad \text{and} \quad y_1 = x, \quad z_1 = x.$$

Further,

$$z^{(1)} = x^2/2 \quad \text{and} \quad z^{(2)} = (1 - x)^2/2.$$

Then

$$y^{(1)} = x^2/2, \quad y^{(2)} = (1 - x)^2/2, \quad C = 1/6.$$

The function G_1 will therefore have the form

$$G_1 = \begin{cases} x^2/2 + (1 - x_0)^2/2 - 1/6, & x_0 \geqq x \\ x_0^2/2 + (1 - x)^2/2 - 1/6, & x \geqq x_0. \end{cases} \tag{20.47}$$

EXAMPLE 3. We consider the same equation (20.41) in the interval $0 \leqq x \leqq 2\pi$ and shall regard x as the length of a circular arc of unit radius measured from some initial point. Points whose coordinates differ by a multiple of 2π coincide. It is clear, therefore, that the required function y must be periodic with period 2π. We have for this function the conditions

$$[y]_{x=2\pi} = [y]_{x=0}, \quad [y']_{x=2\pi} = [y']_{x=0}. \tag{20.48}$$

These conditions express the circumstance that the function y and its first derivative are continuous on the circle. The fact that the point $x = 0$ plays a special part in these conditions is due merely to the choice of the initial point.

We shall not develop the general theory for problems of this sort, but limit ourselves to the analysis of this problem which is entirely similar to our previous work.

It is not difficult to see that the homogeneous equation

$$y'' = 0$$

has, as in the previous problem, the non-trivial solution $y = 1$ satisfying the conditions (20.48). Consequently we shall have to construct the generalized Green's function. This construction is most easily accomplished by using the periodicity of the required function, and also the fact that all points on the circle are equivalent and therefore the function will depend only on the difference $x - x_0$. In view of this we shall put from the start $x_0 = \pi$.

The equation for Green's function will have the form

$$\frac{\mathrm{d}^2 G_1}{\mathrm{d}x^2} = 0,$$

and its solution will obviously be a quadratic form

$$G_1(x, \pi) = c_1 x^2 + c_2 x + c_3, \quad -\pi \leqq x \leqq \pi.$$

From the continuity condition we have:

$$G_1(\pi - 0, \pi) - G_1(\pi + 0, \pi) = G_1(\pi, \pi) - G_1(-\pi, \pi) = 2\pi c_2,$$

and therefore

$$c_2 = 0.$$

Further, from the condition for the jump in the derivative, we see that:

$$G_1'(\pi + 0, \pi) - G_1'(\pi - 0, \pi) = -4\pi c_1 = 1,$$

so that

$$c_1 = -1/4\pi.$$

Moreover, Green's function must be orthogonal to a constant, whence

$$-\frac{1}{4\pi} \int_{-\pi}^{+\pi} x^2 \, dx + 2\pi c_3 = 0 \quad \text{or} \quad c_3 = \pi/12.$$

Finally, we have:

$$G_1(x, \pi) = -\frac{x^2}{4\pi} + \frac{\pi}{12}, \quad -\pi \leqq x \leqq +\pi.$$

By virtue of the periodicity, $G_1(x, \pi) = G_1(x, (2k + 1) \pi)$, and in the general case

$$G_1(x, x_0) = \begin{cases} G_1(x - x_0 + \pi, \pi) = -\dfrac{1}{4\pi} (x - x_0 + \pi)^2 + \dfrac{\pi}{12} \\ \qquad \text{for} \quad x_0 - 2\pi \leqq x \leqq x_0, \\ G_1(x - x_0 - \pi, \pi) = -\dfrac{1}{4\pi} (x - x_0 - \pi)^2 + \dfrac{\pi}{12} \\ \qquad \text{for} \quad x_0 \leqq x \leqq x_0 + 2\pi. \end{cases} \tag{20.49}$$

This Green's function which we have constructed is symmetrical with respect to x and x_0 and has all the properties which were established earlier for the generalized Green's function.

In practical problems we often encounter cases where in the equation (20.1) $p(0) = 0$ or $p(1) = 0$, and sometimes both ends of the interval are roots of $p(x) = 0$. Under these conditions we can sometimes still construct Green's function, albeit with certain additional limitations. We limit ourselves here to the analysis of a single example which will be useful to us in the sequel.

EXAMPLE. We consider the equation

$$Ly \equiv xy'' + y' - \frac{m^2}{x}y = f(x) \tag{20.50}$$

where m is an integer, and we shall seek a solution satisfying the condition

$$[y]_{x=1} = 0. \tag{20.51}$$

It is not difficult to see that

$$y_1 = x^m, \quad y_2 = x^{-m}$$

are integrals of the homogeneous equation

$$Ly = 0.$$

One of these is unbounded at the beginning of the interval. Hence, in order that one of the constants entering into the general solution shall be fully determinate, it is sufficient to require that

$$|y| \leqq \frac{1}{x^{m-1}}.$$

The other constant is determined from the condition (20.51).

In this case the adjoint operator coincides with the original one:

$$L_Z \equiv M_Z$$

It is natural to replace in the definition of Green's function the condition at $x = 0$ by a requirement for boundedness. In this case Green's function for our problem may be expressed in the form

$$G(x, x_0) = \begin{cases} \dfrac{x^m(x_0^m - x_0^{-m})}{2m}, & x \leqq x_0 \\[3mm] \dfrac{x_0^m(x^m - x^{-m})}{2m}, & x \geqq x_0. \end{cases} \qquad (20.52)$$

We assume that $f(x)$ satisfies the inequality

$$|f(x)| < \frac{A}{x^k}, \quad 0 \leqq k \leqq m.$$

Applying formula (20.26) to the required solution, we shall have:

$$y(x_0) = \frac{x_0^m - x_0^{-m}}{2m} \int_0^{x_0} x^m f(x)\, dx + \frac{x_0^m}{2m} \int_{x_0}^1 (x^m - x^{-m}) f(x)\, dx.$$

It is not difficult to find an upper bound for the absolute value of this solution:

$$|y(x_0)| \leqq \frac{x_0^{-m}}{2m} \int_0^{x_0} x^{m-k} A\, dx + \frac{x_0^m}{2m} \int_{x_0}^1 x^{-m-k} A\, dx + \frac{x_0^m}{2m} \int_0^1 A x^{m-k}\, dx$$

$$\leqq A \left(\frac{x_0^{-k+1}}{2m(m-k+1)} + \frac{x_0^{-k+1}}{2m(m+k-1)} + \frac{x_0^m}{2m(m-k+1)} \right),$$

or

$$|y(x_0)| \leqq \frac{A_1}{x_0^{k-1}}.$$

Consequently, since $k \leqq m$, the solution will satisfy the conditions imposed.

If $m = 0$, then $y_1 = 1$, $y_2 = \log_e x$ will be solutions of the homogeneous equation. And in this case, for bounded functions $f(x)$ a bounded solution may be looked for.

It is also possible to construct Green's function for equations of higher order than the second. We shall again restrict ourselves to one example.

We consider the solution of the equation in Example 1 of § 2 subject to the conditions (20.9). In this case Green's function is defined by the formulae:

$$G(x, x_0) = 0, \quad x > x_0,$$

$$MG = 0, \quad x < x_0,$$

and satisfies the conditions

$$[G]_{x=x_0} = [G'_x]_{x=x_0} = [G''_{xx}]_{x=x_0} = \cdots = [G^{(n-2)}_{xx\ldots x}]_{x=x_0} = 0$$

$$[G^{(n-1)}_{xx\ldots x}]_{x=x_0-0} = -\frac{1}{p_0(x_0)}.$$

Using this Green's function, we get the solution of the problem in the form

$$y(x_0) = \int_0^1 G(x, x_0) f(x) \, dx.$$

The reader may verify this for himself immediately. The Green's function defined by these conditions differs, as before, from the Green's function for the adjoint problem only in the interchange of the arguments.

GREEN'S FUNCTION FOR THE LAPLACE OPERATOR

§ 1. Green's Function for the Dirichlet Problem

Having analysed the most important cases of construction of Green's function for various ordinary differential equations, we now proceed to investigate Green's function for various problems connected with Poisson's equation.

We consider in a space with the coordinates x, y, z a domain Ω bounded by a sufficiently smooth surface S. We shall seek a function u which in this domain satisfies the equation

$$\nabla^2 u = f(P) \tag{21.1}$$

and one of the following conditions:

$$[u]_S = F_0(S) \tag{21.2}$$

or

$$\left[\frac{\partial u}{\partial n}\right]_S = F_1(S). \tag{21.3}$$

A problem of this sort is encountered, for example, in seeking the potential of an electric field due to a given distribution of charges. The problem of the equilibrium form of a membrane subjected to specified transverse forces also reduces to the same type; and so on.

The Laplace operator, as we know, is self-adjoint. The homogeneous conditions

$$[u]_S = 0 \quad \text{or} \quad \left[\frac{\partial u}{\partial n}\right]_S = 0,$$

corresponding to (21.2) or (21.3), are also self-adjoint, as is clear from Green's classical formula

$$\iiint_\Omega (v\nabla^2 u - u\nabla^2 v)\, d\Omega = \iint_S \left(u\frac{\partial v}{\partial n} - v\frac{\partial u}{\partial n}\right) dS. \tag{21.4}$$

The arguments which we shall adduce can be carried over without difficulty to the case of more general conditions, but we shall not deal with this question.

DEFINITION. *A function $G(P, P_0)$ of two variable points P and P_0 which satisfies the following conditions will be called Green's function for the equation (21.1) subject to the conditions (21.2), or Green's function for the Dirichlet problem:*

1. $G(P, P_0)$ is a harmonic function of the point P throughout the domain Ω, including the point P_0 itself.
2. As a function of the point P $G(P, P_0)$ satisfies the condition $[G(P, P_0)]_S = 0$.
3. In the domain, Ω, $G(P, P_0)$ can be expressed in the form

$$G(P, P_0) = \frac{1}{4\pi r} + g(P, P_0),$$

where r is the distance between P and P_0, and $g(P, P_0)$ is a regular harmonic function.

The function $G(P, P_0)$ is entirely similar to the Green's function which we constructed earlier for the ordinary linear differential equation.

Instead of the definition just given, we might also have defined this function as an influence function (as in § 2, Lecture 20), but we shall not go into this in detail.

We prove that Green's function for the Dirichlet problem exists.

By property 3

$$G(P, P_0) - \frac{1}{4\pi r} = g(P, P_0)$$

is a harmonic function of the point P throughout the domain Ω:

$$\nabla^2 g = 0. \tag{21.5}$$

Its boundary values on S will be

$$[g(P, P_0)]_S = \left[-\frac{1}{4\pi r} \right]_S \tag{21.6}$$

In view of (21.5) and (21.6) it follows that $g(P, P_0)$ may be constructed by means of the solution of the corresponding Dirichlet problem.

We shall examine one more important property of Green's function. We first prove a lemma.

Lyapunov's Lemma. Consider a domain Ω bounded by a twice-differentiable surface S. Suppose two surfaces S_1 and S_2 are drawn, one on each side of S, at a distance h from S, and that $h < d$, where d is the least radius of curvature of any plane normal section of the surface S.

Suppose that a function $F(x, y, z)$ is continuous and has continuous first-order derivatives in the region included between S_1 and S_2 and that its second-order derivatives are continuous everywhere within this region except on the surface S itself. Suppose further that $\nabla^2 F$ is bounded.

Then a function which is harmonic within Ω and coincides with F on the surface S will have a regular normal derivative on S.

To prove this lemma we apply Green's formula separately to the two layers Ω_1 and Ω_2 included between S_1 and S and between S_2 and S. For any point P_0 lying inside the inner layer Ω_1 we shall have:

$$F(P_0) = \frac{1}{4\pi} \iint_{S+S_1} \left\{ F \frac{\partial}{\partial n}\left(\frac{1}{r}\right) - \frac{1}{r}\frac{\partial F}{\partial n} \right\} dS - \frac{1}{4\pi} \iiint_{\Omega_1} \frac{1}{r} \nabla^2 F \, d\Omega$$

or

$$F(P_0) = \frac{1}{4\pi} \iint_S F \frac{\partial}{\partial n}\left(\frac{1}{r}\right) dS + \frac{1}{4\pi} \iint_{S_1} F \frac{\partial}{\partial n}\left(\frac{1}{r}\right) dS$$
$$- \frac{1}{4\pi} \iint_{S+S_1} \frac{1}{r} \frac{\partial F}{\partial n} dS - \frac{1}{4\pi} \iiint_{\Omega_2} \frac{1}{r} \nabla^2 F \, d\Omega .$$

The integrals standing on the right-hand side

$$\frac{1}{4\pi} \iiint_{\Omega_1} \frac{1}{r} \nabla^2 F \, d\Omega, \quad \frac{1}{4\pi} \iint_{S+S_1} \frac{1}{r} \frac{\partial F}{\partial n} dS, \quad \frac{1}{4\pi} \iint_{S_1} F \frac{\partial}{\partial n}\left(\frac{1}{r}\right) dS$$

have regular normal derivatives; this follows from Theorem 2 of Lecture 15. The left-hand member has continuous first-order derivatives in the neighbourhood of S. Consequently the integral

$$\frac{1}{4\pi} \iint_S F \frac{\partial}{\partial n}\left(\frac{1}{r}\right) dS$$

also has a continuous, regular normal derivative.

Similarly for a point P_0 lying outside the surface S we have

$$0 = \frac{1}{4\pi} \iint_{S+S_2} \left\{ F \frac{\partial}{\partial n}\left(\frac{1}{r}\right) - \frac{1}{r}\frac{\partial F}{\partial n} \right\} dS - \frac{1}{4\pi} \iiint_{\Omega_1} \frac{1}{r} \nabla^2 F \, d\Omega ,$$

whence by the same arguments we see that the integral

$$\frac{1}{4\pi} \iint_S F \frac{\partial}{\partial n}\left(\frac{1}{r}\right) dS .$$

has a regular normal derivative outside S as well.

Hence by Lyapunov's Theorem, Lecture 15, we see that a function which is harmonic inside Ω and takes on S values equal to those of the function F has a regular normal derivative, as was to be shown.

We can now establish the further important property of Green's function.

THEOREM 1. *If a surface S satisfies the conditions of Lyapunov's lemma, then Green's function regarded as a function of the point P will have a regular normal derivative when the point P_0 lies inside Ω.*

The proof follows immediately from Lyapunov's lemma. For we have

$$G = \frac{1}{4\pi r} - g.$$

The function $1/4\pi r$ obviously has a regular normal derivative. The function g takes on the surface S the same values as the function $1/4\pi r$, which is twice-differentiable in the region near S. By Lyapunov's lemma, g has a regular normal derivative, as was to be shown.

THEOREM 2. *A solution of the equation* (21.1) *subject to the conditions* (21.2), *if it exists, can be represented in the form*

$$u(P_0) = \iint_S F_0(S) \frac{\partial G}{\partial n} \, dS - \iiint_\Omega G(P, P_0) f(P) \, dP. \qquad (21.7)$$

For the proof, we apply Green's formula to the domain Ω' obtained from Ω by removing a small sphere with surface σ of radius δ described about the point P_0. We take u to be the unknown solution of equation (21.1) and v to be Green's function G. (We may note that Green's formula can be applied because G has a regular normal derivative.) Both u and G will be continuous and have continuous first-order derivatives inside Ω'. We get:

$$\iiint_{\Omega'} G(P, P_0) \, \nabla^2 u \, dP = \iint_{S'} \left(u \frac{\partial G}{\partial n} - G \frac{\partial u}{\partial n} \right) dS,$$

where S' is the complete surface of Ω'.

We denote by n' the direction of the normal on the surface σ, taken in the direction inwards into Ω'. Let n be the direction of a normal going into the centre of the sphere σ. Taking into account the fact that on S the function G vanishes, we get

$$\iiint_\Omega G(P, P_0) f(P) \, dP = \iint_S F_0(S) \frac{\partial G}{\partial n} \, dS$$
$$- \frac{1}{4\pi} \iint_\sigma \left\{ u \frac{\partial}{\partial n} \left(\frac{1}{r} \right) - \frac{1}{r} \frac{\partial u}{\partial n} \right\} d\sigma$$
$$- \iint_\sigma \left\{ u \frac{\partial g}{\partial n} - g \frac{\partial u}{\partial n} \right\} d\sigma.$$

The minus sign before the last two integrals arises because in the integration over the surface σ the direction of the normal n is opposite to that of the inward normal n'.

The limit of the last integral as $\delta \to 0$ is zero, since u and g and their derivatives are bounded. Further,

$$\lim_{\delta \to 0} \frac{1}{4\pi} \iint_\sigma \left\{ u \frac{\partial}{\partial n} \left(\frac{1}{r} \right) - \frac{1}{r} \frac{\partial u}{\partial n} \right\} d\sigma$$

$$= \lim_{\delta \to 0} \frac{1}{4\pi\delta^2} \iint_\sigma u \, d\sigma - \lim_{\delta \to 0} \frac{1}{4\pi\delta^2} \iint_\sigma \delta \frac{\partial u}{\partial n} \, d\sigma.$$

The limit of the first term is the limit of the mean value of u on the sphere σ and is equal to $u(P_0)$. Using this fact, we at once obtain the required formula (21.7).

THEOREM 3. *The function $G(P, P_0)$ is a symmetrical function of its arguments.*

Proof. We apply Green's formula to the functions $G(P, P_1)$ and $G(P, P_2)$ in the domain Ω'' obtained from Ω by excluding both the points P_1 and P_2 by means of small spheres. Denoting the boundary of Ω'' by S'' we shall obtain

$$\iint_{S''} \left\{ G(P, P_1) \frac{\partial G(P, P_2)}{\partial n} - G(P, P_2) \frac{\partial G(P, P_1)}{\partial n} \right\} dS'' = 0.$$

This integral taken over the surface S is zero, since there both the functions $G(P, P_1)$ and $G(P, P_2)$ vanish. The limit of the integral over the sphere round P_1 will obviously be equal to $G(P_1, P_2)$, and the limit of the integral taken over the sphere round P_2 will be equal to $G(P_2, P_1)$: thus we have shown that $G(P_1, P_2) = G(P_2, P_1)$.

THEOREM 4. *The function $u(P_0)$ defined by formula (21.7) gives the solution of the problem under consideration.*

To prove this, it is sufficient to show that the solution exists, since by Theorem 2 it must be given by formula (21.7) if it does exist.

Let ψ be the Newtonian potential of the domain Ω with the density $f(P)$:

$$\psi(P_0) = -\frac{1}{4\pi} \iiint_\Omega \frac{f(P)}{r} \, dv.$$

The function ψ satisfies Poisson's equation (see p. 169):

$$\nabla^2 \psi = f.$$

We put

$$v = u - \psi.$$

If we can find a harmonic function v satisfying the conditions

$$[v]_S = [u]_S - [\psi]_S,$$

then $u = v + \psi$ will be the solution of the problem. But the existence of such a function v follows from our earlier investigation of the Dirichlet problem. Hence the theorem.

One more point is worth noting. Green's function satisfies the inequality

$$G(P, P_0) \leqq \frac{1}{4\pi r}.$$

This follows from the fact that the function g is negative on the boundary and so is negative everywhere.

Just as we have constructed Green's function for the Dirichlet problem in space, so we could construct the same function in a plane. We should then have to separate out the term $(1/2\pi) \log_e (1/r)$ instead of $1/4\pi r$. The solution would be given by the formula (21.7) if we took Ω to be a plane domain and S its boundary curve.

§ 2. The Concept of Green's Function for the Neumann Problem

When we try to solve the Neumann problem, either in space or in a plane, we meet the difficulty that Green's function, as we should like to construct it, does not exist, since the corresponding homogeneous problem has a non-trivial solution, *viz.*, a constant. We should have to find a generalized Green's function, *i.e.*, to require it to satisfy not Laplace's equation but the equation

$$\nabla^2 G(P, P_0) = C. \tag{21.8}$$

We shall examine this question in more detail. We shall show that a solution of (21.8) exists, having the form

$$G_1 = \frac{1}{4\pi r} + g \tag{21.9}$$

where $[\partial G_1 / \partial n]_S = 0$, and g is a function which is regular within the domain. To do this, it is sufficient to show that, for a proper choice of the constant C, a function exists which satisfies equation (21.8) and the condition

$$\left[\frac{\partial g}{\partial n}\right]_S = -\frac{1}{4\pi} \left[\frac{\partial}{\partial n} \left(\frac{1}{r}\right)\right]_S.$$

We shall try to find g in the form $g = \alpha R^2 + g_1$, where $R^2 = x^2 + y^2 + z^2$, and g_1 is a harmonic function. For the function g_1 we get the boundary condition

$$\left[\frac{\partial g_1}{\partial n}\right]_S = -\alpha \left[\frac{\partial R^2}{\partial n}\right]_S - \frac{1}{4\pi} \left[\frac{\partial}{\partial n} \left(\frac{1}{r}\right)\right]_S. \tag{21.10}$$

We shall show that with a suitable choice of the constant α, the following equation holds:

$$\iint_S \left\{ \alpha \left[\frac{\partial R^2}{\partial n} \right]_S + \frac{1}{4\pi} \frac{\partial}{\partial n} \left(\frac{1}{r} \right) \right\} dS = 0. \qquad (21.11)$$

From this it will follow that a harmonic function exists satisfying (21.10), and moreover we shall have

$$\nabla^2 g = 6\alpha = C.$$

Consequently, it will follow that a function G_1 exists satisfying (21.8) and (21.10). We shall call this function G_1 *Green's function for the Neumann problem*.

We have

$$\iint_S \alpha \frac{\partial R^2}{\partial n} \, dS = \alpha \iiint_\Omega \nabla^2(R^2) \, dS = 6\alpha \cdot m\Omega,$$

where $m\Omega$ is the volume of the domain Ω. Further,

$$\frac{1}{4\pi} \iint_S \frac{\partial}{\partial n} \left(\frac{1}{r} \right) dS = \frac{1}{4\pi} \iint_S d\omega = 1.$$

Hence the condition (21.11) is satisfied if we put $\alpha = -(1/6m\Omega)$.

Using Green's function G_1 a solution of the following problem can be constructed: to find a function u satisfying the equation

$$\nabla^2 u = f(P)$$

and the condition $[\partial u/\partial n]_S = 0$.

The necessary and sufficient condition for this problem to be soluble is that

$$\iiint_\Omega f(P) \, dv = 0.$$

If this condition is satisfied, then

$$u(P_0) = \iiint_\Omega G(P, P_0) f(P) \, dv.$$

We shall leave the reader to verify this assertion.

We consider one example.

EXAMPLE. We shall construct Green's function for the Neumann problem, formulated for the sphere $R \leq 1$, where $R = \sqrt{x^2 + y^2 + z^2}$.

By definition,

$$G = \frac{1}{4\pi r} + g$$

where

$$\nabla^2 G = \text{const.} \left[\frac{\partial g}{\partial n} \right]_{R=1} = -\frac{1}{4\pi} \left[\frac{\partial}{\partial n} \left(\frac{1}{r} \right) \right]_{R=1}.$$

For convenience, we suppose that the pole of Green's function is at the point $(0, 0, z_0)$. Then

$$r = \sqrt{ x^2 + y^2 + (z - z_0)^2 },$$

and

$$\left[\frac{\partial}{\partial n} \left(\frac{1}{r} \right) \right]_{R=1}$$

$$= - \left[\frac{x \cos (n, x) + y \cos (n, y) + (z - z_0) \cos (n, z)}{r^3} \right]_{R=1}.$$

But on the surface $R = 1$ we have $\cos (n, x) = -x$, $\cos (n, y) = -y$, $\cos (n, z) = -z$. Hence

$$\left[\frac{\partial}{\partial n} \left(\frac{1}{r} \right) \right]_{R=1} = \left[\frac{1 - z_0 z}{r^3} \right]_{R=1}. \tag{21.12}$$

We now introduce the number $z_0' = 1/z_0$ and the function

$$r_1 = \sqrt{x^2 + y^2 + (z - z_0')^2}.$$

Clearly,

$$\left[\frac{\partial}{\partial n} \left(\frac{1}{r_1} \right) \right]_{R=1} = \left[\frac{1 - z_0' z}{r_1^3} \right]_{R=1}.$$

Putting $R_0 = z_0$, $\varrho = r_1$, $R = 1$ in (10.8), we have $[r]_{R=1} = [z_0 r_1]_{R=1}$. Using this, we find

$$\left[\frac{\partial}{\partial n} \left(\frac{1}{r_1} \right) \right]_{R=1} = \left[z_0^2 \frac{z_0 - z}{r^3} \right]_{R=1} \tag{21.13}$$

whence

$$\left[\frac{\partial}{\partial n} \left(\frac{1}{r} \right) \right]_{R=1} + \frac{1}{z_0} \left[\frac{\partial}{\partial n} \left(\frac{1}{r_1} \right) \right]_{R=1} = \left[\frac{1 - 2zz_0 + z_0^2}{r^3} \right]_{R=1} = \left[\frac{1}{r} \right]_{R=1}.$$

To finish these preliminary calculations, we consider the function

$$w = \log_e (z_0' - z + r_1).$$

We prove that w is a harmonic function. For,

$$\frac{\partial w}{\partial x} = \frac{x}{r_1(z_0' - z + r_1)}, \qquad \frac{\partial w}{\partial y} = \frac{y}{r_1(z_0' - z + r_1)},$$

$$\frac{\partial w}{\partial z} = \frac{-1 + (z - z_0')/r_1}{z_0' - z + r_1} = -\frac{1}{r_1}$$

$$\frac{\partial^2 w}{\partial x^2} = \frac{(z_0' - z)(r_1^2 - x^2) + r_1^3 - 2r_1 x^2}{r_1^3(z_0' - z + r_1)^2},$$

$$\frac{\partial^2 w}{\partial y^2} = \frac{(z_0' - z)(r_1^2 - y^2) + r_1^3 - 2r_1 y^2}{r_1^3(z_0' - z + r_1)^2},$$

$$\frac{\partial^2 w}{\partial x^2} = \frac{z - z_0'}{r_1^3}, \qquad \frac{\partial^2 w}{\partial x^2} + \frac{\partial^2 w}{\partial y^2} = \frac{z_0' - z}{r_1^3}.$$

Hence $\nabla^2 w = 0$.

We next calculate $[\partial w/\partial n]_{R=1}$. We get

$$\left[\frac{\partial w}{\partial n}\right]_{R=1} = \left[\frac{x \cos(n, x) + y \cos(n, y) - (z_0' - z + r_1) \cos(n, z)}{r_1(z_0' - z + r_1)}\right]_{R=1}$$

$$= \left[-\frac{[r_1^2 - (z_0' - z)^2] - z(r_1 + z_0' - z)}{r_1(z_0' - z + r_1)}\right]_{R=1}$$

$$= -\left[\frac{r_1 - z_0'}{r_1}\right]_{R=1} = -1 + \left[\frac{1}{z_0 r_1}\right]_{R=1}$$

$$= -1 + \left[\frac{1}{r}\right]_{R=1}. \tag{21.14}$$

Using (21.12), (21.13), (21.14), it is easily verified that Green's function must have the form

$$G = \frac{1}{4\pi r} + \frac{1}{z_0}\frac{1}{4\pi r_1} - \frac{w}{4\pi} - \frac{R^2}{8\pi} + C,$$

where C is a certain constant which can be determined from the condition

$$\iiint_{R \le 1} G \, dx \, dy \, dz = 0.$$

The calculation, which we shall not give in detail, leads to

$$C = -\frac{z_0^2}{8\pi} - \frac{1}{4\pi} \log_e z_0.$$

Hence

$$G = \frac{1}{4\pi r} + \frac{1}{z_0}\frac{1}{4\pi r} - \frac{w}{4\pi} - \frac{R^2}{8\pi} - \frac{1}{4\pi} \log_e z_0 - \frac{z_0^2}{8\pi}.$$

Returning to the notation of Lecture 19, we have

$$
\begin{aligned}
G = \frac{1}{4\pi} \Bigg\{ & \frac{1}{\sqrt{R^2 - 2RR_0 \cos y + R_0^2}} \\
& + \frac{1}{\sqrt{R^2 R_0^2 - 2RR_0 \cos y + 1}} - \frac{R^2}{2} - \frac{R_0^2}{2} \\
& - \log_e \left(1 - RR_0 \cos y + \sqrt{R^2 R_0^2 - 2RR_0 \cos y + 1} \right) \Bigg\}.
\end{aligned}
$$

CORRECTNESS OF FORMULATION
OF THE BOUNDARY-VALUE PROBLEMS
OF MATHEMATICAL PHYSICS

§ 1. The Equation of Heat Conduction

With most of the problems which we have so far considered, the method of solution itself answers the question whether the particular problem was correctly formulated. In some cases, however, it is more convenient to determine first of all whether the problem is correctly formulated.

Let us consider the equation of heat conduction in a bounded domain Ω of three-dimensional space, variables x, y, z, with the boundary surface S, when there are internal heat-sources present with density $F(x, y, z)$. Let $u(x, y, z, t)$ be the temperature at the point (x, y, z) at time t. The function u satisfies the equation

$$\nabla^2 u = \frac{\partial u}{\partial t} - F(x, y, z). \tag{22.1}$$

If we regard the heat influx as being non-negative everywhere, we have

$$F(x, y, z) = F(P) \geqq 0. \tag{22.2}$$

Suppose, further, that the following boundary and initial conditions are given:

$$[u]_S = f(Q, t), \quad [u]_{t=0} = \varphi(P) \tag{22.3}$$

where the functions f and φ are continuous, and the values of f for $t = 0$ on the surface S coincide with the values of φ, and where Q is a point of the surface S and P is a point of the domain Ω. We prove the following.

THEOREM 1. *At any instant of time within any finite internal* $0 \leqq t_0 \leqq T$ *the following inequality holds within the domain* Ω:

$$u(P_0, t_0) \geqq \inf_{\substack{P \subset \Omega; Q \subset S \\ 0 \leqq t \leqq t_0}} \min [f(Q, t), \varphi(P)]. \tag{22.4}$$

In other words, the function u attains its lowest values either at $t = 0$ *or on the boundary of the domain* Ω.

Proof. Suppose the theorem false. Let the function u take at a point P_0 and a time t_0 a value which is less than all its other values within the domain Ω at $t = 0$ and less than its values on the surface S during the time-interval $0 \leq t \leq t_0$. Then

$$u(P_0, t_0) - \inf_{\substack{P \subset \Omega, Q \subset S \\ 0 \leq t \leq t_0}} \min\,[f(Q, t), \varphi(P)] = -\varepsilon_0 < 0.$$

We form the function

$$v(x, y, z, t) = u(x, y, z, t) - \frac{\varepsilon_0}{2} \frac{t_0 - t}{t_0}.$$

As before, the function v has the property that in the interval $0 \leq t \leq t_0$, it does not take its minimum value either at $t = 0$ nor on the boundary of the domain. This follows at once from the fact that its value at the point (P_0, t_0) is less, by at least $\varepsilon_0/2$, than the minimum values of $v(x, y, z, t)$ on the boundary of Ω and at $t = 0$, because

$$v(P_0, t_0) = u(P_0, t_0), \quad [v]_S \geq [u]_S - \varepsilon_0/2, \quad [v]_{t=0} = [u]_{t=0} - \varepsilon_0/2.$$

The function v must take its least value either somewhere within the domain $\{\Omega, 0 \leq t \leq t_0\}$ or at $t = t_0$. It is now easy to see that our hypothesis leads to a contradiction. If the minimum v were to lie within Ω for $t < t_0$, then the first derivatives of v with respect to the space coordinates and time would vanish at this point, and the second derivatives $\partial^2 v / \partial x^2$, $\partial^2 v / \partial y^2$, $\partial^2 v / \partial z^2$ would be non-negative, as would $\nabla^2 v$. On the other hand,

$$\nabla^2 v = \nabla^2 u, \quad \frac{\partial v}{\partial t} = \frac{\partial u}{\partial t} + \frac{\varepsilon_0}{2t_0}, \quad \text{and} \quad \nabla^2 v - \frac{\partial v}{\partial t} = \nabla^2 u - \frac{\partial u}{\partial t} - \frac{\varepsilon_0}{2t} < 0.$$

Thus, a non-negative number would have to equal a negative number; and this shows our hypothesis untrue.

We have still to consider the case when v attains its minimum at $t = t_0$. The derivative $\partial v / \partial t$ clearly cannot be positive at the point where v attains its minimum, for otherwise at smaller values of the time v would have lesser values and we should not have found its minimum. On the other hand, as above, the inequality $\nabla^2 v \geq 0$ holds at this point. Repeating the previous argument, we are led to the absurd conclusion that the non-negative expression $\nabla^2 v - \partial v / \partial t$ must be less than zero. Hence this case also is impossible, and the theorem is proved.

By changing the signs of the functions u, f, φ and F we derive:

COROLLARY 1. *If in equation (22.1) the function F satisfies the inequality*

$$F(x, y, z) \leq 0,$$

then the function u attains its maximum value either for $t = 0$ or on the boundary S of the domain Ω.

COROLLARY 2. *A function $u(x, y, z, t)$ which, in the domain Ω and for $0 \leq t \leq T$, satisfies the homogeneous equation of heat conduction will take its maximum value and its minimum value either at $t = 0$ or on the boundary S of Ω.*

From the foregoing arguments follow the uniqueness of solution of the equation of heat conduction under the conditions (22.3) and the continuous dependence of order $(0, 0)$ of this solution on the right-hand members of the boundary and initial conditions, *i.e.,* the correctness of formulation of our boundary-value problem.

For, if there were two solutions of the problem, then their difference, satisfying the homogeneous equation, would vanish for $t = 0$ and on the surface S. But then, by Theorem 1, both the maximum and the minimum of this difference would be zero. Consequently the difference itself would be zero. That is, the problem cannot have two different solutions.

We can prove that the problem is correctly formulated in a similar way. If the difference of the functions which give the initial and boundary conditions does not exceed in absolute value a certain positive number ε, then the difference of the corresponding solutions, considered as the solution of the homogeneous equation of heat conduction with small boundary values, will also not exceed ε in absolute value.

THEOREM 2. *The solution of equation (22.1) depends continuously not only on the conditions (22.3) but also on the free member of (22.1). More precisely, if for $0 \leq t \leq t_0$ the values of the functions f and φ are less than $\varepsilon_0/2$, and the function F satisfies the inequality $F \leq \varepsilon_0/2t_0$, then for $0 \leq t \leq t_0$ the solution of (22.1) satisfies the inequality $u < \varepsilon_0$.*

Proof. Suppose that at a point (P_0, t_0) the solution of our problem takes a value exceeding ε_0. We form the function

$$v = u + \varepsilon_0(t_0 - t)/2t_0.$$

This function must have a maximum within Ω for $0 < t < t_0$. But, by forming the expression $\nabla^2 v - \partial v/\partial t$, we find that it is positive everywhere within Ω for $0 < t < t_0$, and this contradicts the condition that v should have a maximum.

COROLLARY 1. *If the signs in the inequalities for φ, f, and F are changed, i.e., if $\varphi > -\varepsilon_0/2, f > -\varepsilon_0/2, F > -\varepsilon_0/2t_0$, then $u > -\varepsilon_0$.*

Hence, finally, we have:

COROLLARY 2. *If*

$$|\varphi| < \varepsilon_0/2, \quad |f| < \varepsilon_0/2, \quad |F| < \varepsilon_0/2t_0,$$

then in the interval $0 < t < t_0$ for all $P \subset \Omega$ we have $|u| < \varepsilon_0$.

§ 2. The Concept of the Generalized Solution

In many of the problems of mathematical physics which we encounter, the existence of a solution is established only if we lay considerable restrictions on the boundary conditions. We now introduce a concept which will enable us to avoid having to investigate these questions.

Suppose we have three sequences of continuous functions F_n, φ_n, f_n, which tend uniformly to the continuous functions F, φ, f respectively, and let the equation

$$\nabla^2 u_n - \frac{\partial u_n}{\partial t} = F_n$$

have a solution u_n subject to the conditions $[u]_{t=0} = \varphi_n$, $[u]_S = f_n$ (by what has been proved, such a solution will be unique). By Theorem 2 of this lecture, the difference

$$u_m - u_n$$

will be arbitrarily small in absolute magnitude if m and n are sufficiently large. That is, the sequence u_n tends uniformly to a function u which satisfies our boundary conditions. But we know nothing about the convergence of the derivatives of u, and therefore we cannot assert that the limit function satisfies the equation

$$\nabla^2 u - \frac{\partial u}{\partial t} = F.$$

We shall call the function u the generalized solution of the equation (22.5) *with the given initial and boundary conditions.*

Such a generalized solution is unique, for there cannot be two sequences u_n and $u_n^{(1)}$ for which the functions f_n and $f_n^{(1)}$, φ_n and $\varphi_n^{(1)}$, F_n and $F_n^{(1)}$ converge respectively to single limits while the sequences themselves converge to different limits, because in this case the sequence

$$u_1, u_1^{(1)}, u_2, u_2^{(1)}, \ldots, u_n, u_n^{(1)}, \ldots$$

would diverge, and this is impossible.

In practice, instead of posing the problem of finding the true solution, it is sufficient to solve the problem of finding the generalized solution. For, in physical problems we do not know precisely the values of f, φ, and F. The values which we take for them will not be precise, but will differ from the exact values by only small amounts. Hence the generalized solution, even though it is not the true solution, will differ from it but little.

We have just discussed the generalized solution of the equation for heat conduction. Similar considerations apply to Poisson's equation, considered earlier. We can define a generalized solution of the equation

$$\nabla^2 u = \varrho$$

subject to the conditions

$$[u]_S = \varphi \quad \text{or} \quad \left[\frac{\partial u}{\partial n}\right]_S = \psi$$

as the limit of solutions of the equation

$$\nabla^2 u = \varrho_n,$$

subject to the conditions

$$[u]_S = \varphi_n \quad \text{or} \quad \left[\frac{\partial u}{\partial n}\right]_S = \psi_n,$$

provided that

$$\varrho_n \to \varrho, \quad \varphi_n \to \varphi, \quad \psi_n \to \psi$$

uniformly in each case.

We give an example for which such a generalized solution exists.

EXAMPLE 1.

Let

$$\nabla^2 u = \frac{x^2 - y^2}{R^2} \left[\frac{5}{\log_e R} - \frac{1}{(\log_e R)^2} \right] \qquad (22.6)$$

where

$$R = \sqrt{x^2 + y^2 + z^2}.$$

We shall show that, for $R \leqq 1/2$, the function

$$u_0 = (x^2 - y^2) \log_e \left| \log_e R \right|$$

will be a generalized solution of equation (22.6) subject to the boundary conditions

$$[u]_{R=1/2} = [u_0]_{R=1/2}.$$

For, write

$$u_k = (x^2 - y^2) \log_e \left| \log_e R_k \right|,$$

where

$$R_k = \sqrt{R^2 + \delta_k}, \quad \delta_k \to +0 \quad \text{as } k \to \infty.$$

Calculating $\varDelta u_k$ we get:

$$\frac{\partial^2 u_k}{\partial x^2} = 2 \log_e \left| \log_e R_k \right| + \frac{4x^2}{R_k^2 \log_e R_k}$$

$$+ (x^2 - y^2) \left[\frac{1}{R_k^2 \log_e R_k} - \frac{2x^2}{R_k^4 \log_e R_k} - \frac{x^2}{R_k^4 (\log_e R_k)^2} \right],$$

$$\frac{\partial^2 u_k}{\partial y^2} = -2 \log_e \left| \log_e R_k \right| - \frac{4\,y^2}{R_k^2 \log_e R_k}$$

$$+ (x^2 - y^2) \left[\frac{1}{R_k^2 \log_e R_k} - \frac{2\,y^2}{R_k^4 \log_e R_k} - \frac{y^2}{R_k^4 (\log_e R_k)^2} \right]$$

$$\frac{\partial^2 u_k}{\partial z^2} = (x^2 - y^2) \left[\frac{1}{R_k^2 \log_e R_k} - \frac{2\,z^2}{R_k^4 \log_e R_k} - \frac{z^2}{R_k^4 (\log_e R_k)^2} \right],$$

whence

$$\nabla^2 u_k = \frac{x^2 - y^2}{R_k^2} \left[\frac{5}{\log_e R_k} - \frac{1}{(\log_e R_k)^2} \right].$$

The sequence u_k obviously converges uniformly to u_0 as $k \to \infty$. Moreover, for $R \leqq 1/2$, the sequence $\nabla^2 u_k$ also converges uniformly to the right-hand side of equation (22.6). For, if $R < r$ and if δ_k is sufficiently small, then both $\nabla^2 u$ and the function

$$\frac{x^2 - y^2}{R^2} \left[\frac{5}{\log_e R} - \frac{1}{(\log_e R)^2} \right]$$

will be as small as we please if r is sufficiently small, and if $r \leqq R \leqq 1/2$ the uniform convergence is obvious.

Consequently, u_0 is the generalized solution of equation (22.6). However, at the point $(0, 0, 0)$, the second derivatives of u_0 have no meaning.

A solution of equation (22.6) with the boundary conditions

$$[u]_{R=1/2} = [u_0]_{R=1/2}$$

does not exist.

We notice, preliminarily, that u_0 satisfies equation (22.6) in the usual sense everywhere in the circle under consideration except at the point $(0, 0, 0)$.

If u were a solution of the proposed problem, then the difference $u - u_0$ would have to:

(i) be a harmonic function everywhere except perhaps at the origin,

(ii) be continuous everywhere,

(iii) vanish on the boundary of the domain.

But we have already seen that a continuous function which is harmonic everywhere except perhaps at one point must be harmonic at that point also.

Hence the difference $u - u_0$ is a harmonic function which is zero on the surface of the sphere, and is therefore zero everywhere. Hence the only possible solution of our problem is the function u_0. Since u_0 does not satisfy equation (22.6) at the origin, our problem has no solution at all.

If we now take the equation for heat conduction with the same right-hand member as in (22.6):

$$\nabla^2 u - \frac{\partial u}{\partial t} = \frac{x^2 - y^2}{R^2}\left[\frac{5}{\log_e R} - \frac{1}{(\log_e R)^2}\right]$$

and try to solve it with the conditions

$$[u]_{R=1/2} = [u_0]_{R=1/2}, \quad [u]_{t=0} = u_0,$$

we shall be unable to find a solution.

The function u_0 can again serve as the generalized function in this case. It can be shown that a continuous function which satisfies the homogeneous equation of heat conduction everywhere except perhaps at a point (x_0, y_0, z_0) for all values of time must also satisfy the equation at this point too. Using this fact and repeating the argument used in investigating Poisson's equation, we can prove that the problem has no solution.

§ 3. The Wave Equation

We now investigate the wave equation

$$\nabla^2 u - \frac{\partial^2 u}{\partial t^2} = F \tag{22.7}$$

and consider its solution in a domain Ω bounded by a surface S, with the initial conditions

$$[u]_{t=0} = \varphi(P), \quad \left[\frac{\partial u}{\partial t}\right]_{t=0} = \psi(P) \tag{22.8}$$

and the boundary conditions

$$\left[\frac{\partial u}{\partial n}\right]_S = f(S). \tag{22.9}$$

Our argument will apply without change if, instead of its normal derivative, the unknown function itself is specified on the boundary.

As regards the function u, we shall assume that within Ω it has continuous derivatives up to the second order inclusive and that its first-order derivatives are continuous in the closed domain $\overline{\Omega}$.

Without loss of generality we can always suppose that

$$\varphi = 0, \quad \psi = 0, \quad f = 0.$$

For if this is not initially true, by taking a new function v defined by

$$v = u - w,$$

where w is any function whatever which satisfies (22.8) and (22.9), we at once derive the homogeneous conditions for v.

We consider the integral

$$K_1(t) = \iiint_\Omega \left\{ \left(\frac{\partial u}{\partial x}\right)^2 + \left(\frac{\partial u}{\partial y}\right)^2 + \left(\frac{\partial u}{\partial z}\right)^2 + \left(\frac{\partial u}{\partial t}\right)^2 \right\} dx\, dy\, dz.$$

Calculating $dK_1(t)/dt$, we use (22.7) to find

$$\frac{dK_1}{dt} = 2 \iiint_\Omega \left\{ \frac{\partial u}{\partial x} \frac{\partial^2 u}{\partial x\, \partial t} + \frac{\partial u}{\partial y} \frac{\partial^2 u}{\partial y\, \partial t} + \frac{\partial u}{\partial z} \frac{\partial^2 u}{\partial z\, \partial t} + \frac{\partial u}{\partial t} \frac{\partial^2 u}{\partial t^2} \right\} dx\, dy\, dz$$

$$= 2 \iiint_\Omega \left\{ \left(\frac{\partial u}{\partial x} \frac{\partial^2 u}{\partial x\, \partial t} + \frac{\partial u}{\partial t} \frac{\partial^2 u}{\partial x^2} \right) + \left(\frac{\partial u}{\partial y} \frac{\partial^2 u}{\partial y\, \partial t} + \frac{\partial u}{\partial t} \frac{\partial^2 u}{\partial y^2} \right) \right.$$

$$\left. + \left(\frac{\partial u}{\partial z} \frac{\partial^2 u}{\partial z\, \partial t} + \frac{\partial u}{\partial t} \frac{\partial^2 u}{\partial z^2} \right) - F \frac{\partial u}{\partial t} \right\} dx\, dy\, dz$$

$$= 2 \iiint_\Omega \left\{ \frac{\partial}{\partial x}\left(\frac{\partial u}{\partial t} \frac{\partial u}{\partial x} \right) + \frac{\partial}{\partial y}\left(\frac{\partial u}{\partial t} \frac{\partial u}{\partial y} \right) + \frac{\partial}{\partial z}\left(\frac{\partial u}{\partial t} \frac{\partial u}{\partial z} \right) - F \frac{\partial u}{\partial t} \right\} dx\, dy\, dz$$

$$= 2 \iint_S \frac{\partial u}{\partial t} \frac{\partial u}{\partial n} dS - 2 \iiint_\Omega F \frac{\partial u}{\partial t} dx\, dy\, dz.$$

Since we have assumed $[\partial u/\partial n]_S = 0$, we get

$$\frac{dK_1}{dt} = -2 \iiint_\Omega F \frac{\partial u}{\partial t} dx\, dy\, dz. \tag{22.10}$$

Using the obvious inequality, $|ab| \le a^2/2 + b^2/2$, we can write (22.10) as

$$\frac{dK_1}{dt} \le \iiint_\Omega F^2 dx\, dy\, dz + \iiint_\Omega \left(\frac{\partial u}{\partial t}\right)^2 dx\, dy\, dz$$

or, making use of the same inequality again,

$$\frac{dK_1}{dt} \le \iiint_\Omega F^2 dx\, dy\, dz + K_1.$$

If we put $\displaystyle\iiint_\Omega F^2 dx\, dy\, dz = A(t)$, then $\displaystyle\frac{dK_1}{dt} - K_1 \le A(t)$,

or $$\frac{d(e^{-t}K_1)}{dt} \le e^{-t}A(t). \tag{22.11}$$

Consequently,

$$\left.\begin{aligned}
e^{-t}K_1(t) &\leq K_1(0) + \int_0^t e^{-t_1}A(t_1)\,dt_1 \\
K_1(t) &\leq e^t K_1(0) + \int_0^t e^{t-t_1}A(t_1)\,dt_1
\end{aligned}\right\}. \tag{22.12}$$

It follows at once that, provided $K_1(0) = 0$, for any fixed, finite interval of t the value of $K_1(t)$ will be as small as we please if $A(t)$ is also sufficiently small.

We note also that, putting $K_0(t) = \iiint_\Omega u^2\,dx\,dy\,dz$, we shall have

$$\frac{dK_0(t)}{dt} = 2\iiint_\Omega u\frac{\partial u}{\partial t}\,dx\,dy\,dz \leq K_0(t) + K_1(t),$$

i.e.,

$$\frac{dK_0(t)}{dt} - K_0(t) \leq K_1(t). \tag{22.13}$$

From this inequality, in the same way as above, we get

$$K_0(t) \leq e^t K_0(0) + \int_0^t e^{t-t_1}K_1(t_1)\,dt_1. \tag{22.14}$$

If $K_0(0) = 0 = K_1(0)$, then $K_0(t)$, like $K_1(t)$, will be as small as we please if $A(t)$ is sufficiently small.

A number of conclusions can be drawn from these results (22.12) and (22.14).

THEOREM 3. *Let the sequence of functions u_n satisfy the equations*

$$\nabla^2 u_n - \frac{\partial^2 u_n}{\partial t^2} = F_n$$

and the conditions

$$[u_n]_{t=0} = \left[\frac{\partial u_n}{\partial t}\right]_{t=0} = 0 \tag{22.15}$$

$$\left[\frac{\partial u}{\partial n}\right]_S = 0. \tag{22.16}$$

Let the functions F_n satisfy the condition

$$\lim_{n\to\infty}\int_0^T\left\{\iiint_\Omega F_n^2\,dx\,dy\,dz\right\}dt = 0.$$

Then for all values of t such that $0 \leq t \leq T$,

$$\lim_{n \to \infty} K_0(t) = 0$$

$$\lim_{n \to \infty} K_1(t) = 0.$$

Proof. From (22.12) and (22.14) we have *a fortiori*

$$K_1(t) \leq e^T \int_0^T A(t_1)\, dt_1 \tag{22.17}$$

and

$$K_0(t) \leq e^T \int_0^T K_1(t_1)\, dt_1 \leq e^{2T}T \int_0^T A(t_1)\, dt_1, \tag{22.18}$$

and our theorem follows at once from (22.17) and (22.18).

By Theorem 3, the solution is continuously dependent in the mean to order $(0, 1)$ on the right-hand member of the equation (see p. 29). If we compare the solutions of two equations

$$\nabla^2 u_1 - \frac{\partial^2 u_1}{\partial t^2} = F_1 \quad \text{and} \quad \nabla^2 u_2 - \frac{\partial^2 u_2}{\partial t^2} = F_2, \tag{22.19}$$

with the homogeneous conditions (22.15) and (22.16), the difference $(F_1 - F_2)$ being sufficiently small in the mean, *i.e.*, the integral $\iiint_\Omega (F_1 - F_2)^2\, dx\, dy\, dz$ is sufficiently small for all t, then although we cannot assert that $|u_1 - u_2|$ will be everywhere small, nevertheless the difference $(u_1 - u_2)$ and its derivatives of the first order will be arbitrarily small in the mean at any instant, *i.e.*,

$$\iiint_\Omega (u_1 - u_2)^2\, dx\, dy\, dz < \varepsilon,$$

$$\iiint_\Omega \left\{ \left(\frac{\partial u_1}{\partial x} - \frac{\partial u_2}{\partial x} \right)^2 + \left(\frac{\partial u_1}{\partial y} - \frac{\partial u_2}{\partial y} \right)^2 + \left(\frac{\partial u_1}{\partial z} - \frac{\partial u_2}{\partial z} \right)^2 \right\} dx\, dy\, dz < \varepsilon.$$

Suppose we wish to compare the solutions of the two equations

$$\nabla^2 u_1 - \frac{\partial^2 u_1}{\partial t^2} = F_1, \quad \nabla^2 u_2 - \frac{\partial^2 u_2}{\partial t^2} = F_2 \tag{22.20}$$

subject to the conditions

$$[u_1]_{t=0} = \varphi_1, \quad [u_2]_{t=0} = \varphi_2,$$

$$\left[\frac{\partial u_1}{\partial t} \right]_{t=0} = \psi_1, \quad \left[\frac{\partial u_2}{\partial t} \right]_{t=0} = \psi_2,$$

$$\left[\frac{\partial u_1}{\partial n} \right]_S = f_1, \quad \left[\frac{\partial u_2}{\partial n} \right]_S = f_2,$$

where φ_1, φ_2 have continuous derivatives up to the second order, and $\psi_1, \psi_2,$ f_1, f_2 have continuous first-order derivatives. Suppose that these functions are near to one another so that the following inequalities hold:

$$\left| \varphi_1 - \varphi_2 \right| < \delta, \quad \left| \frac{\partial \varphi_1}{\partial x_i} - \frac{\partial \varphi_2}{\partial x_i} \right| < \delta, \quad \left| \frac{\partial^2 \varphi_1}{\partial x_i \, \partial x_j} - \frac{\partial^2 \varphi_2}{\partial x_i \, \partial x_j} \right| < \delta,$$

$$\left| \psi_1 - \psi_2 \right| < \delta, \quad \left| \frac{\partial \psi_1}{\partial x_i} - \frac{\partial \psi_2}{\partial x_i} \right| < \delta, \quad \left| f_1 - f_2 \right| < \delta, \quad \left| \frac{\partial f_1}{\partial x_i} - \frac{\partial f_2}{\partial x_i} \right| < \delta,$$

$$\left| F_1 - F_2 \right| < \delta, \quad (x_1 = x, \quad x_2 = y, \quad x_3 = z).$$

It is easy to see that, if we reduce these problems to problems with homogeneous conditions, we again obtain two equations of the form (22.20) whose right-hand members are close together, and consequently their solutions will be close together in the mean for any values of t in a finite interval.

In exactly the same way as for the equation of heat conduction, it is easy to show that the solution of equation (22.7) with arbitrary initial conditions (22.8) is unique. To do this, it is sufficient to show that the homogeneous equation with homogeneous conditions has only a trivial, identically zero, solution. And this follows because, as we have seen, the integral of the square of such a solution is zero.

Finally, just as in the previous problems, the question of the existence of a solution presents very considerable difficulties, which even exceed those arising with Laplace's equation and the equation of heat conduction. These difficulties may be avoided by introducing again the concept of a generalized solution.

§ 4. The Generalized Solution of the Wave Equation

We make a few preliminary remarks. Consider a function $u(x, y, z, t)$ of four independent variables, x, y, z being the coordinates of an arbitrary point in a certain domain Ω and t the time; suppose that u^2 is integrable with respect to x, y, z for any value of t in a certain interval. We shall say that the function $u(x, y, z, t)$ is *continuous in the mean* with respect to the variable t at the point t_0 if the magnitude of

$$\left[\iiint_{\Omega} \{ u(x, y, z, t_0 + h) - u(x, y, z, t) \}^2 \, dx \, dy \, dz \right]^{1/2}$$

can be made arbitrarily small for sufficiently small $|h|$. We shall say that a function which is continuous in the mean at every instant of an interval $\alpha \leq t \leq \beta$ is *continuous in the mean in this interval*. It is convenient to introduce an abbreviated notation. For any function of the variables x, y, z

which is specified in the domain Ω and whose square is integrable, we shall call the quantity

$$\left[\iiint_\Omega u^2(x, y, z) \, dx \, dy \, dz \right]^{1/2}$$

the norm of the function u and denote it by $\|u\|$.

By Minkovski's inequality (see § 6 below)

$$\|u + v\| \leqq \|u\| + \|v\|.$$

If $\|u\| = 0$, then, as shown in § 7, Lecture 6, the function u vanishes almost everywhere in Ω.

If a is an arbitrary constant, then obviously $\|au\| = |a| . \|u\|$.

We shall need these properties of the norm later. Using this new notation, the definition of continuity in the mean can be re-formulated thus: the function $u(x, y, z, t)$ is continuous in the mean at the point t_0 if, given any positive number ε, a positive number $\eta(\varepsilon)$ can be found such that

$$\|u(t_0 + h) - u(t_0)\| < \varepsilon$$

provided only that $|h| < \eta(\varepsilon)$. (The arguments x, y, z of u have been left out for brevity.)

We shall say that a sequence of functions $u_n(x, y, z, t)$, $(n = 1, 2, \ldots)$, *converges uniformly in the mean* to the function $u_0(x, y, z, t)$ if

$$\|u_n - u_0\| = \left[\iiint_\Omega \{u_n(x, y, z, t) - u_0(x, y, z, t)\}^2 \, dx \, dy \, dz \right]^{\frac{1}{2}} < \varepsilon$$

for all t in the given interval, provided only that $n \geqq N(\varepsilon)$.

THEOREM 4. *If the sequence of functions*

$$u_n(x, y, z, t), \quad n = 1, 2, \ldots,$$

converges uniformly in the mean, and if each function is continuous in the mean, then the limit function is continuous in the mean.

The theorem is proved in the same way as that for ordinary continuous functions in elementary mathematical analysis. Let ε be a given positive number. Choose N so large that

$$\|u_n(t) - u_0(t)\| < \frac{\varepsilon}{3} \tag{22.21}$$

for all t in the given interval, provided only that $n \geqq N$. The function $u_n(x, y, z, t)$ is, by the conditions of the theorem, continuous in the mean at the point t_0. Consequently, for $|h| < \eta$,

$$\left\| u_n(t + h) - u_n(t) \right\| < \frac{\varepsilon}{3}. \tag{22.22}$$

From (22.21), (22.22) and Minkovski's inequality we obtain

$$\left\| u_0(t + h) - u_0(t) \right\| \leqq \left\| u_0(t + h) - u_n(t + h) \right\| + \left\| u_0(t) - u_n(t) \right\|$$

$$+ \left\| u_n(t + h) - u_n(t) \right\| < \frac{\varepsilon}{3} + \frac{\varepsilon}{3} + \frac{\varepsilon}{3} = \varepsilon.$$

Thus for $|h| < \eta$, $\| u_0(t + h) - u_0(t) \| < \varepsilon$, and the theorem is proved.

We shall say that the function u, whose square is integrable in the domain Ω and which satisfies

$$\lim_{n \to \infty} \iiint_{\Omega} (u_n - u)^2 \, \mathrm{d}x \, \mathrm{d}y \, \mathrm{d}z = 0$$

is the *limit in the mean* of the sequence u_n.

We define the *generalized solution* of equation (22.7) subject to the conditions (22.15) and (22.16) to be the function u which is the limit in the mean of the sequence of functions u_n which satisfy the equations

$$\nabla^2 u_n - \frac{\partial^2 u_n}{\partial t^2} = F_n$$

and the conditions (22.15) and (22.16), where the sequence F_n converges in the mean to F.

An important result follows from the inequalities established in Theorem 3: if the sequence $F_n(x, y, z, t)$ converges in the mean uniformly with respect to t, then the sequence of solutions $u_n(x, y, z, t)$ also converges in the mean uniformly with respect to t.

Just as we did earlier (see, for example, Lecture 18), we shall regard two functions $u^{(1)}$ and $u^{(2)}$ as *equivalent* if

$$\iiint_{\Omega} (u^{(1)} - u^{(2)})^2 \, \mathrm{d}x \, \mathrm{d}y \, \mathrm{d}z = 0.$$

This means that the two functions can differ from each other only over a set of measure zero (see Theorem 21, Lecture 6).

The generalized solution is unique, for the sequence u_n cannot have two limits in the mean; otherwise we should have for the two different limits the inequality

$$\iiint_{\Omega} (u^{(1)} - u^{(2)})^2 \, \mathrm{d}x \, \mathrm{d}y \, \mathrm{d}z \leqq \iiint_{\Omega} [(u^{(1)} - u_n) + (u_n - u^{(2)})]^2 \, \mathrm{d}x \, \mathrm{d}y \, \mathrm{d}z$$

$$\leqq 2 \iiint_{\Omega} (u^{(1)} - u_n)^2 \, \mathrm{d}x \, \mathrm{d}y \, \mathrm{d}z + 2 \iiint_{\Omega} (u^{(2)} - u_n)^2 \, \mathrm{d}x \, \mathrm{d}y \, \mathrm{d}z \leqq 4\varepsilon,$$

so that $u^{(1)}$ and $u^{(2)}$ must coincide.

The existence of the generalized solution can be established by using a theorem in the theory of functions of a real variable, known as the *Riesz–Fischer Theorem*. This theorem states that:

If the integral of u_i^2 exists for all i, and if the sequence of functions u_1, u_2, ..., u_n, ... has the property that, for all sufficiently large m and n,

$$\iiint_{\Omega} (u_m - u_n)^2 \ dx \ dy \ dz < \varepsilon,$$

then the sequence has a limit in the mean, i.e., there is a function u such that

$$\lim_{n \to \infty} \iiint_{\Omega} (u_n - u)^2 \ dx \ dy \ dz = 0.$$

We shall prove this theorem at the end of this lecture.

We now show how the Riesz–Fischer theorem can be used to prove the existence of the generalized solution of the wave equation. Let $u_n(x, y, z, t)$ be a solution of equation (22.7) satisfying the conditions (22.15) and (22.16). Consider the expression

$$\left\| u_m(t) - u_n(t) \right\|.$$

The function $v_{m,n} \equiv u_m(t) - u_n(t)$ is a solution of the equation

$$\Delta v_{m,n} - \frac{\partial^2 v_{m,n}}{\partial t^2} = F_m - F_n.$$

By virtue of the uniform convergence in the mean of the function F_n to F we shall have $\|F_m(t) - F_n(t)\| < \eta$ for sufficiently large m and n and for any $\eta > 0$. As we have seen, it follows that $\|v_{m,n}\| < \varepsilon$, where ε is an arbitrary positive number. By the Riesz–Fischer theorem, the sequence $u_n(x, y, z, t)$ converges uniformly in the mean to a limit function, which, by Theorem 4, will be continuous in the mean, as was to be shown.

In the examples considered in the previous sections it happened that the generalized solutions of Laplace's equation, Poisson's equation, and the equation for heat conduction, had continuous first-order derivatives. As we shall soon see this circumstance is not fortuitous.

In contrast to these problems, the generalized solution of the wave equation and its first-order derivatives may be discontinuous. It would take too much time to explain the circumstances in which this can happen, and we shall not deal with the question here.

To conclude this section we give a simple example.

EXAMPLE 2. We take as the domain Ω a sphere of radius 1. Let $\psi(\xi)$ be a certain function which is specified in the interval $-\infty < \xi < \infty$. Consider the function

$$u_0 = \frac{\psi(t + r) - \psi(t - r)}{r} \qquad (22.23)$$

where r is the distance of the variable point from the origin. If $r \neq 0$, the function u_0 has the same number of derivatives as the function ψ. If $r \neq 0$, u_0 has derivatives of order one less than the order of the highest derivatives that ψ has. It is not difficult to establish this by differentiation of the derivatives with respect to the space coordinates x, y, z, and for the derivatives with respect to time it is immediately obvious. For,

$$\frac{\partial^k u}{\partial t^k} = \frac{1}{r} [\psi^{(k)}(t + r) - \psi^{(k)}(t - r)].$$

This function is determinate for $r = 0$ only if

$$\lim_{r \to 0} \frac{1}{r} [\psi^{(k)} (t + r) - \psi^{(k)}(t - r)] = 2\psi^{(k+1)} (t)$$

exists. Consequently the necessary (and sufficient) condition for the existence of the derivative $\partial^k u / \partial t^k$ at $r = 0$ is that the continuous derivative $\psi^{(k+1)}$ shall exist. It is easy to see by differentiation that u_0 is a solution of the equation

$$\nabla^2 u_0 - \frac{\partial^2 u_0}{\partial t^2} = 0,$$

if ψ has continuous third-order derivatives. Suppose now that the function ψ from (22, 23) is differentiable only once or not at all; then it can be shown that u_0 is still the generalized solution.

The sequence of solutions

$$u_n = \frac{\psi_n(r + t) - \psi_n(r - t)}{r}$$

will tend to this function u_0 if the sequence of thrice-differentiable functions ψ_n tends to ψ.

If at some point $\xi = t_0$ the function $\psi(\xi)$ has no derivative, then u_0 will have a discontinuity at the point $r = 0$, $t = t_0$. In this case there will be no function u satisfying the conditions

$$[u]_S = [u_0]_S, \quad [u]_{t=0} = [u_0]_{t=0}, \quad \left[\frac{\partial u}{\partial t}\right]_{t=0} = \left[\frac{\partial u_0}{\partial t}\right]_{t=0}$$

and the equation

$$\nabla^2 u - \frac{\partial^2 u}{\partial t^2} = 0.$$

For, if such a solution existed, then, since a solution depends continuously on the initial conditions, the sequence u_n would have to converge towards this solution, and this is impossible because the sequence converges to u_0.

There is another approach to the generalized solutions by means of which they can be defined immediately without having recourse to a passage to the limit.

If u_n is some solution of the equation $Lu_n = F_n$, and if ψ is a completely arbitrary function having continuous derivatives up to the second order and which is different from zero only in a certain interior part σ of the domain Ω, then

$$\iiint_\Omega (\psi F_n - u_n M\psi)\, d\Omega = 0.$$

The terms outside the integral drop out since ψ and its derivatives vanish outside σ.

We now require to use an important inequality known as *Bunyakovski's inequality (Schwarz's inequality)*. Let φ_1, φ_2 be two functions of n variables $x_1, x_2, \ldots, x_n$ which are defined in an open set Ω. If the integrals

$$\int_\Omega \varphi_1^2\, dv \quad \text{and} \quad \int_\Omega \varphi_2^2\, dv$$

exist, then the integral

$$\int_\Omega \varphi_1\varphi_2\, dv$$

also exists and

$$\left\{\int_\Omega \varphi_1\varphi_2\, dv\right\}^2 \leqq \left\{\int_\Omega \varphi_1^2\, dv\right\}\left\{\int_\Omega \varphi_2^2\, dv\right\}.$$

We shall prove this inequality later.

Minkovski's and Bunyakovski's inequalities give, for any generalized solution of equation (22.7), for any $\varepsilon > 0$ and for sufficiently large n,

$$\left|\iiint_\Omega (\psi F - uM\psi)\, d\Omega\right| = \left|\iiint_\Omega [\psi(F - F_n) - (u - u_n)\, M\psi]\, d\Omega\right|$$

$$\leqq \sqrt{\iiint_\Omega \psi^2\, d\Omega \cdot \iiint_\Omega (F - F_n)^2\, d\Omega}$$

$$+ \sqrt{\iiint_\Omega (u - u_n)^2\, d\Omega \cdot \iiint_\Omega (M\psi)^2\, d\Omega} \leqq \varepsilon.$$

This means that for any generalized solution, the integral equation

$$\iiint_\Omega uM\psi\, d\Omega = \iiint_\Omega \psi F\, d\Omega \qquad (22.24)$$

holds for any function ψ. The differential equation may be replaced by this integral relation. We can say that the generalized solution of the equation

$$Lu = F$$

is any function which satisfies the integral relation (22.24) where ψ is any function which has continuous derivatives up to the second order and which is different from zero only in a certain internal part σ of the domain Ω.

If we recall the proof of the existence of the solution which we obtained by Kirchhoff's method, we see that the formula (22.24) was an important link in the argument. We have proved, in this way, the existence only of the generalized solution. Now from (22.24), supposing that u has derivatives, and integrating by parts, we deduce the following formula:

$$\iiint_\Omega \psi(Lu - F)\, d\Omega = 0, \tag{22.25}$$

and hence conclude that u is a solution of the equation $Lu = F$. This last step can be taken only if u is a solution of the problem in the classical sense of the word.

§ 5. A Property of Generalized Solutions of Homogeneous Equations

The following general theorem holds.

THEOREM 5. *For Laplace's equation, the equation $\nabla^2 u + \lambda u = 0$, and the homogeneous equation for heat conduction, any generalized solution in the sense of the integral relation (22.24) is necessarily differentiable as often as we please and is a solution in the ordinary sense.*

This property sharply distinguishes equations of elliptic or parabolic type from equations of hyperbolic type, for which this property does not hold.

We shall prove the theorem first for the equation of heat conduction.

We first construct certain auxiliary functions. We define a function $\Psi(\xi)$ by the formulae

$$\Psi(\xi) = \begin{cases} 0 & \xi \leq \tfrac{1}{4}, \\ \left\{ 1 + \exp\left[\dfrac{-\xi + \tfrac{1}{2}}{(\xi - \tfrac{1}{4})(1 - \xi)} \right] \right\}^{-1}, & \tfrac{1}{4} < \xi < 1, \\ 1 & \xi \geq 1. \end{cases}$$

The function $\Psi(\xi)$ has certain obvious properties:

(a) It is continuous in the interval $0 \leq \xi \leq \infty$.

For, the fraction $(-\xi + 1/2)/[(\xi - 1/4)(1 - \xi)]$ attains its limit $+\infty$

when $\xi \to (1/4) + 0$ and its limit $-\infty$ when $\xi \to 1 - 0$: consequently,

$$\lim_{\xi \to 1/4} \exp\left[\frac{-\xi + \frac{1}{2}}{(\xi - \frac{1}{4})(1 - \xi)}\right] = \infty, \quad \lim_{\xi \to 1} \exp\left[\frac{-\xi + \frac{1}{2}}{(\xi - \frac{1}{4})(1 - \xi)}\right] = 0$$

and the continuity of $\Psi(\xi)$ follows.

(b) The function $\Psi(\xi)$ has continuous derivatives of all orders.

Proof. It is sufficient to show that the limit values of the derivatives of $\Psi(\xi)$ of any order when $\xi \to 1/4$ or $\xi \to 1$ are zero. We have

$$\Psi'(\xi) = \frac{-\exp\left[\dfrac{-\xi + \frac{1}{2}}{(\xi - \frac{1}{4})(1 - \xi)}\right]}{\left(1 + \exp\left[\dfrac{-\xi + \frac{1}{2}}{(\xi - \frac{1}{4})(1 - \xi)}\right]\right)^2} \frac{d}{d\xi}\left[\frac{-\xi + \frac{1}{2}}{(\xi - \frac{1}{4})(1 - \xi)}\right]$$

$$= \frac{\dfrac{d}{d\xi}\left[\dfrac{\xi - \frac{1}{2}}{(\xi - \frac{1}{4})(1 - \xi)}\right]}{\left(1 + \exp\left[\dfrac{-\xi + \frac{1}{2}}{(\xi - \frac{1}{4})(1 - \xi)}\right]\right)\left(1 + \exp\left[\dfrac{\xi - \frac{1}{2}}{(\xi - \frac{1}{4})(1 - \xi)}\right]\right)}.$$

The expression $\exp\left[\dfrac{-\xi + \frac{1}{2}}{(\xi - \frac{1}{4})(1 - \xi)}\right]$ tends to infinity, when $\xi \to \frac{1}{4}$, faster than any rational function of ξ, and the expression $\exp\left[\dfrac{\xi - \frac{1}{2}}{(\xi - \frac{1}{4})(1 - \xi)}\right]$ also tends to infinity when $\xi \to 1$ faster than any rational function of ξ. Consequently,

$$\Psi'(\tfrac{1}{4} + 0) = 0; \quad \Psi'(1 - 0) = 0. \tag{22.26}$$

Any derivative $\Psi^m(\xi)$ may be expressed in the form

$$\Psi^m(\xi) = \sum_{\substack{p \geq 1 \\ q \geq 1}} \frac{R_{m, p, q}(\xi)}{\left(1 + \exp\left[\dfrac{-\xi + \frac{1}{2}}{(\xi - \frac{1}{4})(1 - \xi)}\right]\right)^p \left(1 + \exp\left[\dfrac{\xi - \frac{1}{2}}{(\xi - \frac{1}{4})(1 - \xi)}\right]\right)^q}$$

$$\tag{22.27}$$

where $R_{m,p,q}(\xi)$ are certain rational functions.
This formula may be proved by induction.

(22.27) shows that

$$\Psi^m(\tfrac{1}{4} + 0) = 0, \quad \Psi^m(1 - 0) = 0 \tag{22.28}$$

as was to be proved.

We now bring into consideration the function

$$w_n(x - x_0, y - y_0, z - z_0, t - t_0)$$

$$= \begin{cases} 0 & , \quad t_0 \leqq t \\ \dfrac{-1}{8\pi^{\frac{3}{2}}(t - t_0)^{\frac{3}{2}}} \left(\exp\left[\dfrac{-r^2}{4(t_0 - t)} \right] \right) \Psi[n^2(r^2 + t_0 - t)], & t < t_0 \end{cases}$$

where

$$r = \sqrt{(x - x_0)^2 + (y - y_0)^2 + (z - z_0)^2}, \quad 0 \leqq r < +\infty.$$

We note one or two properties of the function w_n.

(a) It is clear that

$$w_n(x - x_0, y - y_0, z - z_0, t_0 - t)$$
$$= n^3 w_1[n(x - x_0), n(y - y_0), n(z - z_0), n^2(t_0 - t)]. \qquad (22.29)$$

(b) We form the expression

$$\nabla^2 w_n + \frac{\partial w_n}{\partial t} = \Phi_n(x - x_0, y - y_0, z - z_0, t_0 - t).$$

Then it follows from (22.29) that

$$\Phi_n(x - x_0, y - y_0, z - z_0, t_0 - t)$$
$$= n^5 \Phi_1[n(x - x_0), n(y - y_0), n(z - z_0), n^2(t_0 - t)]. \qquad (22.30)$$

(c) The function Φ_n is different from zero only in the domain D_n:

$$\frac{1}{4n^2} \leqq r^2 + t_0 - t \leqq \frac{1}{n^2}; \quad t < t_0,$$

which shrinks to the point x_0, y_0, z_0 as $n \to \infty$.

For, in the domain where $\Psi[n^2(r^2 + t_0 - t)]$ is equal to unity, *i.e.*, where $r^2 + t_0 - t \geqq 1/n^2$, we have $w_n = -v$, where v is a particular solution of the equation $\nabla^2 v + \partial v/\partial t = 0$ (see Lecture 8).
That is, $\nabla^2 w_n + \partial w_n/\partial t = 0$ for $r^2 + t_0 - t \geqq 1/n^2$.
That Φ_n vanishes for $r^2 + t_0 - t \leqq 1/4n^2$ is obvious.

(d) The following formula holds:

$$\iiiint_{D_n} \Phi_n \, dx \, dy \, dz \, dt = \int_{t_0 - 1/n^2}^{t_0} \left\{ \iiint_{-\infty}^{+\infty} \Phi_n \, dx \, dy \, dz \right\} dt = 1. \qquad (22.31)$$

For, by Green's formula (21.4) we have

$$\int_{t_0-1/n^2}^{t_0} \left\{ \iiint_{-\infty}^{+\infty} \Phi_n \, dx \, dy \, dz \right\} dt$$

$$= \int_{t_0-1/n^2}^{t_0} \left\{ \iiint_{-\infty}^{+\infty} \left(\nabla^2 w_n + \frac{\partial w_n}{\partial t} \right) dx \, dy \, dz \right\} dt$$

$$= \iiint_{-\infty}^{+\infty} \left[w_n \right]_{t_0-1/n^2}^{t} dx \, dy \, dz = \iiint_{-\infty}^{+\infty} [v]_{t=t_0-1/n^2} \, dx \, dy \, dz,$$

where v is a principal solution of the equation of heat conduction, as defined in Lecture 8. The last integral, as was shown in Lemma 2 of that lecture, is equal to unity, and this proves (22.31).

(e) Let $f(x, y, z, t)$ be an arbitrary continuous function in the domain Ω of the variables x, y, z, t.

We select a domain Ω_n of values of x_0, y_0, z_0, t_0 so that for points x_0, y_0, z_0, t_0 of Ω_n the domain D_n lies wholly within Ω. We construct in Ω_n the function

$$f_n(x_0, y_0, z_0, t_0)$$

$$= \iiiint_\Omega \Phi_n(x - x_0, y - y_0, z - z_0, t - t_0) f(x, y, z, t) \, dx \, dy \, dz \, dt.$$

Then the sequence of functions $f_n(x_0, y_0, z_0, t_0)$ converges to the function $f(x_0, y_0, z_0, t_0)$, and the convergence is uniform in any domain interior to Ω.

We write the function f in the form

$$f(x, y, z, t) = f(x_0, y_0, z_0, t_0) + \eta.$$

It is evident that in the domain D_n for sufficiently large n we shall have $|\eta| < \varepsilon$ because f is continuous. We write

$$\int_{-\infty}^{t_0} \left\{ \iiint_{-\infty}^{+\infty} |\Phi_1| \, dx \, dy \, dz \right\} dt = M.$$

By virtue of formula (22.30)

$$\int_{-\infty}^{t_0} \left\{ \iiint_{-\infty}^{+\infty} |\Phi_n| \, dx \, dy \, dz \right\} dt = \int_{-\infty}^{t_0} \left\{ \iiint_{-\infty}^{+\infty} |\Phi_1| \, dx \, dy \, dz \right\} dt = M.$$

We have:

$$f_n(x_0, y_0, z_0, t_0) = \iiiint_\Omega \Phi_n f \, dx \, dy \, dz \, dt$$

$$= \iiiint_\Omega \Phi_n f(x_0, y_0, z_0, t_0) \, dx \, dy \, dz \, dt$$

$$+ \iiiint_\Omega \Phi_n \eta \, dx \, dy \, dz \, dt$$

$$= f(x_0, y_0, z_0, t_0) + \iiiint_{D_n} \Phi_n \eta \, dx \, dy \, dz \, dt$$

whence

$$|f_n(x_0, y_0, z_0, t_0) - f(x_0, y_0, z_0, t_0)| \leqq \varepsilon M,$$

as was to be shown.

(f) The function f_n which we have constructed will be differentiable without limit. This follows because Φ_n is differentiable without limit.

After these remarks we can now prove our Theorem 5.

Let u be some generalized solution of the equation of heat conduction. We set up $u_n(x_0, y_0, z_0, t_0)$. Then all the $u_n(x_0, y_0, z_0, t_0)$ for any values of n are equal and do not depend on n. For,

$$u_n(x_0, y_0, z_0, t_0) - u_{n+p}(x_0, y_0, z_0, t_0)$$

$$= \iiiint_\Omega u(x, y, z, t)\,(\Phi_n - \Phi_{n+p}) \, dx \, dy \, dz \, dt$$

$$= \iiiint_\Omega u(x, y, z, t)\left[\nabla^2(w_n - w_{n+p}) + \frac{\partial}{\partial t}(w - w_{n+p})\right] \times dx \, dy \, dz \, dt.$$

But $w_n - w_{n+p} = \psi$ is different from zero only in the domain

$$\frac{1}{4(n + p)^2} \leqq r^2 + t_0 - t \leqq \frac{1}{n^2},$$

since if $r^2 + t_0 - t \geqq 1/n^2$ the function w_n coincides with w_{n+p}.

The function ψ satisfies all the conditions for formula (22.24) to be applicable. In this case we can take

$$M = \nabla^2 + \frac{\partial}{\partial t}$$

$$L = \nabla^2 - \frac{\partial}{\partial t},$$

whence, noting that our equation has the form $Lu = 0$, $i.e.$, $F = 0$, we have

$$u_n - u_{n+p} = \iiiint_{\Omega} \psi F \, dx \, dy \, dz \, dt = 0.$$

Hence it follows that $u = u_n$; but u_n has an unlimited number of derivatives. Consequently, u also may be differentiated any number of times. Hence the theorem.

Suppose now a function satisfies the equation $\nabla^2 u + \lambda u = 0$. We put

$$v = e^{-\lambda t} u.$$

Then

$$\nabla^2 v = e^{-\lambda t} \nabla^2 u = -\lambda e^{-\lambda t} u.$$

Consequently,

$$\nabla^2 v = \frac{\partial v}{\partial t}.$$

By what we have just proved, the function v will have an unlimited number of derivatives. Consequently, the function u will also have the same property, as was to be shown.

It can also be shown that all solutions of the equation $\nabla^2 u + \lambda u = 0$ will, in fact, be analytical throughout the domain Ω, in contrast to solutions of the equation of heat conduction, for which this property does not hold.

§ 6. Bunyakovski's Inequality and Minkovski's Inequality

Let $\varrho(P)$ be a non-negative function, which we shall call the *weight*.

Let $\varphi(P)$ and $\psi(P)$ be two functions, specified in the domain Ω, and such that the product of the weight $\varrho(P)$ and the squares of their moduli are integrable, $i.e.$,

$$\int_{\Omega} |\varphi(P)|^2 \varrho(P) \, dP < \infty$$

$$\int_{\Omega} |\psi(P)|^2 \varrho(P) \, dP < \infty. \tag{22.32}$$

Then the following inequality holds:

$$\left| \int_{\Omega} \varphi(P) \, \psi(P) \, \varrho(P) \, dP \right|^2 \le \int_{\Omega} |\varphi(P)|^2 \varrho(P) \, dP \int_{\Omega} |\psi(P)|^2 \varrho(P) \, dP. \tag{22.33}$$

It is clearly sufficient to prove the theorem only for the case when $\psi(P)$ and $\varphi(P)$ are real, non-negative functions, since

$$\left| \int_{\Omega} \varphi(P) \, \psi(P) \, \varrho(P) \, dP \right| \le \int_{\Omega} |\varphi(P)| \, |\psi(P)| \, \varrho(P) \, dP. \tag{22.34}$$

We have, for any two non-negative functions whatsoever,

$$\varphi\psi \leqq \varphi^2/2 + \psi^2/2.$$

Consequently the integral $\int_\Omega \varphi(P)\psi(P)\varrho(P)\,\mathrm{d}P$ is meaningful.

Consider the following (also meaningful) integral:

$$\int_\Omega (\varphi - \lambda\psi)^2 \,\varrho\, \mathrm{d}P = \int_\Omega \varphi^2\varrho\, \mathrm{d}P - 2\lambda \int_\Omega \varphi\psi\varrho\, \mathrm{d}P + \lambda^2 \int_\Omega \psi^2\varrho\, \mathrm{d}P$$

$$= a\lambda^2 - 2b\lambda + c. \tag{22.35}$$

The parabola defined by $y = a\lambda^2 - 2b\lambda + c$ in the plane of the variables y and λ cannot have a single point below the λ-axis, since the magnitude of (22.35) is greater than or equal to zero. This implies that the equation

$$a\lambda^2 - 2b\lambda + c = 0$$

cannot have different real roots, and it can have a repeated root only when there is a λ_0 such that

$$\int (\varphi - \lambda_0\psi)^2 \,\varrho\, \mathrm{d}P = 0,$$

i.e., when the expression $(\varphi - \lambda_0\psi)\varrho$ is equivalent to zero,
i.e., when it is different from zero only on a set of measure zero.

In the case when the weight-function $\varrho(P)$ vanishes, if at all, only on a set of measure zero, this means that the expression $\varphi - \lambda_0\psi$ is equivalent to zero.

Consequently,

$$b^2 \leqq ac,$$

and the equality sign can hold only when φ and ψ are proportional. This proves Bunyakovski's inequality (22.33).

From Bunyakovski's inequality it follows in an obvious manner that, for any functions φ and ψ such that the products of their squares with the weight-function are each integrable,

$$\left| \int_\Omega (\varphi^2 + 2\varphi\psi + \psi^2)\varrho\, \mathrm{d}P \right|$$

$$\leqq \int_\Omega |\varphi|^2\varrho\, \mathrm{d}P + \int_\Omega |\psi|^2\varrho\, \mathrm{d}P + 2\sqrt{\int_\Omega |\varphi|^2\varrho\, \mathrm{d}P \cdot \int_\Omega |\psi|^2\varrho\, \mathrm{d}P}$$

$$= \left[\sqrt{\int_\Omega |\varphi|^2\varrho\, \mathrm{d}P} + \sqrt{\int_\Omega |\psi|^2\varrho\, \mathrm{d}P} \right]^2,$$

or

$$\sqrt{\left|\int_{\Omega} (\varphi + \psi)^2 \varrho \, dP\right|} \leqq \sqrt{\int_{\Omega} |\varphi|^2 \varrho \, dP} + \sqrt{\int_{\Omega} |\psi|^2 \varrho \, dP}. \quad (22.36)$$

(22.36) is *Minkovski's Inequality*. The left-hand side exists if the right-hand side exists.

§ 7. The Riesz–Fischer Theorem

The Riesz–Fischer Theorem. If $\varphi_1, \varphi_2, \ldots, \varphi_k, \ldots$ *is a sequence of functions such that* φ_k^2 *is integrable in a bounded domain* Ω *for all* k, *and if, given any* ε, *a number* $N(\varepsilon)$ *can be found such that*

$$\int_{\Omega} (\varphi_k - \varphi_s)^2 \, dv < \varepsilon \quad if \quad k > N(\varepsilon), \quad s > N(\varepsilon), \quad (22.37)$$

then there is a function φ_0 *such that* φ_0^2 *is integrable in* Ω *and*

$$\lim_{k \to \infty} \int_{\Omega} (\varphi_k - \varphi_0)^2 \, dv = 0 \quad (22.38)$$

It is clear that such a function would be "unique with accuracy up to equivalence"; that is to say, if there were two such functions, they would necessarily be *equivalent*. For, if there were two such limit functions, say φ_0 and φ_*, then we should have by Minkovski's inequality

$$\sqrt{\int_{\Omega} (\varphi_0 - \varphi_*)^2 \, dv} \leqq \sqrt{\int_{\Omega} (\varphi_0 - \varphi_n)^2 \, dv} + \sqrt{\int_{\Omega} (\varphi_n - \varphi_*)^2 \, dv} < \varepsilon$$

for an arbitrary small $\varepsilon > 0$, and hence

$$\int_{\Omega} (\varphi_0 - \varphi_*)^2 \, dv = 0.$$

We shall call φ_0 *the limit in the mean square for the sequence* φ_k.

We now prove the theorem. We first show that the sequence φ_k converges in the mean, *i.e.*, that there is a summable function φ_0 such that

$$\lim_{k \to \infty} \int_{\Omega} |\varphi_0 - \varphi_k| \, dv = 0$$

To do this, we note that by Bunyakovski's inequality

$$\int_{\Omega} |\varphi_m - \varphi_p| \, d\Omega = \int_{\Omega} |\varphi_m - \varphi_p| . 1 . d\Omega$$

$$\leqq \left[\int_{\Omega} |\varphi_m - \varphi_p|^2 \, d\Omega\right]^{\frac{1}{2}} \left[\int_{\Omega} 1 . d\Omega\right]^{\frac{1}{2}}$$

and consequently

$$\int_{\Omega} |\varphi_m - \varphi_p|\, d\,\Omega < \eta,$$

provided only that

$$m > N(\eta), \quad p > N(\eta).$$

Applying Theorem 23 of Lecture 6, we see that a summable function φ_0 exists which is the limit in the mean of the sequence φ_k.

It follows from the above method of proof that there is a sequence φ_{k_i}, belonging to the sequence φ_k, which converges to φ_0 almost everywhere and which does so uniformly in the closed set F_δ on which all the functions are continuous. The set F_δ can be chosen so that its measure is as close as we please to that of Ω. We have

$$\int_{F_\delta} \varphi_0^2\, dv = \lim_{i \to \infty} \int_{F_\delta} \varphi_{k_i}^2\, dv \leqq A < \infty$$

whence

$$\int_{\Omega} \varphi_0^2\, dv \leqq A$$

and the integral on the left-hand side exists. It is not difficult to see that the magnitude of

$$\int_{\Omega} (\varphi_0 - \varphi_{k_i})^2\, d\Omega$$

can be made arbitrarily small by taking i sufficiently large. For brevity, we write $\psi_i = \varphi_{k_i}$. Then for any closed set F_δ and for any given positive number ε, we have, for a sufficiently large l (which will depend, in general, on F_δ):

$$\int_{F_\delta} (\varphi_0 - \psi_i)^2\, d\Omega = \int_{F_\delta} [(\psi_l - \psi_i) + (\varphi_0 - \psi_l)]^2\, d\Omega$$

$$\leqq 2\int_{F_\delta} (\psi_l - \psi_i)^2\, d\Omega + 2\int_{F_\delta} (\varphi_0 - \psi_l)^2\, d\Omega$$

$$\leqq 2\int_{\Omega} (\psi_l - \psi_i)^2\, d\Omega + \varepsilon.$$

By the hypothesis, for a sufficiently large l and i, which are independent of F_δ

$$\int_{\Omega} (\psi_l - \psi_i)^2\, d\Omega \leqq \varepsilon.$$

Hence

$$\int_{F_\delta} (\varphi_0 - \psi_i)^2 \, d\Omega \leqq 3\varepsilon$$

and consequently

$$\int_\Omega (\varphi_0 - \psi_i)^2 \, d\Omega \leqq 3\varepsilon.$$

We now show that φ_0 is also the limit in the mean square for the original sequence φ_k.

For s, $k_i > N(\varepsilon)$ we have:

$$\int_\Omega (\varphi_0 - \varphi_s)^2 \, d\Omega \leqq 2 \int_\Omega (\varphi_0 - \psi_i)^2 \, d\Omega + 2 \int_\Omega (\psi_i - \varphi_s)^2 \, d\Omega$$

$$\leqq 6\varepsilon + 2\varepsilon = 8\varepsilon.$$

Hence the theorem.

FOURIER'S METHOD

§ 1. Separation of the Variables

Boundary-value problems in mathematical physics for equations of para-bolic and hyperbolic type can conveniently be solved by a method put for-ward by Fourier which we shall refer to as *separation of the variables*.

We shall illustrate the essence of the method by means of particular ex-amples. The reader who has mastered the arguments set out in the previous lectures will have no difficulty in understanding immediately how and in what circumstances the Fourier method enables the solution of a problem to be found.

Suppose a solution is required of the equation

$$\nabla^2 u = \frac{1}{a} \frac{\partial u}{\partial t} \tag{23.1}$$

in a domain Ω of the space x, y, z bounded by a surface S, and for t satis-fying $0 \leq t \leq T$, the solution being subject to the conditions

$$[u]_S = 0 \tag{23.2}$$

and

$$[u]_{t=0} = \varphi(x, y, z). \tag{23.3}$$

For the time being we shall disregard the condition (23.3) and shall try to find particular solutions of equation (23.1) which satisfy condition (23.2). We seek these solutions in the form of a product of two functions

$$u = U(x, y, z) \, T(t) \tag{23.4}$$

Substituting (23.4) into (23.1) and dividing both parts by u, we get

$$\frac{\nabla^2 U(x, y, z)}{U(x, y, z)} = \frac{1}{a} \frac{T'(t)}{T(t)}. \tag{23.5}$$

In equation (23.5) the variables x, y, z and t are separated; the left-hand side does not depend on t nor the right-hand side on x, y, z. The equality is possible only if both the left-hand side and the right-hand side are equal to one and the same constant, say $-\lambda$:

$$\frac{\nabla^2 U}{U} = -\lambda, \quad \frac{1}{a} \frac{T'(t)}{T(t)} = -\lambda, \tag{23.6}$$

or

$$\nabla^2 U + \lambda U = 0 \tag{23.7}$$

$$T(t) = C e^{-\lambda a t}. \tag{23.8}$$

In order that our solution shall satisfy the condition (23.2) it is necessary for the function $U(x, y, z)$ to satisfy this condition. The values of λ for which the equation (23.7) has a solution satisfying the boundary condition (23.2) are called the *eigenvalues* (or *characteristic values*) of the boundary-value problem for this equation subject to the condition (23.2). By no means every value of λ is an eigenvalue. For, transferring the term λU in (23.7) to the right-hand side and considering it as a free member, we have, by applying formula (21.7)

$$U(P_0) = \frac{\lambda}{4\pi} \iiint_{\Omega} G(P, P_0)\, U(P)\, \mathrm{d}P \tag{23.9}$$

where $G(P, P_0)$ is Green's function for the Laplacian in the domain Ω.

Equation (23.9) is a linear, homogeneous, integral equation of the second Fredholm type with a kernel satisfying the requirements of § 7 of Lecture 18. By Fredholm's fourth theorem, it can have a non-zero solution only for certain discrete values of λ. Let these values of λ be $\lambda_1, \lambda_2, ..., \lambda_n, ...$ and let $U_1, U_2, ..., U_n, ...$ be the corresponding solutions of equation (23.9); these functions are called *eigenfunctions* (or *characteristic functions*). Then we have a whole set of particular solutions of equation (23.1) of the required type:

$$u_i = U_i\, e^{-\lambda_i a t}.$$

We may mention that the kernel of equation (23.9) is a symmetric function of the coordinates of the points P and P_0.

We shall prove shortly (in Lecture 24), in the theory of integral equations with symmetric kernels, that there are infinitely many such solutions, and that for any function φ whose square is integrable a series

$$u = \sum_{i=1}^{\infty} a_i U_i e^{-\lambda_i a t} \tag{23.10}$$

may be constructed which will satisfy the conditions (23.2) and (23.3) and which in the mean will form a generalized solution of equation (23.1). It follows from Theorem 5 of Lecture 22 that any generalized solution of (23.1) is a solution in the usual sense of the word.

We point out, for the moment without proof, five important properties of the system of functions $U_i(x, y, z)$ and the numbers λ_i:

1. Of the infinite number of numbers λ_i, all are real and positive. (In certain other problems, this circumstance becomes: only a finite number of the λ_i are negative.)

2. The set of functions U_i is orthogonal and can be normalized, *i.e.*,

$$\iiint_\Omega U_i(x, y, z)\, U_j(x, y, z)\, \mathrm{d}x\, \mathrm{d}y\, \mathrm{d}z = \begin{cases} 1, & i = i \\ 0, & i \neq j. \end{cases}$$

3. The functions U_i form a so-called *complete* system, *i.e.*, any continuous function $\varphi(x, y, z)$ can be expressed as a series

$$\varphi(x, y, z) = \sum_{i=1}^{\infty} a_i U_i(x, y, z) \tag{23.11}$$

which converges in the mean, where the numbers a_i are the so-called *Fourier coefficients* for the function φ. If, further, φ satisfies the condition

$$[\varphi]_S = 0$$

and has second-order derivatives which are continuous everywhere, including the boundary, then the series (23.11) converges uniformly.

4. In addition to the conditions for orthogonality, written above, the following equations hold:

$$\iiint_\Omega \left\{ \frac{\partial U_i}{\partial x} \frac{\partial U_j}{\partial x} + \frac{\partial U_i}{\partial y} \frac{\partial U_j}{\partial y} + \frac{\partial U_i}{\partial z} \frac{\partial U_j}{\partial z} \right\} \mathrm{d}x\, \mathrm{d}y\, \mathrm{d}z = \begin{cases} \lambda_i, & i = j \\ 0, & i \neq j \end{cases}.$$

5. If a function φ which is continuously twice-differentiable satisfies the condition (23.2), then the series (23.11) not only converges uniformly to φ, but the series obtained from it by termwise differentiation also converges in the mean to the corresponding derivative of φ. In other words, if we put

$$\varphi_N = \sum_{i=1}^{N} a_i U_i.$$

then the integral

$$\iiint_\Omega \left\{ \left(\frac{\partial(\varphi - \varphi_N)}{\partial x} \right)^2 + \left(\frac{\partial(\varphi - \varphi_N)}{\partial y} \right)^2 + \left(\frac{\partial(\varphi - \varphi_N)}{\partial z} \right)^2 \right\} \mathrm{d}x\, \mathrm{d}y\, \mathrm{d}z$$

tends to zero.

Consider now the problem of finding the Fourier coefficients for a function $\varphi(x, y, z)$. If we multiply both sides of (23.11) by U_j and integrate over the domain Ω, integrating the series term by term, we get

$$\iiint_{\Omega} U_j(x, y, z)\, \varphi(x, y, z)\, dx\, dy\, dz$$

$$= \sum_{i=1}^{\infty} \left[a_i \iiint_{\Omega} U_i(x, y, z)\, U_j(x, y, z)\, dx\, dy\, dz \right] = a_j.$$

$$(23.12)$$

Formula (23.12) gives us, in fact, the value of the Fourier coefficient a_j.

In order that the function expressed in the form of the series (23.10) shall satisfy the condition (23.3), we naturally require equation (23.11) to be satisfied. If the sum of the series (23.10) is continuous for $t \geqq 0$, then the conditions (23.2) and (23.3) will be satisfied for the values of a_i which are the Fourier coefficients for the function φ.

It is not difficult to establish that the series (23.10) converges uniformly and gives the generalized solution of equation (23.11). For, let

$$\varphi_N = \sum_{i=1}^{N} a_i U_i(x, y, z).$$

It is clear that

$$u_N = \sum_{i=1}^{N} a_i U_i(x, y, z)\, e^{-\lambda_i a t}$$

gives the solution of equation (23.1) subject to the conditions (23.2) and

$$[u_N]_{t=0} = \varphi_N.$$

But we proved in the last lecture that under these conditions it follows from the convergence of the sequence φ_N that the sequence U_N converges uniformly in any finite interval of the time variable to the generalized solution u; and this is what we had to show.

In exactly the same way the problem of integrating the wave equation may be solved by the method of separation of variables. Suppose we require a solution of the equation

$$\nabla^2 u - \frac{\partial^2 u}{\partial t^2} = 0 \qquad (23.13)$$

with the initial conditions

$$[u]_{t=0} = \varphi_0(x, y, z); \quad \left[\frac{\partial u}{\partial n} \right]_{t=0} = \varphi_1(x, y, z) \qquad (23.14)$$

and the boundary conditions, say, of the form

$$\left[\frac{\partial u}{\partial n} \right]_s = 0, \qquad (23.15)$$

where S is the surface bounding the domain Ω in the space of the variables x, y, z.

The particular solutions of this equation which satisfy the conditions (23.15) may, as before, be sought in the form

$$u = U(x, y, z) \, T(t). \tag{23.16}$$

Substituting (23.16) into (23.13), we find:

$$T(t)\nabla^2 U(x, y, z) = U(x, y, z) \, T''(t),$$

or

$$\frac{T''(t)}{T(t)} = \frac{\nabla^2 U(x, y, z)}{U(x, y, z)} = -\lambda^2$$

whence

$$T'' + \lambda^2 T = 0 \tag{23.17}$$

$$\nabla^2 U + \lambda^2 U = 0. \tag{23.18}$$

A solution of (23.18) subject to the conditions (23.15) can be found by the use of Green's function again. In this case the function $G_1(P, P_0)$ satisfies not Laplace's equation but the equation

$$\nabla^2 G_1 = -\frac{4\pi}{D}$$

where D is the volume of the domain Ω. The constant $-4\pi/D$ serves in this case as the solution of the adjoint homogeneous problem, *i.e.*,

$$\nabla^2\left(-\frac{4\pi}{D}\right) = 0, \quad \frac{\partial}{\partial n}\left(-\frac{4\pi}{D}\right) = 0,$$

and is chosen so as to ensure the existence of Green's function. This follows from the earlier investigation of the Neumann problem (see § 2, Lecture 21). G_1 will thus be the generalized Green's function.

We shall seek that solution U of the equation $\nabla^2 U = -\lambda^2 U$ which is orthogonal to a constant in our domain: $\iiint_\Omega \lambda^2 \, U \, dx \, dy \, dz = 0$. We shall solve equation (23.18), regarding $\lambda^2 U$ as the free term. In this case, by the results of the previous lectures, the solution of (23.18) which is itself orthogonal to a constant must have the form

$$U(P_0) = \lambda^2 \iiint_\Omega \frac{G_1(P, P_0) \, U(P) \, dP}{4\pi}. \tag{23.19}$$

We have again obtained an integral equation with symmetric kernel for the eigenfunctions of the problem. For this integral equation all our previous assertions are valid except No. 3, which is now modified thus:

3(a). Any function $\varphi(P)$ which has continuous second-order derivatives in and on the boundary of Ω, and which is orthogonal to a constant:

$$\iiint_\Omega \frac{4\pi}{D} \varphi(P) \, dP = 0,$$

and which satisfies the boundary conditions, may be expressed as a uniformly convergent series of eigenfunctions of equation (23.19).

Equation (23.17) has two linearly independent solutions:

$$T_1 = \cos \lambda t, \quad T_2 = \sin \lambda t.$$

If $U_i(x, y, z) = U_i(P)$ is an eigenfunction of equation (23.19) and λ_i is its eigenvalue, then the required particular solutions of equation (23.13) will be:

$$U_i(x, y, z) \cos \lambda_i t, \quad U_i(x, y, z) \sin \lambda_i t.$$

We shall seek a solution of the problem in which we are interested in the form of series

$$u = \sum_{i=1}^\infty a_i U_i(x, y, z) \cos \lambda_i t + \sum_{i=1}^\infty b_i U_i(x, y, z) \sin \lambda_i t + c_0 + c_1 t.$$

$$(23.20)$$

If the series (23.20) and its derivative with respect to time both converge uniformly in the mean, then u and $\partial u/\partial t$ will be functions of time which are continuous in the mean. This was proved in Lecture 22, Theorem 4. We shall soon see that the conditions for uniform convergence in the mean are satisfied.

We require the series (23.20) to satisfy the initial conditions:

$$[u]_{t=0} = \sum_{i=1}^\infty a_i U_i(x, y, z) + c_0, \quad \left[\frac{\partial u}{\partial t}\right]_{t=0} = \sum_{i=1}^\infty \lambda_i b_i U_i(x, y, z) + c_1$$

$$(23.21)$$

If we choose the coefficients a_i and b_i so that

$$\sum_{i=1}^\infty a_i U_i(x, y, z) = \varphi_0(x, y, z) - c_0$$

$$\sum_{i=1}^\infty \lambda_i b_i U_i(x, y, z) = \varphi_1(x, y, z) - c_1,$$

then the initial conditions (23.14) will be satisfied in the mean. Again assuming the system of functions U_i to be orthogonal and normalized

$$\iiint_\Omega U_i(P) U_j(P) \, dP = \begin{cases} 1, & i = j \\ 0, & i \neq j \end{cases},$$

we can again determine all the coefficients of the series (23.21), having first chosen the constants c_0 and c_1 so that $\varphi_0 - c_0$ and $\varphi_1 - c_1$ are orthogonal to a constant. Repeating the former arguments, we get

$$a_i = \iiint_\Omega \varphi_0 U_i \, dP, \quad \lambda_i b_i = \iiint_\Omega \varphi_1 U_i \, dP.$$

We pass on now to the proof of the convergence in the mean of the series (23.20). Exactly as before, we first verify that

$$u_N = \sum_{j=1}^N a_i U_i \cos \lambda_i t + \sum_{i=1}^N b_i U_i \sin \lambda_i t$$

is a solution of the problem with approximate initial conditions.

We consider two finite sequences of terms of the series (23.20), u_n and u_m, and suppose that $n > m$. The difference of these two sequences

$$v_{nm} = u_n - u_m = \sum_{i=m+1}^N a_i U_i \cos \lambda_i t + \sum_{i=m+1}^N b_i U_i \sin \lambda_i t$$

satisfies the equation

$$\Delta v_{mn} - \frac{\partial^2 v_{nm}}{\partial t^2} = 0,$$

the boundary conditions (23.15) and the initial conditions

$$[v_{nm}]_{t=0} = \sum_{j=m+1}^n a_i U_i, \quad \left[\frac{\partial v_{nm}}{\partial t}\right]_{t=0} = \sum_{i=m+1}^n \lambda_i b_i U_i.$$

It is clear that

$$\left\{\iiint_\Omega [v_{nm}^2]_{t=0} \, d\Omega\right\}^{\frac{1}{2}} < \varepsilon, \quad \left\{\iiint_\Omega \left[\left[\frac{\partial v_{nm}}{\partial t}\right]_{t=0}\right]^2 d\Omega\right\}^{\frac{1}{2}} < \varepsilon$$

provided only that n and m are sufficiently large.

If we now make use of Property 5 of the system of eigenfunctions, we have, in the notation of the previous lecture,

$$K_0(0) < \varepsilon^2, \quad K_1(0) < \varepsilon^2, \quad A(t) \equiv 0.$$

It follows from the inequalities (23.12) and (23.14) that, under these conditions $K_0(t)$ and $K_1(t)$ will also be arbitrarily small in any finite interval of t. Thus the sequences v_{nm} and $\partial v_{nm}/\partial t$ satisfy the conditions of the Riesz–Fischer theorem and consequently converge in the mean uniformly with respect to t in any finite interval $a \leqq t \leqq b$. Hence the series (23.29) converges in the mean to a certain generalized solution. We shall not investigate in more detail the circumstances under which this solution will be a solution in the ordinary sense of the word.

We see that each term of the series (23.20) represents a so-called harmonic oscillation; the frequencies λ_i of the oscillations form a discrete sequence.

§ 2. The Analogy between the Problems of Vibrations of a Continuous Medium and Vibrations of Mechanical Systems with a Finite Number of Degrees of Freedom

An analogy may easily be pursued between the problem considered for equation (23.13) and the problem of free, small vibrations of mechanical systems having a finite number of degrees of freedom. The latter problem may be formulated as a problem of integrating a system of equations

$$\frac{d^2q}{dt^2} = \sum_{k=1}^{m} a_{jk}q_k \quad (j = 1, 2, \ldots, m) \tag{23.22}$$

subject to the conditions

$$[q_j]_{t=0} = (q_j)_0, \quad \left[\frac{dq_j}{dt}\right]_{t=0} = (q_j')_0$$

and it is also assumed that $a_{ij} = a_{ji}$.

In such a system, instead of a function $u(P, t)$ depending on t and on a variable point of space, we have a quantity depending on t and a discrete number j. If we construct a grid of a finite number of points $P_1, P_2, \ldots, P_m$ in the domain Ω, and consider instead of the function $u(P, t)$ the finite number of quantities $q_j = u(P_j, t)$, and replace the derivatives in equation (23.13) by finite differences, then we shall obtain a system similar to (23.22).

We know from the theory of ordinary differential equations that the solution of the system (23.22) has the form

$$q_j = \sum_{r=1}^{m} (c_{jr} \cos \lambda_r t + d_{jr} \sin \lambda_r t),$$

where the λ_r are the frequencies of vibration determined from the so-called *characteristic equation* (or *frequency equation* or *secular equation*)

$$\begin{vmatrix} a_{11} - \lambda^2 & a_{12} & \cdots & a_{1m} \\ a_{21} & a_{22} - \lambda^2 & \cdots & a_{2m} \\ \cdot & \cdot & \cdots & \cdot \\ a_{m1} & a_{m2} & \cdots & a_{mm} - \lambda^2 \end{vmatrix} = 0.$$

The difference between our earlier problem and the problem of finding solutions for the system (23.22) consists only in the fact that the system

(23.22) has a finite number of characteristic frequencies of vibration (eigenvalues) while our problem of integrating the wave equation has infinitely many. The analogy goes even deeper.

If we interpret the set of numbers $q_1, q_2, \ldots, q_m$ as a point in m-dimensional space, or, more precisely, as a vector q joining the origin to a certain point in this space, then the system of equations (23.22) may be written in the form

$$\frac{d^2 q}{dt^2} = Aq \qquad (23.23)$$

where A is the matrix of a linear substitution on the vector q. It is known from the theory of ordinary differential equations that the integration of (23.23) reduces, in fact, to an *orthogonal* change of variables (a transformation of coordinates in the space q) such that the matrix A of the linear substitution is reduced to diagonal form. If these new coordinates are denoted by $r_1, r_2, \ldots, r_m$, then after the change of variables

$$q_j = \sum_{s=1}^{m} \beta_{js} r_s \qquad (23.24)$$

or

$$r_s = \sum_{i=1}^{m} \beta_{sj} q_j,$$

we shall have the system

$$\frac{d^2 r_j}{dt^2} = \sum_{h=1}^{m} \bar{a}_{jk} r_k,$$

where

$$\bar{a}_{jk} = \begin{cases} -\lambda_j^2 & j = k \\ 0 & j \neq k. \end{cases}$$

The consideration in the case of the wave equation of the eigenfunctions

$$U_1, U_2, \ldots, U_m$$

is exactly similar to the choice of new coordinates just described. Instead of the values of some function $u(P, t)$ taken over all possible points P we shall regard this function as specified by its coefficients $f_i(t)$ in a series expansion

$$f(P, t) = \sum_{i=1}^{\infty} f_i(t) \, U_i(P). \qquad (23.25)$$

The formula (23.25) and the formula

$$f_i(t) = \iint\limits_{\Omega} \int f(P, t) \, U_i(P) \, dP \qquad (23.26)$$

are similar to (23.24) except that the suffix i in (23.25) and (23.26), which corresponds to the suffix s in (23.24) goes not from 1 to m but from 1 to infinity, and the role of the suffix j varying from 1 to m in (23.24) is now played by the point P varying over the domain Ω.

This point of view gives a new insight into the Fourier method itself of solving the problem of integration of the wave equation (23.13) with given initial and boundary conditions. The analogy which we have noted does not, of course, arise fortuitously; both problems are really particular cases of a more general problem which may be formulated in terms of the abstract theory of equations in functional spaces.

§ 3. The Inhomogeneous Equation

Without going more deeply into the question, we shall use the existing analogy which has been pointed out in § 2 and shall extend it further. We illustrate this analogy by the problem of integrating the wave equation with a free member and subject to zero initial conditions. As we have seen, the general case can be reduced to this one.

We consider the equation

$$\nabla^2 u - \frac{\partial^2 u}{\partial t^2} = F$$

and try to find a solution satisfying the conditions

$$[u]_{t=0} = \left[\frac{\partial u}{\partial t}\right]_{t=0} = 0,$$

and

$$\left[\frac{\partial u}{\partial n}\right]_s = 0.$$

Seeking u in the form of a series

$$u = \sum_{i=1}^{\infty} a_i(t)\, U_i + c_0(t) \qquad (23.27)$$

(any twice-differentiable function u which satisfies the condition $[\partial u/\partial n]_s = 0$ can be expanded in such a series), and expressing F in the form of a similar series†, we obtain:

$$c_0''(t) + \Delta\left(\sum_{i=1}^{\infty} a_i(t)\, U_i\right) - \frac{\partial^2}{\partial t^2}\left(\sum_{i=1}^{\infty} a_i(t)\, U_i\right) = \sum_{i=1}^{\infty} F_i(t)\, U_i + F_0(t).$$

† The function F will not, generally speaking, satisfy the boundary conditions. However, it can obviously be replaced by a function F' which does satisfy these conditions and also $\iiint_\Omega (F' - F)^2 \, dx \, dy \, dz < \varepsilon$. Then, by the argument of the previous lecture, a substitution of this sort will introduce into the solution only an error which may be made as small as we please.

We carry out the differentiation under the summation signs. This is permissible, if we suppose the series of derivatives converge uniformly. Then we have:

$$c_0''(t) - F_0(t) + \sum_{i=1}^{\infty} U_i[-\lambda_i^2 \, a_i(t) - a_i''(t) - F_i(t)] = 0.$$

We multiply both sides of this equation by U_j and integrate, noting that all terms except those having the suffix j then drop out; hence we have the differential equation

$$a_j''(t) + \lambda_j^2 \, a_j(t) + F_j(t) = 0 \tag{23.28}$$

for the determination of $a_j(t)$. We also have

$$c_0''(t) - F_0(t) = 0.$$

Using known formulae for the solution of ordinary differential equations we get

$$a_i(t) = \frac{1}{\lambda_i} \int_0^t \sin\left[\lambda_i(t - t_1)\right] F_i(t_1) \, dt_1,$$

$$c_0(t) = \int_0^t (t - t_1) \, F_0(t_1) \, dt_1.$$

For such values of $a_i(t)$ the formula (23.27) gives us the required solution, provided only that the series (23.27) and the one obtained from it by differentiating twice converge uniformly. To avoid having to investigate the convergence, we can again replace the free member F by a function F_N which is a finite sequence of the Fourier series. Then, passing to the limit, and using the Riesz–Fischer theorem, we obtain a solution which, if it is not a solution in the usual sense of the word, is a generalized solution.

Our substitution (23.27) is the analogue of the change of variables in the system (23.23) which brought the latter to canonical form. Just as for (23.23), it quickly solves the problem.

It is also easy to indicate the way to solve the problem of integrating the equation of heat conduction when it has a free member and when the conditions on the boundary are inhomogeneous:

$$\nabla^2 u - \frac{1}{a} \frac{\partial u}{\partial t} = F(P, t),$$

$$[u]_S = f(S, t),$$

$$[u]_{t=0} = \varphi(P).$$

It is sufficient to remark that this problem can be reduced to that of integrating the same equation under the conditions

$$[u]_S = 0,$$

$$[u]_{t=0} = 0.$$

We expand the free member as a series of the form

$$F(P, t) = \sum_{i=1}^{\infty} U_i(P) \, F_i(t). \tag{23.29}$$

Such an expansion is possible, since for a fixed value of t the function $F(P, t)$ is developable in a series of the form (23.29). The coefficients of this series depend, in general, on t:

$$F_i(t) = \iiint_{\Omega} U_i(P) \, F(P, t) \, dP. \tag{23.30}$$

The $F_i(t)$ are continuous, differentiable functions if the partial derivative of the function $F(P, t)$ with respect to t is continuous, as is seen from the formula (23.30).

Seeking a solution of the equation

$$\nabla^2 u_N - \frac{1}{a} \frac{\partial u_N}{\partial t} = \sum_{i=1}^{N} F_i(t) \, U_i(P) \tag{23.31}$$

in the form

$$u_N = \sum_{i=1}^{N} a_i(t) \, U_i(P).$$

we get

$$\sum_{i=1}^{N} U_i(P) \left\{ F_i(t) + \frac{1}{a} a_i'(t) + \lambda_i a_i(t) \right\} = 0,$$

whence, multiplying by U_j and integrating, we see that the coefficient $a_i(t)$ must satisfy the equation

$$\frac{1}{a} a_i'(t) + \lambda_i a_i(t) + F_i(t) = 0. \tag{23.32}$$

Taking as $a_i(t)$ that solution of (23.32) which vanishes at $t = 0$, i.e.,

$$a_i = \int_0^t e^{a\lambda_i(t_1 - t)} \, F_i(t) \, dt,$$

we see that $u_N(t)$ will satisfy equation (23.31) and also the required initial and boundary conditions.

Passing then to the limit as $N \to \infty$ and noting that the right-hand member of (23.31) tends to the function F, we obtain, by the previous argument, the generalized solution, which for a sufficiently smooth F will be the solution in the usual sense of the word (see § 2, Lecture 22).

In practical problems our greatest interest in solving the wave equation is to determine all the frequencies λ_i or, as we say, the *spectrum of natural frequencies of vibration*. A knowledge of this spectrum enables undesirable resonances to be avoided. Resonance generally arises when an external force varies according to a sinusoidal law and its frequency coincides with a natural frequency of vibration of the system.

Suppose, for example, that in formula (23.28)

$$F_i(t) = \sin \lambda_i t;$$

then

$$a_i(t) = \frac{1}{\lambda_i} \int_0^t \sin \lambda_i(t - t_1) \sin \lambda_i t_1 \, dt_1$$

$$= \frac{1}{2\lambda_i} \int_0^t [\cos \lambda_i(2t_1 - t) - \cos \lambda_i t] \, dt_1$$

$$= -\frac{t}{2\lambda_i} \cos \lambda_i t + \frac{1}{2\lambda_i^2} \sin \lambda_i t.$$

We see from this that $a_i(t)$, and with it the amplitude of the vibration, increases without limit as t increases.

§ 4. Longitudinal Vibrations of a Bar

Many other cases of the use of the Fourier method could be given besides those which we have examined. It would be natural to study in this way, for example, the equation

$$p(x) \frac{\partial^2 u}{\partial x^2} + q(x) \frac{\partial u}{\partial x} + r(x) \, u = \varrho(x) \frac{\partial u}{\partial t} + F(x, t),$$

or

$$p(x) \frac{\partial^2 u}{\partial x^2} + q(x) \frac{\partial u}{\partial x} + r(x) \, u = \varrho(x) \frac{\partial^2 u}{\partial t^2} + F(x, t),$$

subject to conditions on u at the instant $t = 0$ and at the ends of the interval $0 \leq x \leq 1$; and a host of similar problems.

Without going into details, we shall examine one more simple application of the general theory. We examine the equation

$$\frac{\partial^2 u}{\partial x^2} - \frac{\partial^2 u}{\partial t^2} = 0$$

with the conditions

$$[u]_{t=0} = \varphi_0(x), \qquad \left[\frac{\partial u}{\partial t}\right]_{t=0} = \varphi_1(x), \tag{23.33}$$

$$\left[\frac{\partial u}{\partial x}\right]_{x=0} = 0, \qquad \left[\frac{\partial u}{\partial x}\right]_{x=1} = 0. \tag{23.34}$$

This problem arises, for example, in studying the longitudinal vibrations of a bar, free at both ends.

In accordance with the method already explained, we seek a solution in the form of a series

$$U = \sum_{j=1}^{\infty} (a_j \cos \lambda_j t + b_j \sin \lambda_j t) \, U_j(x) + c_0 + c_1 t,$$

where the $U_j(x)$ are solutions of the differential equation

$$\frac{d^2 U_j}{dx^2} + \lambda_j^2 U_j = 0. \tag{23.35}$$

In this case we do not need to bring in an integral equation in order to find solutions of (23.35) satisfying the boundary conditions

$$\left[\frac{dU_j}{dx}\right]_{x=0} = 0 \quad \text{and} \quad \left[\frac{dU_j}{dx}\right]_{x=1} = 0.$$

The general solution of (23.35) will be

$$U_j = c_j \cos \lambda_j x + d_j \sin \lambda_j x \quad \text{if } \lambda_j^2 > 0,$$

or

$$U_j = c_j \cosh i\lambda_j x + d_j \sinh i\lambda_j x \quad \text{if } \lambda^2 < 0.$$

Correspondingly,

$$\frac{dU_j}{dx} = \lambda_j(-c_j \sin \lambda_j x + d_j \cos \lambda_j x),$$

or

$$\frac{dU_j}{dx} = \lambda_j(c_j \sinh i\lambda_j x + d_j \cosh i\lambda_j x).$$

The first of the boundary conditions then shows that we must take $d_j = 0$ in both cases, and the second condition leads to the conclusion that imaginary values of λ_j (i.e., negative values of λ_j^2) are impossible. Hence

$$U_j = c_j \cos \lambda_j x, \qquad \frac{dU_j}{dx} = -\lambda_j c_j \sin \lambda_j x.$$

Using the second boundary condition, we conclude that $\sin \lambda_j = 0$, whence

$$\lambda_j = j\pi \quad \text{and} \quad U_j = c_j \sin j\pi x.$$

It is well known that

$$\int_0^1 \cos j\pi x \cos k\pi x \, dx = \begin{cases} \frac{1}{2} & \text{if } j = k, \\ 0 & \text{if } j \neq k. \end{cases}$$

If we now choose $c_j = \sqrt{2}$, then we obtain the required system of orthogonal, normalized functions U_j in the form

$$U_j = \sqrt{2} \cos j\pi x.$$

The coefficients a_j and b_j are obtained in the form

$$a_j = \sqrt{2} \int_0^1 \varphi_0(x) \cos j\pi x \, dx,$$

$$b_j = \frac{\sqrt{2}}{\pi j} \int_0^1 \varphi_1(x) \cos j\pi x \, dx,$$

and we derive the final answer in the form of the series

$$u = \sqrt{2} \sum_{j=1}^{\infty} (a_j \cos j\pi x + b_j \sin j\pi x) + c_0 + c_1 t,$$

where

$$c_0 = \int_0^1 \varphi_0(x) \, dx, \quad c_1 = \int_0^1 \varphi_1(x) \, dx.$$

We see that the natural frequencies of vibration of such a bar will have the form

$$\pi j, \quad j = 1, 2, \ldots, n, \ldots$$

To conclude this exposition of the Fourier method we make one or two further remarks of a general nature. An essential feature of the argument has been that the numbers λ_j are nowhere dense. Consequently, in the formulae (23.25) and (23.26), which we considered as the analogue of linear transformations of n numbers, one of the variables — with the suffix i — could take only a denumerable set of values. A more detailed investigation would show that this circumstance is intimately connected with the fact that the domain Ω is bounded. Those properties of integral equations with symmetric kernels upon which we have relied may be lost if the domain is unbounded. In such a case it can happen, for example, that the required orthogonal, normalized eigenfunctions do not exist. They would be replaced by a whole set of functions $U(P, \xi)$ depending on a continuously varying parameter ξ. If the problem is not self-adjoint, then the eigenvalues λ are not necessarily real but may be complex.

INTEGRAL EQUATIONS WITH REAL, SYMMETRIC KERNELS

§ 1. Elementary Properties. Completely Continuous Operators

We have already seen that the problem of finding the eigenvalues and eigenfunctions for many of the problems of mathematical physics is reducible, with the aid of Green's function, to the problem of finding the eigenvalues and eigenfunctions for some integral equation of the second Fredholm type with a real, symmetric kernel, *i.e.*, with a kernel such that

$$K(P, P_0) = K(P_0, P).$$

We shall examine a rather more general integral equation, *viz.*,

$$\varphi(P_0) = f(P_0) + \lambda \int_\Omega K(P_0, P)\, \varphi(P)\, \varrho(P)\, \mathrm{d}P \qquad (24.1)$$

and the corresponding homogeneous equation

$$\varphi(P_0) = \lambda \int_\Omega K(P_0, P)\, \varphi(P)\, \varrho(P)\, \mathrm{d}P, \qquad (24.2)$$

where $K(P_0, P)$ is a symmetrical function of the coordinates of the points P_0 and P, and $\varrho(P)$ is a non-negative, measurable function, called the *weight*. If $\varrho = 1$ we get integral equations with a symmetric kernel.

For integral equations of the type (24.1) and (24.2), and, in particular, for equations with a symmetric kernel, a whole series of important propositions hold good, and to the investigation of these we now turn.

We shall say that equation (24.1) has a *weighted symmetric kernel* or a *symmetric kernel with weight* $\varrho(P)$. In order not to complicate the argument, we shall consider only the case when $\varrho(P)$ is bounded. We shall further suppose that $\varrho(P)$ vanishes only on a set of measure zero.

LEMMA 1. *Let $\varphi(P_0)$ and $\psi(P)$ be two arbitrary functions, real or complex†, such that their moduli are quadratically integrable with weight $\varrho(P)$ in the*

---

† By a complex function of a real argument (or real arguments) we mean a function which can be expressed as $\varphi(P) = \varphi_1(P) + i\varphi_2(P)$ where $\varphi_1(P)$ and $\varphi_2(P)$ are real functions.

bounded domain Ω, *i.e.*,

$$\int_\Omega |\varphi(P)|^2 \varrho(P)\, \mathrm{d}P < \infty, \quad \int_\Omega |\psi(P_0)|^2 \varrho(P_0)\, \mathrm{d}P_0 < \infty.$$

Further, let the real kernel† $K(P, P_0)$ *satisfy the inequality*

$$|K(P, P_0)| \leq \frac{A}{r^a},$$

where r *is the distance between the points* P *and* P_0, *and* $a < n$. *Then the integral*

$$\int_\Omega \int_\Omega |K(P, P_0)\, \varphi(P)\, \psi(P_0)|\, \varrho(P)\, \varrho(P_0)\, \mathrm{d}P\, \mathrm{d}P_0 \tag{24.3}$$

converges.

By Bunyakovski's inequality,

$$\int_\Omega \int_\Omega |K(P, P_0)\, \varphi(P)\, \psi(P_0)|\, \varrho(P)\, \varrho(P_0)\, \mathrm{d}P\, \mathrm{d}P_0$$

$$\leq \sqrt{\int_\Omega |\varphi(P)|^2\, \varrho(P)\, \mathrm{d}P} \cdot \sqrt{\int_\Omega \left\{\int_\Omega |K(P, P_0)\, \psi(P_0)|\, \varrho(P_0)\, \mathrm{d}P_0\right\}^2 \varrho(P)\, \mathrm{d}P}$$

If we establish the existence of the integral

$$\int_\Omega |K(P, P_0)\, \psi(P_0)|\, \varrho(P_0)\, \mathrm{d}P_0 \tag{*}$$

for almost all P, and also the convergence of the integral

$$\int_\Omega \left\{\int_\Omega |K(P, P_0)\, \psi(P_0)|\, \varrho(P_0)\, \mathrm{d}P_0\right\}^2 \varrho(P)\, \mathrm{d}P, \tag{24.4}$$

then our lemma will follow.

We have:

$$\int_\Omega |K(P, P_0)\, \psi(P_0)|\, \varrho(P_0)\, \mathrm{d}P_0 \leq \int_\Omega \frac{A}{r^a} |\psi(P_0)|\, \varrho(P_0)\, \mathrm{d}P_0$$

$$= A \int_\Omega \frac{1}{r^{\frac{a}{2}}} \frac{1}{r^{\frac{a}{2}}} |\psi(P_0)|\, \varrho(P_0)\, \mathrm{d}P_0$$

$$\leq A \sqrt{\int_\Omega \frac{\varrho(P_0)}{r^a}\, \mathrm{d}P_0} \times$$

$$\times \sqrt{\int_\Omega \frac{1}{r^a} |\psi(P_0)|^2\, \varrho(P_0)\, \mathrm{d}P_0};$$

† The kernel $K(P, P_0)$ is not assumed to be symmetric.

but

$$A^2 \int_\Omega \frac{\varrho(P_0)\,\mathrm{d}P_0}{r^a} \leqq M, \qquad (**)$$

where M is a certain constant.

Consequently the integral (*) exists for all those points P for which the integral

$$\int_\Omega \frac{1}{r^a} |\psi(P_0)|^2 \varrho(P_0)\,\mathrm{d}P_0 \qquad (24.5)$$

exists. Thus

$$\int_\Omega \left[\int_\Omega |K(P, P_0)\,\psi(P_0)|\varrho(P_0)\,\mathrm{d}P_0 \right]^2 \varrho(P)\,\mathrm{d}P$$

$$\leqq M \int_\Omega \left[\int_\Omega \frac{1}{r^a} |\psi(P_0)|^2 \varrho(P_0)\,\mathrm{d}P_0 \right] \varrho(P)\,\mathrm{d}P.$$

We shall establish simultaneously both the existence almost everywhere of the integral (24.5) and the convergence of the integral on the right-hand side of the last inequality, by using the Lebesgue–Fubini theorem.

We shall show that the $2n$-dimensional integral

$$\int_\Omega \int_\Omega \frac{1}{\varrho^a} |\psi(P_0)|^2 \varrho(P_0)\,\varrho(P)\,\mathrm{d}P_0\,\mathrm{d}P \qquad (24.6)$$

converges. Hence it will follow that the multiple integral

$$\int_\Omega \left[\int_\Omega \frac{1}{r^a} |\psi(P_0)|^2 \varrho(P_0)\,\mathrm{d}P_0 \right] \varrho(P)\,\mathrm{d}P$$

is meaningful and convergent, as well as the integral (24.4).

In order to establish the convergence of the integral (24.6) we shall carry out the integration in the other order:

$$\int_\Omega \left[|\psi(P_0)|^2 \varrho(P_0) \int_\Omega \frac{1}{r^a} \varrho(P)\,\mathrm{d}P \right] \mathrm{d}P_0. \qquad (24.7)$$

By virtue of the inequality (**) the function under the integral sign is less than

$$B|\psi(P_0)|^2 \varrho(P_0)$$

where B is a constant; consequently the integral (24.7) is less than

$$B \int_\Omega |\psi(P_0)|^2 \varrho(P_0)\,\mathrm{d}P_0,$$

which converges by the conditions of the lemma. This implies that the integral (24.6) also converges; hence the lemma.

COROLLARY. *The following equation holds:*

$$\int_{\Omega} \varphi(P) \left\{ \int_{\Omega} K(P, P_0)\, \psi(P_0)\, \varrho(P_0)\, \mathrm{d}P_0 \right\} \varrho(P)\, \mathrm{d}P$$

$$= \int_{\Omega} \psi(P_0) \left\{ \int_{\Omega} K(P, P_0)\, \varphi(P)\, \varrho(P)\, \mathrm{d}P \right\} \varrho(P_0)\, \mathrm{d}P_0.$$

The proof follows from the remark made after the Lebesgue–Fubini theorem (see p. 125).

We now introduce the notation:

$$\int_{\Omega} K(P_0, P)\, \varphi(P)\, \varrho(P)\, \mathrm{d}P = A\varphi \tag{24.8}$$

$$\int_{\Omega} K(P_0, P)\, \psi(P_0)\, \varrho(P_0)\, \mathrm{d}P_0 = A^*\psi. \tag{24.9}$$

From what we have just proved it follows that the integrals

$$\int_{\Omega} |A\varphi|^2 \varrho(P)\, \mathrm{d}P \quad \text{and} \quad \int_{\Omega} |A^*\psi|^2 \varphi(P)\, \mathrm{d}P \quad \text{exist}.$$

We notice that if the function $K(P_0, P)$ is symmetrical, then the operator A coincides with the operator A^*.

Let also

$$(\varphi, \psi) = \int_{\Omega} \varphi(P)\, \overline{\psi(P)}\, \varrho(P)\, \mathrm{d}P, \tag{24.10}$$

where $\overline{\psi}$ denotes the complex function conjugate to ψ. Then clearly

$$(\varphi, \overline{\psi}) = \int_{\Omega'} \varphi(P)\, \psi(P)\, \varrho(P)\, \mathrm{d}P.$$

We shall call the expression (φ, ψ) *the scalar product* of the functions φ and ψ.

We note some properties of these symbols:

1. $\qquad A(a_1\varphi_1 + a_2\varphi_2) = a_1 A\varphi_1 + a_2 A\varphi_2.$

2. $\qquad (a_1\varphi_1 + a_2\varphi_2, \psi) = a_1(\varphi_1, \psi) + a_2(\varphi_2, \psi).$

3. $\qquad (\varphi, a_1\psi_1 + a_2\psi_2) = \bar{a}_1(\varphi, \psi_1) + \bar{a}_2(\varphi, \psi_2).$

4. $\qquad (\varphi, \psi) = \overline{(\psi, \varphi)}.$

5. For any quadratically integrable function $\varphi, (\varphi, \varphi) \geqq 0$, where the equality sign holds if and only if the function φ is equivalent to zero.

In the equalities 1, 2, 3, a_1 and a_2 are constants. All these equalities can be verified directly.

Our corollary may be written in the form

$$\left(\varphi, \overline{A^*\psi}\right) = (A\varphi, \overline{\psi}). \qquad (24.11)$$

If the function $K(P, P_0)$ is real and symmetrical and if φ and ψ are real, we have $(\varphi, A\psi) = (A\varphi, \psi)$.

We now pass on to the investigation of the integral equations, and we shall deal first with the homogeneous equation.

Let us agree to consider in future only those eigenfunctions such that their moduli are quadratically integrable with weight ϱ.

THEOREM 1. *If λ_1 and λ_2 are two different characteristic numbers of the equations*

$$\varphi = \lambda_1 A\varphi$$

and

$$\psi = \lambda_2 A^*\varphi,$$

then the eigenfunctions φ and ψ of these equations satisfy the relation

$$(\varphi, \overline{\psi}) = 0. \qquad (24.12)$$

For,

$$\left(\varphi, \overline{A^*\psi}\right) = \left(\varphi, \overline{\frac{1}{\lambda_2}\psi}\right) = \frac{1}{\lambda_2}(\varphi, \overline{\psi}),$$

$$(A\varphi, \overline{\psi}) = \left(\frac{1}{\lambda_1}\varphi, \overline{\psi}\right) = \frac{1}{\lambda_1}(\varphi, \overline{\psi});$$

hence, from (24.11)

$$\lambda_2(\varphi, \overline{\psi}) = \lambda_1(\varphi, \overline{\psi}),$$

and this is possible only if $(\varphi, \overline{\psi}) = 0$, as was to be shown.

We shall call functions φ and ψ which satisfy (24.12) *orthogonal with weight ϱ*, or simply orthogonal, if this will not lead to ambiguity.

COROLLARY. *The fundamental functions of an integral equation with a weighted symmetric kernel which correspond to different characteristic numbers are orthogonal.*

Proof. For a weighted symmetric kernel, all the fundamental functions of the equation $\psi = \lambda A^*\psi$ simply become the fundamental functions of the equation $\varphi = \lambda A\varphi$, since $A \equiv A^*$.

THEOREM 2. *A real, weighted, symmetric kernel cannot have complex characteristic numbers.*

Proof. Let λ_0 be a characteristic number and φ_0 a fundamental function of our equation, *i.e.*,

$$\varphi_0 = \lambda_0 A\varphi_0.$$

Taking the complex quantities conjugate to both sides of this equation, and taking into account that for a symmetric kernel $K(P, P_0)$

$$\overline{A\varphi} = A\overline{\varphi}.$$

we get

$$\overline{\varphi}_0 = \overline{\lambda}_0 A \overline{\varphi}_0.$$

Hence it follows that $\overline{\lambda}_0$ is also a characteristic number, and $\overline{\varphi}_0$ a fundamental solution of our equation.

If $\lambda_0 \neq \overline{\lambda}_0$, then, by the corollary to Theorem 1, we see that φ_0 and $\overline{\varphi}_0$ must be orthogonal with weight ϱ, $i.e.$,

$$(\varphi_0, \varphi_0) = \int_{\Omega'} \varphi_0(P) \, \overline{\varphi_0(P)} \, \varrho(P) \, \mathrm{d}P = \int_{\Omega'} |\varphi_0(P)|^2 \, \varrho(P) \, \mathrm{d}P = 0;$$

this implies that $\varphi_0 = 0$, which contradicts the hypothesis that φ_0 is a nontrivial solution of the equation $\varphi = \lambda A \varphi$.

THEOREM 3. *All fundamental functions of a real symmetric kernel are themselves real (or, more precisely, can be chosen to be real).*

Let $\varphi_0(P) = \alpha(P) + i\beta(P)$ be a fundamental function. Substituting it in the equation, we get

$$\varphi_0 = \alpha + i\beta = \lambda A \, \varphi_0 = \lambda A\alpha + i\lambda A\beta;$$

and separating real and imaginary parts,

$$\alpha = \lambda A\alpha, \quad \beta = \lambda A\beta.$$

Consequently, α and β are themselves fundamental functions, and in place of φ_0 we may consider either of these functions, or a linear combination of them.

LEMMA 2. *All fundamental functions of a weighted kernel may be considered orthogonal with weight ϱ.*

We remark that for an equation with symmetric kernel of the type described, all the Fredholm theory obviously holds, since the integrals

$$\int_{\Omega} |K(P, P_0)| \, \varrho(P_0) \, \mathrm{d}P_0, \quad \int_{\Omega} |K(P, P_0)| \, \varrho(P_0) \, \mathrm{d}P$$

are bounded. In particular, to each eigenvalue λ corresponds only a finite number of linearly independent functions.

Turning now to the proof of the lemma, any non-orthogonal functions could only be those which correspond to one and the same characteristic number λ. Suppose these functions are, for example, $\varphi_1, \varphi_2, \ldots, \varphi_q$. In place of them we consider the linear combinations of them:

$$\psi_1 = \varphi_1,$$

$$\psi_2 = \varphi_2 - \frac{(\psi_1, \overline{\varphi}_2)}{(\psi_1, \overline{\psi}_1)} \psi_1,$$

$$\psi_3 = \varphi_3 - \frac{(\psi_1, \overline{\varphi}_3)}{(\psi_1, \overline{\psi}_1)} \psi_1 - \frac{(\psi_2, \overline{\varphi}_3)}{(\psi_2, \overline{\psi}_2)} \psi_2,$$

$$\cdot \qquad \cdot \qquad \cdot \qquad \cdot \qquad \cdot \qquad \cdot \qquad \cdot$$

$$\psi_k = \varphi - \frac{(\psi_1, \overline{\varphi}_k)}{(\psi_1, \overline{\psi}_1)} \psi_1 - \frac{(\psi_2, \overline{\varphi}_k)}{(\psi_2, \overline{\psi}_2)} \psi_2 - \cdots - \frac{(\psi_{k-1}, \overline{\varphi}_k)}{(\psi_{k-1}, \overline{\psi}_{k-1})} \psi_{k-1}$$

It is easily proved by induction that each function ψ_s is orthogonal to all the preceding functions, since

$$\psi_s, \overline{\psi}_t) = (\varphi_s, \overline{\psi}_t) - \frac{(\psi_1, \overline{\varphi}_s)}{(\psi_1, \overline{\psi}_1)} (\psi_1, \overline{\psi}_t) - \cdots - \frac{(\psi_{s-1}, \overline{\varphi}_s)}{(\psi_{s-1}, \overline{\psi}_{s-1})} (\psi_{s-1}, \overline{\psi}_t).$$

By the hypothesis of the induction all the terms on the right-hand side are zero except $(\varphi_s, \overline{\psi}_t)$ and $-\frac{(\psi_t, \overline{\varphi}_s)}{(\psi_t, \overline{\psi}_t)} (\psi_t, \overline{\psi}_t)$ because $t < s$, and these two cancel each other out.

The ψ_k are obviously solutions of the homogeneous equation, as was to be proved.

DEFINITION. *We shall say that the sequence $\{\varphi_n\}$ of functions whose moduli are quadratically integrable converges in the mean with weight ϱ to the function φ with a quadratically integrable modulus if the relation*

$$\lim_{n \to \infty} \int_{\Omega} |\varphi_n - \varphi|^2 \varrho \, dP = 0$$

holds.

We note that if the sequence $\{\varphi_n\}$ of functions whose squares are integrable converges uniformly to φ, then this sequence converges in the mean to φ. But the converse proposition does not hold. If we reject the requirement for uniform convergence, then examples can be constructed of sequences which converge everywhere but which do not converge in the mean.

We encountered the concept of convergence in the mean earlier in Lecture 22, § 7, where we proved that a sequence cannot converge in the mean to two different functions. Here we should remind ourselves again that, if we are supposing integrability in the Lebesgue sense, then in the proposition mentioned we should regard functions as different only if their values differ on a set of positive measure.

DEFINITION. *We shall say that a set of functions is compact if, from any infinite subset of these functions, a convergent sequence can be selected.*

With different conditions imposed on the convergence, which may, for example, be convergence in the mean, or uniform convergence, etc., we get different conditions for compactness.

Compact sets of functions are, by definition, strongly reminiscent of bounded sets of points. It is clear that any infinite sub-set of a bounded set of points is itself a bounded, infinite set and therefore has at least one limit point and consequently also contains a sequence converging to this point.

In mathematical analysis, the concept is introduced of the *equicontinuity* of a set of functions. We recall the definition.

DEFINITION. *A set of functions* $\{\varphi\}$ *is said to be equicontinuous if, given any positive* ε, *a number* $\eta(\varepsilon)$ *can be found such that*

$$\left| \varphi(P) - \varphi(P_1) \right| < \varepsilon,$$

provided only that the distance between the points P and P_1 *is less than* $\eta(\varepsilon)$, *the number* $\eta(\varepsilon)$ *being the same for all functions belonging to the set* $\{\varphi\}$.

One of the most important applications of the idea of equicontinuity is the so-called Arzela's theorem†, *viz.*

From any set $\{\varphi\}$ *of functions which are uniformly bounded and equicontinuous, a uniformly convergent sequence of functions can be chosen.*

(A set of functions is uniformly bounded if every function satisfies the inequality $|\varphi| < A$, where A is independent of φ.)

Using the concept of compactness, we can formulate this result thus: A set consisting of uniformly bounded, equicontinuous functions will be compact if uniform convergence, or, *a fortiori*, if convergence in the mean is considered.

For, by Arzela's theorem, such an infinite set has at least one limit function, *i.e.*, it contains a uniformly convergent sequence. It is clear that in such case there will also be convergence in the mean if the domain in which the functions are specified is bounded, as we shall suppose. A set of functions which is compact for uniform convergence will evidently be compact for convergence in the mean.

We shall say that a set of functions $\{\varphi\}$ is bounded in the mean with weight $\varrho(P)$ if it satisfies the condition.

$$\int_{\Omega} |\varphi|^2 \varrho \; dP \leqq A.$$

DEFINITION. We shall say that the operator A is *completely continuous* if, when it is applied to all functions of some set $\{\varphi\}$ which is *bounded in the mean*, it transforms it into *a compact set in the sense of convergence in the mean*. If the set $\{A\varphi\}$ is *compact in the sense of uniform convergence*, we shall say that A is *a strong, completely continuous operator.*

THEOREM 4. *The integral operator A defined by*

$$A\varphi = \int_{\Omega} K(P_0, P) \; \varphi(P) \; \varrho(P) \; dP,$$

---

† See V. V. Stepanov, *Course of Differential Equations*, 4th edition, Chapter II, § 2 page 64, or I. G. Petrovskii, *Lectures on the Theory of Ordinary Differential Equations*, § 11.

where the function $K(P_0, P)$ is continuous, and $\varrho(P)$ is an integrable function, is a strong, completely continuous operator.

Let $\{\varphi\}$ be a set of functions bounded in the mean: $\int_\Omega |\varphi|^2 \varrho \, dP < A$.

Using Bunyakovski's inequality we can easily see that the family of functions $\{A\varphi\}$ is uniformly bounded. Writing $\omega = A\varphi$, we have

$$|\omega|^2 \leq \int_\Omega |\varphi|^2 \varrho \, dP \cdot \int_\Omega |K(P_0, P)|^2 \varrho(P) \, dP \leq AM^2 B,$$

if

$$|K(P_0, P)| \leq M, \quad B = \int_\Omega \varrho(P) \, dP.$$

We shall prove that the family $\{\omega\}$ is equicontinuous. We take the difference

$$\omega(P_1) - \omega(P_2) = \int_\Omega [K(P, P_1) - K(P, P_2)] \, \varphi(P) \, \varrho(P) \, dP:$$

we have

$$|\omega(P_1) - \omega(P_2)|^2 \leq \int_\Omega |\varphi(P)|^2 \varrho(P) \, dP \cdot \int_\Omega |K(P, P_1) - K(P, P_2)|^2 \varrho(P) \, dP.$$

We choose the point P_2 to be sufficiently close to P_1 so that

$$|K(P, P_1) - K(P, P_2)| < \varepsilon.$$

This is possible because the kernel $K(P, P_0)$ is continuous in the closed domain of variation of the variables P, P_0 and, by a known theorem, will be uniformly continuous therein. Then

$$|\omega(P_1) - \omega(P_2)|^2 \leq A \cdot \int_\Omega \varepsilon^2 \varrho(P) \, dP = \varepsilon^2 AB$$

where

$$B = \int_\Omega \varrho(P) \, dP.$$

We see, then, that the difference $|\omega(P_1) - \omega(P_2)|$ can be made less than any previously assigned number for P_1 sufficiently close to P_2, simultaneously for all $\omega = A\varphi$, since we did not use any individual property of φ in obtaining the last inequality. Consequently the family $\{A\varphi\}$ is equicontinuous. Hence follows the compactness of this family in the sense of uniform convergence and, *a fortiori*, in the sense of convergence in the mean.

Suppose now that the kernel $K(P_0, P)$ is real and continuous everywhere in the domain $\Omega \times \Omega$ in both variables with the exception of the set $\{P = P_0\}$

and that it satisfies everywhere in $\Omega \times \Omega$ the inequality

$$|K(P_0, P)| \leqq \frac{A}{r^{\frac{n}{2}-a}}, \quad a > 0 \tag{24.13}$$

We shall say that such a kernel is *almost regular*.

THEOREM 5. *The integral operator A defined by*

$$A\varphi = \int_\Omega K(P_0, P)\, \varphi(P)\, \varrho(P)\, \mathrm{d}P,$$

where the kernel $K(P_0, P)$ is almost regular, and the function $\varrho(P)$ is bounded, is a strong, completely continuous operator.

The proof of this theorem is very similar to that of the preceding one.

Let $\{\varphi\}$ be a family of functions uniformly bounded in the mean. We shall show that the family $\{A\varphi\}$ is uniformly bounded and equicontinuous.

As before, the uniform boundedness follows from Bunyakovski's inequality. For, if

$$\int_\Omega |\varphi(P)|^2 \varrho(P)\, \mathrm{d}(P) \leqq M,$$

then

$$\left[\int_\Omega K(P_0, P)\, \varphi(P)\, \varrho(P)\, \mathrm{d}P\right]^2$$
$$\leqq \left[\int_\Omega |K(P_0, P)|^2 \varrho(P)\, \mathrm{d}P\right]\left[\int_\Omega |\varphi(P)|^2 \varrho(P)\, \mathrm{d}P\right].$$

By the condition (24.13) we have

$$\int_\Omega |K(P_0, P)|^2 \varrho(P)\, \mathrm{d}P \leqq \int_\Omega \frac{A^2}{r^{n-2a}} \varrho(P)\, \mathrm{d}P \leqq C$$

where C is a certain constant. Using this result, we get

$$\left[\int_\Omega K(P_0, P)\, \varphi(P)\, \varrho(P)\, \mathrm{d}P\right]^2 \leqq MC.$$

Hence the family $\{A\varphi\}$ is bounded.

We now write $\omega = A\varphi$ and set up the difference

$$\omega(P_1) - \omega(P_2) = \int_\Omega [K(P_1, P) - K(P_2, P)]\, \varphi(P)\, \varrho(P)\, \mathrm{d}P.$$

We have

$$[\omega(P_1) - \omega(P_2)]^2$$

$$\leqq \int_\Omega |\varphi(P)|^2 \varrho(P)\, dP \cdot \int_\Omega |K(P_1, P) - K(P_2, P)|^2 \varrho(P)\, dP$$

$$\leqq M \int_\Omega |K(P_1, P) - K(P_2, P)|^2 \varrho(P)\, dP = MI,$$

where

$$I = \int_\Omega |K(P_1, P) - K(P_2, P)|^2 \varrho(P)\, dP. \tag{24.14}$$

We shall estimate the magnitude of this last integral.

The function $|K(P_1, P) - K(P_2, P)|^2$ is a function of the $3n$ coordinates of the points P, P_1, P_2 which is continuous everywhere in the domain $\Omega \times \Omega \times \Omega$ except on the set of points $\{P = P_1$ and $P = P_2\}$. Let r_1 be the distance from the point P to the point P_1, and r_2 be the distance from the point P to the point P_2. In the space of the $3n$ variables we exclude from the domain $\Omega \times \Omega \times \Omega$ the open sets of points which satisfy the conditions $r_1 < \eta$ and $r_2 < \eta$, where η is any previously specified positive number. In the remaining closed set, the function $|K(P_1, P) - K(P_2, P)|^2$ will be uniformly continuous, by Weierstrass's theorem. For $P_1 = P_2$ it will vanish. Consequently, given any $\varepsilon > 0$, we can find a positive number $\delta(\varepsilon, \eta)$ such that

$$|K(P_1, P) - K(P_2, P)|^2 < \varepsilon,$$

provided that the distance r^* between P_1 and P_2 is less than δ.

Suppose then that ε is a given positive number. In the domain of variation of P we surround both the singular points P_1 and P_2 by small spheres of radius η, and we divide the integral (24.14) into two parts

$$I = I_1 + I_2$$

where

$$I_1 = \int_{\substack{r_1 < \eta \\ r_2 < \eta}} |K(P_1, P) - K(P_2, P)|^2 \varrho(P)\, dP$$

and

$$I_2 = \int_{\substack{r_1 \geqq \eta \\ \text{or} \\ r_2 \geqq \eta}} |K(P_1, P) - K(P_2, P)|^2 \varrho(P)\, dP.$$

It is not difficult to see that $\eta(\varepsilon)$ can always be chosen so small that

$$I_1 \leqq \frac{\varepsilon}{2}.$$

For,

$$[K(P_1, P) - K(P_2, P)]^2 \leqq [K(P_1, P) - K(P_2, P)]^2$$

$$+ [K(P_1, P) + K(P_2, P)]^2 = 2[K(P_1, P)]^2 + 2[K(P_2, P)]^2$$

and consequently

$$I_1 \leqq 2 \int_{\substack{r_1 < \eta \\ r_2 < \eta}} [K(P_1, P)]^2 \, \varrho(P) \, dP + 2 \int_{\substack{r_1 < \eta \\ r_2 < \eta}} [K(P_2, P)]^2 \, \varrho(P) \, dP.$$

Hence, using the inequality (24.13) for $K(P_1, P)$, $K(P_2, P)$, we get $I_1 \leqq \varepsilon/2$.

Having chosen the value of $\eta(\varepsilon)$, we can further find $\delta(\varepsilon)$ so that everywhere in the domain under the integral sign I_2 we have

$$[K(P_1, P) - K(P_2, P)]^2 \leqq \frac{\varepsilon}{2 \sup \varrho(P) \, m\Omega}$$

where $m\Omega$ is the volume of the domain Ω, provided only that $r^* < \delta$. We shall then have

$$I_2 \leqq \frac{\varepsilon}{2m\Omega} \int_\Omega \frac{\varrho(P)}{\sup \varrho} \, dP \leqq \frac{\varepsilon}{2}.$$

Hence

$$I \leqq \varepsilon.$$

We see that the value of $\delta(\varepsilon)$ depends only on the distance between the points P_1 and P_2 and does not depend on the function φ. Consequently, the set of functions $\{A\varphi\}$ is equicontinuous. And we proved earlier that it was uniformly bounded. Hence, by Arzela's theorem, the set of functions $\{A\varphi\}$ is compact in the sense of uniform convergence. Thus, the operator A is a strong, completely continuous operator, as was to be shown.

As we shall explain later, this property of an operator of being completely continuous is of the greatest importance, for from it follow the main theorems for equations with weighted symmetric kernels.

It can be shown that the whole qualitative aspect of the Fredholm theory for asymmetrical kernels, i.e., the Fredholm alternative, the conditions for the solubility of equations, etc., is carried over completely for the equation

$$\varphi = \lambda A\varphi + f$$

with a completely continuous operator A. We shall not, however, go into this question.

In future, if we wish to establish the complete continuity of an operator A, we shall show that it is a strong, completely continuous operator.

§ 2. Proof of the Existence of an Eigenvalue

THEOREM 6. *If A is a completely continuous operator, the integral equation*

$$\varphi = \lambda A \varphi \qquad (24.15)$$

with a real, symmetric kernel has at least one eigenvalue.

Proof. We notice first of all that, for our purpose, it suffices to show that the equation

$$\varphi = \mu A^2 \varphi$$

has at least one characteristic number and fundamental function. For, let us write the last equation in the form

$$\varphi - \mu A^2 \varphi = 0$$

and let λ_0 be its eigenvalue and φ_0 the corresponding eigenfunction. Then, putting $\mu_0^2 = \lambda_0$, we have, in the notation of Lecture 18,

$$(E - \lambda_0 A)\, [(E + \lambda_0 A)\, \varphi_0] = 0.$$

The left-hand member of the last equation can vanish only if the function

$$\psi_0 = (E + \lambda_0 A)\, \varphi_0$$

is a solution of the equation

$$(E - \lambda A)\, \psi = 0 \quad \text{for} \quad \lambda = \lambda_0.$$

Consequently, either $\psi_0 = 0$, or the equation

$$(E - \lambda_0 A)\, \psi_0 = 0$$

has a nontrivial solution. In the first case we have

$$(E + \lambda_0 A)\, \varphi_0 = 0,$$

and this implies that φ_0 is an eigenfunction of equation (24.15) corresponding to the eigenvalue $\lambda = \lambda_0$ of (24.15). In the second case we also arrive at the existence of an eigenvalue $\lambda = \lambda_0$ for equation (24.15).

Consider the expression

$$\frac{(A\varphi, A\varphi)}{(\varphi, \varphi)} = \frac{(\varphi, A^2\varphi)}{(\varphi, \varphi)}$$

for all possible functions φ which are real, quadratically integrable, and not equivalent to zero. This ratio cannot be zero for all φ, for otherwise $A\varphi$

would be identically zero. It is obviously not negative. Moreover, it cannot take unboundedly large values. For, this ratio does not change when the function is multiplied by any constant. Hence it is sufficient to establish the boundedness of this ratio for functions which are uniformly bounded in the mean, and this follows in an obvious way from what has been proved above. Consequently, the expression $(A\varphi, A\varphi)/(\varphi, \varphi)$ has a least upper bound, which we denote by $\varkappa_1$.

It is then clear that

$$\frac{(A^2\varphi, A^2\varphi)}{(\varphi, \varphi)} = \frac{(A^2\varphi, A^2\varphi)}{(A\varphi, A\varphi)} \cdot \frac{(A\varphi, A\varphi)}{(\varphi, \varphi)} \leq \varkappa_1^2. \tag{24.16}$$

Let φ_n be a sequence of functions such that

$$(\varphi_n, \varphi_n) = 1 \tag{24.17}$$

$$\lim_{n \to \infty}(A\varphi_n, A\varphi_n) = \varkappa_1. \tag{24.18}$$

Such a sequence can always be constructed. By the property of the upper bound there is a sequence φ_n^* such that

$$\lim_{n \to \infty} \frac{(A\varphi_n^*, A\varphi_n^*)}{(\varphi_n^*, \varphi_n^*)} = \varkappa_1.$$

Hence putting

$$\varphi_n = \frac{\varphi_n^*}{\sqrt{(\varphi_n^*, \varphi_n^*)}},$$

we obtain the required sequence.

Consider the sequence

$$\psi_n = \varkappa_1\varphi_n - A^2\varphi_n.$$

We show that

$$\lim_{n \to \infty} (\psi_n, \psi_n) = 0. \tag{24.19}$$

For,

$$(\psi_n, \psi_n) = \varkappa_1^2(\varphi_n, \varphi_n) - 2\varkappa_1(\varphi_n, A^2\varphi_n) + (A^2\varphi_n, A^2\varphi_n).$$

By (24.16) we have

$$(\psi_n, \psi_n) \leq 2\varkappa_1^2 - 2\varkappa_1(\varphi_n, A^2\varphi_n) = 2\varkappa_1(\varkappa_1 - (A\varphi_n, \varphi_n)).$$

Hence, using (24.18), we obtain (24.19).

Thus the sequence of functions $\varkappa_1\varphi_n - A^2\varphi_n$ tends to zero in the mean. Since the operator A^2 is completely continuous, we can choose from the sequence $A^2\varphi_n$ a sequence $A^2\varphi_{n_k}$ which converges in the mean to a certain continuous function.

But

$$\lim A^2 \varphi_{n_k} = \varkappa_1 \lim \varphi_{n_k} \tag{24.20}$$

where the limit on the right is to be understood as in the mean. This implies that the sequence φ_{n_k} also has as its limit in the mean a certain function, which we denote by φ_0. Passing to the limit in the relation (24.20), we get

$$\varkappa_1 \varphi_0 - A^2 \varphi_0 = 0.$$

Or, putting $\varkappa_1 = 1/\mu_1$, we have

$$\varphi_0 = \mu_1 A^2 \varphi_0 = 0.$$

Consequently equation (24.15) has the eigenvalue μ_1, as was to be shown.

Comparing Theorems 4, 5 and 6, we see that the existence has been proved of at least one eigenfunction and one eigenvalue for integral equations with weighted, real, symmetric kernels in two cases—for continuous kernels, and for almost regular kernels.

It is important to note that the eigenfunction, whose existence we have proved, is always continuous. This follows from the strong, complete continuity of the operator A.

THE BILINEAR FORMULA
AND THE HILBERT–SCHMIDT THEOREM

§ 1. The Bilinear Formula

In the last lecture we proved that an integral equation with a weighted symmetric kernel

$$\varphi = \lambda A\varphi \tag{25.1}$$

always has a fundamental function φ_1 and a characteristic number λ_1. By multiplying this function by a constant, we can arrange to make

$$\int_\Omega [\varphi_1(P)]^2 \, \varrho(P) \, \mathrm{d}P = 1.$$

We now introduce a new integral operator B_1 defined by the conditions

$$\psi(P_0) = B_1\varphi \equiv \frac{1}{\lambda_1} \, \varphi_1(P_0) \int_\Omega \varphi_1(P) \, \varphi(P) \, \varrho(P) \, \mathrm{d}P.$$

Obviously

$$L(P, P_0) = \frac{\varphi_1(P_0) \, \varphi_1(P)}{\lambda_1}$$

is the kernel of the operation B_1.

The equation

$$\varphi = \lambda B_1 \varphi \tag{25.2}$$

obviously has an eigenvalue λ_1 and a fundamental solution φ_1. For, the function $B_1\varphi_1$ differs from φ_1 only by a (constant) factor. Consequently, for any λ, the solution of equation (25.2) can be only the function φ_1. Substituting $\varphi = \varphi_1$ in (25.2), we get

$$\varphi_1(P) = \frac{\lambda}{\lambda_1} \, \varphi_1(P),$$

whence $\lambda = \lambda_1$.

LEMMA. *All fundamental functions of the integral equation*

$$\varphi = \lambda(A - B_1) \, \varphi \tag{25.3}$$

with the kernel

$$K - L = K(P, P_0) - \frac{\varphi_1(P)\, \varphi_1(P_0)}{\lambda_1} \tag{25.4}$$

are fundamental functions of equation (25.1) for the same values of λ. Conversely, all fundamental functions of equation (25.1) which are orthogonal to φ_1 serve as fundamental functions of equation (25.3) for the same values of λ.

Proof. Let $\varphi_k (k \neq 1)$ be some solution of equation (25.1), corresponding to the eigenvalue λ_k. Then $B_1 \varphi_k = 0$, since all fundamental functions of equation (25.1) are orthogonal. This implies

$$(A - B_1)\, \varphi_k = A\varphi_k = \frac{1}{\lambda_k}\, \varphi_k$$

i.e.,

$$\varphi_k = \lambda_k (A - B_1)\, \varphi_k.$$

Further,

$$\lambda(A - B_1)\, \varphi_1 = 0.$$

Consequently, all solutions of equation (25.1) except φ_1 satisfy equation (25.3), and, moreover, with the same eigenvalues.

We now prove that any eigenfunction of equation (25.3) satisfies the equation (25.1). We first show that all solutions of (25.3) for any λ are orthogonal to φ_1. For, from (25.3) we have

$$(\varphi, \varphi_1) = \lambda[(A - B_1)\, \varphi, \varphi_1] = \lambda(\varphi, (A - B_1)\, \varphi_1) = 0.$$

Now let φ_k^* be any solution whatever of (25.3) corresponding to $\lambda = \lambda^*$. Since φ_k^* is orthogonal to φ_1, we have $B_1 \varphi_k^* = 0$. Thus $(A - B_1)\, \varphi_k^* = A\varphi_k^*$, and consequently $\varphi_k^* = \lambda^*(A - B_1)\, \varphi_k^* = \lambda^* A \varphi_k^*$, as was to be shown.

For brevity we now write

$$A - B_1 = A_1.$$

The operator A_1 is again a symmetrical integral operator. Two possibilities now present themselves: either its kernel is identically zero, or it has at least one more fundamental function $\varphi_2(P)$ corresponding to an eigenvalue λ_2 (λ_2 may sometimes be equal to λ_1). But in the latter case we can set up an operation B_2 with the kernel

$$\frac{\varphi_2(P)\, \varphi_2(P_0)}{\lambda_2}$$

and, repeating the foregoing argument, arrive at an operator

$$A_2 = A - B_1 - B_2,$$

for which the integral equation $\varphi = \lambda A_2 \varphi$ will have the same fundamental functions as (25.1) has, except φ_1 and φ_2, and only those fundamental functions. We continue this process.

Then if the kernel of the operator A has only m eigenfunctions, the operator

$$A_m = A - B_1 - B_2 - \cdots - B_m \qquad (25.5)$$

will have no eigenfunction, *i.e.*, it will be identically zero. Hence we have:

THEOREM 1. *A symmetric weighted kernel which has a finite number of fundamental functions can be expressed in the form*

$$K(P, P_0) = \sum_{i=1}^{m} \frac{\varphi_i(P)\, \varphi_i(P_0)}{\lambda_i} \qquad (25.6)$$

and consequently is degenerate.

If the kernel $K(P, P_0)$ has an infinite set of eigenfunctions, then, by arranging all the λ_m in order of increasing absolute value, we can obtain operators

$$A_m = A - B_1 - B_2 - \cdots - B_m, \quad m = 1, 2, 3, \ldots,$$

whose eigenvalues are arbitrarily large in absolute value for sufficiently large values of m; for, by Fredholm's 4th theorem, the sequence $\{\lambda_m\}$ must, if it is infinite, be unbounded. Let

$$\varkappa_m = \frac{1}{\lambda_m^2}.$$

Then for all φ such that $(\varphi, \varphi) = 1$,

$$\max (A_m\varphi, A_m\varphi) = \varkappa_m.$$

For, if $(A_m\varphi, A_m\varphi)$ could take values greater than $\varkappa_m$, the equation $\varphi = \lambda A_m\varphi$ would have an eigenvalue λ_{m+1} with $|\lambda_{m+1}| < |\lambda_m|$, and this is impossible.

We thus see that

$$\lim_{m \to \infty} (A_m\varphi, A_m\varphi) = 0, \qquad (25.7)$$

and that this holds uniformly for all φ which are uniformly bounded in the mean.

In a certain sense the equation (25.7) means that $A_m\varphi$ tends to zero; consequently the kernel of the operator $A - \sum_{i=1}^{m} B_i$ tends to zero. Hence the series

$$\sum_{i=1}^{\infty} \frac{\varphi_i(P)\, \varphi_i(P_0)}{\lambda_i}$$

in a certain sense represents the kernel $K(P, P_0)$.

If it happens that the series

$$\sum \frac{\varphi_i(P) \; \varphi_i(P_0)}{\lambda_i}$$

converges uniformly, then the kernel of the operator $A - \sum_{i=1}^{\infty} B_i$ will have no eigenvalues, *i.e.*, it will vanish. Hence we have:

THEOREM 2. *(The Bilinear Formula) If the series*

$$\sum_{i=1}^{\infty} \frac{\varphi_i(P) \; \varphi_i(P_0)}{\lambda_i} \tag{25.8}$$

converges uniformly in both variables, then its sum is equal to the kernel $K(P, P_0)$:

$$K(P, P_0) = \sum_{i=1}^{\infty} \frac{\varphi_i(P) \; \varphi_i(P_0)}{\lambda_i}.$$

In general, however, the series (25.8) may not converge uniformly: in which case there is a further question of interest to us.

We shall say that a function $f(P_0)$ which has the form

$$f(P_0) = \int_{\Omega} K(P, P_0) \, h(P) \, \varrho(P) \, \mathrm{d}P,$$

where h is a quadratically integrable function, is a sourcewise-representable function by h with the aid of the kernel K.

If we substituted in place of $K(P, P_0)$ its proposed representation by (25.8) we should obtain for f the formula

$$f(P_0) = \sum_{i=1}^{\infty} \varphi_i(P_0) \int_{\Omega} \frac{\varphi_i(P) \, h(P) \, \varrho(P) \, \mathrm{d}P}{\lambda_i} = \sum_{i=1}^{\infty} \frac{h_i}{\lambda_i} \varphi_i(P_0), \tag{25.9}$$

where

$$h_i = \int_{\Omega} h(P) \, \varphi_i(P) \, \varrho(P) \, \mathrm{d}P.$$

We now prove a theorem.

THEOREM 3. *If $f(P_0)$ is a function which is sourcewise representable by h, and if the integral*

$$\int_{\Omega} h^2(P) \, \varrho(P) \, \mathrm{d}P$$

converges, then the series

$$\sum_{i=1}^{\infty} \frac{h_i}{\lambda_i} \varphi_i(P_0) \tag{25.10}$$

converges in the mean and its sum is equal to $f(P_0)$. *In other words,*

$$\lim_{N \to \infty} \int_\Omega \left[f(P_0) - \sum_{i=1}^{N} \frac{h_i}{\varphi_i} \varphi_i(P_0) \right]^2 \varrho(P_0) \, \mathrm{d}P_0 = 0. \qquad (25.11)$$

This theorem is an obvious consequence of formula (25.7). For,

$$f(P_0) - \sum_{i=1}^{N} \frac{h_i}{\lambda_i} \varphi_i(P_0) = \int_\Omega \left(K - \sum_{i=1}^{N} B_i \right) h(P) \varrho(P) \, \mathrm{d}P = A_N h,$$

whence, applying formula (25.7), we at once obtain our theorem.

The theorems already proved will enable us to make clear later the significance of the argument by which we prove the existence of eigenvalues. But first we prove a lemma.

LEMMA 2. *(Bessel's Inequality)*

Let $\varphi_1, \varphi_2, \ldots, \varphi_n, \ldots$ *be a finite or infinite sequence of real, orthogonal functions which have been normalized with weight* ϱ:

$$\int_\Omega \varphi_i(P) \, \varphi_j(P) \, \varrho(P) \, \mathrm{d}P = \begin{cases} 1, & i = j \\ 0, & i \neq j \end{cases}$$

and let f be a certain function whose square is integrable with weight ϱ:

$$\int_\Omega f^2 \varrho \, \mathrm{d}P = A.$$

We shall call the numbers

$$f_i = \int_\Omega f \varphi_i \varrho \, \mathrm{d}P$$

the Fourier coefficients for the function f.

Then the series $\sum_{i=1}^{\infty} f_i^2$ *converges and its sum does not exceed* A:

$$\sum_{i=1}^{\infty} f_i^2 \leq A. \qquad (25.12)$$

If for a certain function f the inequality (25.12) becomes an equality then the system of functions is said to be *closed* relative to f. A system of functions which is closed relative to all functions whose squares are integrable is called simply a *closed system*.

Equation (25.12) is known as *Bessel's Inequality*.

To prove the lemma, we first show that

$$\sum_{i=1}^{N} f_i^2 \leqq A,$$

from which, when $N \to \infty$, we at once obtain the proof of the lemma. We have

$$0 \leqq \int_{\Omega} \left(f - \sum_{i=1}^{N} \varphi_i f_i \right)^2 \varrho \, dP$$

$$= \int_{\Omega} f^2 \varrho \, dP - 2 \sum_{i=1}^{N} f_i \cdot \int_{\Omega} f \varphi_i \varrho \, dP + \sum_{i=1}^{N} f_i^2 \cdot \int_{\Omega} \varphi_i^2 \varrho \, dP$$

$$= \int_{\Omega} f^2 \varrho \, dP - \sum_{i=1}^{N} f_i^2 = A - \sum_{i=1}^{N} f_i^2.$$

Hence the lemma.

The fact that the system $\{\varphi_i\}$ is closed relative to f has the following significance. If

$$\lim_{N \to \infty} \sum_{i=1}^{N} f_i^2 = A, \quad \text{then} \quad \lim_{N \to \infty} \left[A - \sum_{i=1}^{N} f_i^2 \right] = 0$$

and this implies

$$\lim_{N \to \infty} \int_{\Omega} \left(f - \sum_{i=1}^{N} \varphi_i f_i \right)^2 \varrho \, dP = 0,$$

i.e., the function $f(P)$ is represented by the series $\sum_{i=1}^{\infty} f_i \varphi_i(P)$ which converges in the mean.

Thus, Theorem 3 shows that the system of fundamental functions is closed relative to any sourcewise representable function.

LEMMA 3. *Let $u_1(P), u_2(P), \ldots, u_N(P)$ be any system of orthogonal, normalized functions, and $f(P)$ be an arbitrary, quadratically integrable function. We consider the minimum value of the integral*

$$I_N = \int_{\Omega} \left[f(P) - \sum_{i=1}^{N} a_i u_i(P) \right]^2 dP$$

for all possible values of a_i.

This minimum is attained when $a_i = f_i$, where $f_i = \int_{\Omega} f(P) u_i(P) \, dP$, and is equal to

$$\int_{\Omega} [f(P)]^2 \, dP - \sum_{i=1}^{N} f_i^2.$$

For, let

$$f(P) = \sum_{i=1}^{n} f_i u_i(P) + R_N(P):$$ (25.13)

multiplying both sides of (25.13) by $u_i(P)$ and integrating, we get

$$\int_{\Omega} R_N(P)\, u_i(P)\, \mathrm{d}P = 0, \quad (i = 1,2, \ldots, N).$$

Then

$$I_N = \int_{\Omega} \left[\sum_{i=1}^{N} (f_1 - a_i)\, u_i(P) + R_N(P) \right]^2 \mathrm{d}P$$

$$= \int_{\Omega} R_N^2(P)\, \mathrm{d}P + \sum_{i=1}^{N} (f_i - a_i)^2.$$

It is clear that this expression will have a minimum for $a_i = f_i$. Then

$$\int_{\Omega} R_N^2(P)\, \mathrm{d}P = \int_{\Omega} \left[f(P) - \sum_{i=1}^{N} f_i u_i(P) \right]^2 \mathrm{d}P = \int_{\Omega} f^2(P)\, \mathrm{d}P - \sum_{i=1}^{N} f_i^2,$$

as was to be proved.

We make two more small observations.

THEOREM 4. *For any quadratically integrable function ψ,*

$$(\psi, A\psi) = \sum_{i=1}^{\infty} \frac{\psi_i^2}{\lambda_i},$$ (25.14)

where ψ_i are the Fourier coefficients of the function ψ, i.e.,

$$\psi_i = \int_{\Omega} \psi(P)\, \varphi_i(P)\, \varrho(P)\, \mathrm{d}P.$$

We note that

$$A_m\psi = A\psi - \sum_{i=1}^{m} \frac{\psi_i \varphi_i(P)}{\lambda_i},$$

and consequently

$$(\psi, A_m\psi) = (\psi, A\psi) - \sum_{i=1}^{m} \left(\psi, \frac{\psi_i \varphi_i(P)}{\lambda_i} \right) = (\psi, A\psi) - \sum_{i=1}^{m} \frac{\psi_i^2}{\lambda_i}.$$

But

$$\left| (\psi, A_m\psi) \right| = \left| \int_{\Omega} \psi A_m\psi \varrho\, \mathrm{d}P \right| \le \sqrt{\int_{\Omega} \psi^2\, \mathrm{d}P \cdot \int_{\Omega} (A_m\psi)^2\, \varrho\, \mathrm{d}P}$$

$$= \sqrt{|\varkappa_m| \int_{\Omega} \psi^2 \varrho\, \mathrm{d}P}.$$

Since $\varkappa_m \to 0$ as $m \to \infty$, so also the quantity $(\psi, A_m \psi)$ tends to zero. We have

$$\lim_{m \to \infty} \left| (\psi, A\psi) - \sum_{i=1}^{m} \frac{\psi_i^2}{\lambda_i} \right| = 0,$$

as was to be proved.

COROLLARY. *If all the eigenvalues λ_i are positive, then $(\psi, A\psi) \geqq 0$ for any ψ, and conversely, if $(\psi, A\psi)$ nowhere takes negative values, then all the λ_j are positive.*

Kernels which have only positive eigenvalues are said to be *positive definite*.

THEOREM 5. *The least eigenvalue of a positive definite kernel is determined by the equation*

$$\frac{1}{\lambda_0} = \sup_{(\psi,\psi)=1} (\psi, A\psi). \tag{25.15}$$

For,

$$(\psi, A\psi) = \frac{1}{\lambda_0} \quad \text{for} \quad \psi = \varphi_0.$$

On the other hand,

$$\sum_{i=1}^{\infty} \frac{\psi_i^2}{\lambda_i} \leqq \sum_{i=1}^{\infty} \frac{\psi_i^2}{\lambda_0}$$

where λ_0 is the least positive of the λ_i,
and since $(\psi, \psi) = 1$, we have from (25,12),

$$\sum_{i=1}^{\infty} \frac{\psi_i^2}{\lambda_i} \leqq \frac{1}{\lambda_0},$$

as was to be shown.

We recall that it was, in fact, by determining the upper bound of the expression $(\psi, A\psi)$ subject to the condition $(\psi, \psi) = 1$ that we established the existence of an eigenvalue for the kernel K_2 of the operator A^2.

§ 2. The Hilbert–Schmidt Theorem

To conclude our investigation, we prove further that, under certain conditions, the convergence of the series (25.8) will be uniform.

THEOREM 6. *(Hilbert–Schmidt)*
If a real symmetric kernel $K(P_0, P)$ is quadratically integrable with respect to each variable, and if each of these integrals of its square is uniformly bounded with respect to the other variable, i.e., if

$$\int_{\Omega} |K(P_0)|^2 \varrho(P) \, dP = A(P_0) \leqq A, \tag{25.16}$$

then the series (25.10) converges uniformly to the function $f(P_0)$.

For, by Bessel's inequality, the series

$$\sum_{i=1}^{\infty} \frac{\varphi_i^2(P_0)}{\lambda_i^2}$$

converges and the inequality

$$\sum_{i=1}^{\infty} \frac{\varphi_i^2(P_0)}{\lambda_i^2} \leqq A \tag{25.17}$$

holds. This follows from the fact that $\varphi_i(P_0)/\lambda_i$ serve as the Fourier coefficients for the function $f = K(P_0, P)$:

$$\frac{\varphi_i(P_0)}{\lambda_i} = \int_\Omega K(P_0, P)\, \varphi_i(P)\, \varrho(P)\, \mathrm{d}P. \tag{25.18}$$

Moreover, the series with constant terms

$$\sum_{i=1}^{\infty} h_i^2 \tag{25.19}$$

also converges, again by virtue of Bessel's inequality. From the convergence of the series (25.19) it follows that

$$\sum_{i=m}^{m+p} h_i^2 < \varepsilon_m, \quad \text{where} \quad \varepsilon_m \to 0 \quad \text{as} \quad m \to \infty.$$

By Bunyakovski's inequality,

$$\left(\sum_{i=m}^{m+p} \left| \frac{h_i \varphi_i}{\lambda_i} \right| \right)^2 \leqq \left(\sum_{i=m}^{m+p} h_i^2 \right) \left(\sum_{i=m}^{m+p} \frac{\varphi_i^2}{\lambda_i^2} \right) \leqq \varepsilon_m A:$$

consequently, the sum

$$\sum_{i=m}^{m+p} \left| \frac{h_i \varphi_i(P_0)}{\lambda_i} \right| \tag{25.20}$$

is as small as we please, and this implies that the series (25.10) converges uniformly.

Writing

$$\gamma(P_0) = \sum_{i=1}^{\infty} \frac{h_i \varphi_i(P_0)}{\lambda_i},$$

and passing to the limit in formula (25.11), we get

$$\int_\Omega (f - \gamma)^2 \varrho \, \mathrm{d}P_0 = 0.$$

Consequently,

$$f(P_0) = \gamma(P_0) = \sum_{i=1}^{\infty} \frac{h_i \varphi_i(P_0)}{\lambda_i}. \tag{25.21}$$

Hence the Hilbert–Schmidt theorem.

COROLLARY. (Bilinear Series for an Iterated Kernel)

If the kernel $K(P_0, P)$ satisfies the condition of the theorem, then for an iterated kernel the bilinear series converges and moreover does so uniformly relative to P for a fixed P_0.

We can take $\varphi_i(P_0)/\lambda_i^2$ as the Fourier coefficients for the iterated kernel. For,

$$\int_{\Omega} K_2(P_0, P) \, \varphi_i(P) \, \varrho(P) \, dP$$

$$= \int_{\Omega} K(P_0, P_1) \left[\int_{\Omega} K(P_1, P) \, \varphi_i(P) \, \varrho(P) \, dP \right] \varrho(P_1) \, dP_1$$

$$= \frac{1}{\lambda_i} \int_{\Omega} K(P_0, P_1) \, \varphi_i(P_1) \, \varrho(P_1) \, dP_1 = \frac{1}{\lambda_i^2} \, \varphi_i(P_0).$$

Hence

$$K_2(P_0, P) = \sum_{i=1}^{\infty} \frac{\varphi_i(P_0) \, \varphi_i(P)}{\lambda_i^2}. \tag{25.22}$$

The convergence of the series for kernels K_m, where $m \geq 2$, will be even more rapid; for these kernels we shall obviously have

$$K_m(P_0, P) = \sum_{i=1}^{\infty} \frac{\varphi_i(P_0) \, \varphi_i(P)}{\lambda_i^m}.$$

It can be shown, though we shall not bother to do so here, that the convergence of the bilinear series (25.22) is uniform relative to both variables.

The Hilbert–Schmidt theorem is obviously valid for continuous kernels with any bounded measurable weight. It is not difficult to see that it also holds for almost regular kernels, since for such kernels

$$\int_{\Omega} \{K(P_0, P)\}^2 \, \varrho(P) \, dP \leq \int_{\Omega} \frac{A^2}{r^{n-2a}} \, \varrho(P) \, dP < M,$$

as is required for the hypothesis of the Hilbert–Schmidt theorem.

The Hilbert–Schmidt theorem can be extended immediately for all kernels of the type of Green's function with two or three independent variables. For, such kernels, which have a singularity $\log_e (1/r)$ or $(1/r)$, are clearly almost regular.

Consider the integral

$$d_N = \int_\Omega \int_\Omega \left[K(P, P_1) - \sum_{i=1}^N \frac{\varphi_i(P)\,\varphi_i(P_1)}{\lambda_i} \right]^2 \varrho(P)\,\varrho(P_1)\,\mathrm{d}P\,\mathrm{d}P_1.$$

Simple transformations give

$$d_N = \int_\Omega \int_\Omega [K(P, P_1)]^2 \varrho(P)\,\varrho(P_1)\,\mathrm{d}P\,\mathrm{d}P_1$$

$$- 2 \sum_{i=1}^N \int_\Omega \int_\Omega K(P, P_1) \frac{\varphi_i(P)\,\varphi_i(P_1)}{\lambda_i} \varrho(P)\,\varrho(P_1)\,\mathrm{d}P\,\mathrm{d}P_1$$

$$+ \int_\Omega \int_\Omega \sum_{i=1}^N \frac{\varphi_i^2(P)\,\varphi_i^2(P_1)}{\lambda_i^2} \varrho(P)\,\varrho(P_1)\,\mathrm{d}P\,\mathrm{d}P_1$$

$$= \int_\Omega \left[\int_\Omega [K(P, P_1)]^2 \varrho(P_1)\,\mathrm{d}P_1 - 2 . \sum_{i=1}^N \frac{\varphi_i^2(P)}{\lambda_i^2} + \sum_{i=1}^N \frac{\varphi_i^2(P)}{\lambda_i^2} \right] \varrho(P)\,\mathrm{d}P$$

$$= \int_\Omega \left[K_2(P, P) - \sum_{i=1}^N \frac{\varphi_i^2(P)}{\lambda_i^2} \right] \varrho(P)\,\mathrm{d}P.$$

Obviously, $d_N > 0$. By what we have proved above, the series $\sum_{i=1}^\infty \varphi_i^2(P)/\lambda_i^2$ converges to the function $K_2(P, P)$.

By Lemma 7 of Lecture 6 we can assert that the non-decreasing sequence of functions $\sum_{i=1}^N \varphi_i^2(P)/\lambda_i^2$, which has a bounded integral, converges almost everywhere to a limit function, and we can pass to the limit under the integral sign. This limit function can be none other than $K_2(P, P)$; whence it follows that $\lim d_N = 0$. The result just obtained gives the important.

THEOREM 7. *The bilinear series for a continuous kernel converges in the mean to this kernel with respect to both variables.*

There is also a theorem (which we shall not prove) due to *Mercer: The bilinear series for any continuous, positive definite kernel, i.e., a kernel which has only positive eigenvalues, converges uniformly.*

The uniform convergence with respect to both variables of the series (25.22) follows from Mercer's theorem.

§ 3. Proof of the Fourier Method for the Solution of the Boundary-value Problems of Mathematical Physics

We can now return to the assertions which we left unproved in Lecture 23 in our exposition of the Fourier method. We shall prove the five basic properties of the system of eigenfunctions which were enunciated in Lecture 23 (p. 329).

We saw that eigenfunctions for boundary-value problems of equation (23.7) were solutions of the integral equation (23.9) with a real, symmetric kernel and with an almost regular kernel. Hence it follows that the eigenvalues λ_i will be real. We shall prove shortly that there are infinitely many eigenvalues and that none of them is negative.

By what we have already proved, the eigenfunctions can be regarded as orthogonal and normalized, *i.e.*, Property 2 (p. 329) holds.

Certain properties of the eigenfunctions are best examined separately for the different boundary-value problems. We shall consider first the boundary-value problem of the first kind.

We have proved that the functions U_i which are the solutions of the equation

$$\nabla^2 U_i + \lambda_i U_i = 0 \tag{25.23}$$

satisfying the conditions

$$[U_i]_S = 0 \tag{25.24}$$

are solutions of the integral equation

$$U_i(P_0) = \lambda_i \iiint_\Omega G U_i \, dP. \tag{25.25}$$

We shall now prove the converse proposition:

Any solution of the integral equation (25.25) has continuous derivatives up to the second order inclusive, is continuous right up to the boundary, and satisfies the differential equation (25.23) and the boundary condition (23.24).

As a preliminary, we show that the U_i have continuous first-order derivatives. Because of the strong, complete continuity of the integral operator with almost regular kernel (25.25), the function U_i is bounded (it is uniformly continuous in a bounded domain). Green's function G may be written in the form

$$G(P, P_0) = \frac{1}{4\pi r} + g(P, P_0),$$

where g is regular in the domain Ω relative to either argument of the function when the other argument has a fixed value lying within the domain Ω. Hence

$$U_i(P_0) = \frac{1}{4\pi} \iiint_\Omega \frac{1}{r} U_i(P) \, dP + \iiint_\Omega g U_i(P) \, dP.$$

Both terms on the right-hand side of this equation have continuous first-order derivatives at any internal point P_0 of the domain Ω, since differentiation with respect to the parameter under the integral sign gives integrals which are uniformly convergent at the point P_0. Hence the continuous differentiability of U_i is proved.

We now pass on to the proof of the proposition already enunciated. Consider Poisson's equation

$$\nabla^2 u = -\lambda_i U_i, \tag{25.26}$$

where a solution of the integral equation (25.25) stands on the right-hand side. We shall seek a solution of this equation satisfying the condition

$$[u]_S = 0. \tag{25.27}$$

Such a solution can be constructed in two stages. Consider first of all a Newtonian potential with the continuously differentiable density $\lambda_i U_i / 4\pi$:

$$Q(P_0) = \frac{\lambda_i}{4\pi} \iiint_\Omega \frac{U_i}{r} \, dP. \tag{25.28}$$

We can show, without essential alteration of the proof given in Lecture 11, that the potential $Q(P_0)$ is a solution of equation (25.26). A function u which satisfies equation (25.26) and the condition (25.27) can be put into the form

$$u(P_0) = Q(P_0) + w(P_0),$$

where $w(P_0)$ is a function which is harmonic in the domain Ω and takes the values $Q(S)$ on the boundary S. Such a function exists. By Lyapunov's lemma (see Lecture 21, p. 292), it has a regular normal derivative. Consequently a solution u exists of equation (25.26) satisfying the condition (25.27) and having a regular normal derivative.

But such a solution, as was shown in Lecture 21, can be expressed in the form

$$u(P_0) = \lambda_i \iiint_\Omega G(P_0, P) \, U_i(P) \, dP. \tag{25.29}$$

By (25.25) the right-hand side of the last equation is $U_i(P_0)$. This implies that the function $U_i(P_0)$ coincides with the function $U(P_0)$, from which follows immediately the proposition that was to be proved.

We now go on to investigate the eigenfunctions for the second kind of boundary-value problem. We proved that any function V_i which is bounded and which satisfies the equation

$$\nabla^2 V_i + \lambda_i V_i = 0 \tag{25.30}$$

and the boundary condition

$$\left[\frac{\partial V_i}{\partial n} \right]_S = 0, \tag{25.31}$$

where the derivative is to be understood as the regular normal derivative, will be a solution of the integral equation

$$V_i(P_0) = \lambda_i \iiint_\Omega G^*(P_0, P) \, V_i(P) \, dP, \tag{25.32}$$

where $G^*(P_0, P)$ is the generalized Green's function for the Neumann problem. We now prove the converse proposition.

We shall seek a solution of the equation

$$\nabla^2 v = -\lambda_i V_i \tag{25.33}$$

subject to the condition

$$\left[\frac{\partial v}{\partial n}\right]_s = 0. \tag{25.34}$$

Such a solution exists, since the function V_i is orthogonal to a constant, *i.e.*, to the solution of the homogeneous Neumann problem. In order to find the required solution, we first set up the Newtonian potential

$$R(P_0) = \frac{\lambda_i}{4\pi} \iiint_\Omega \frac{V_i}{r} \, d\Omega. \tag{25.35}$$

Since V_i is bounded, this potential will have everywhere continuous first-order derivatives satisfying the conditions of Theorem 4 of Lecture 15 (the Lyapunov conditions).

Moreover, the function v will be expressible in the form

$$v(P_0) = R(P_0) + s(P_0), \tag{25.36}$$

where $s(P_0)$ is a harmonic function satisfying the condition

$$\left[\frac{\partial s}{\partial n}\right]_s = -\left[\frac{\partial R}{\partial n}\right]_s.$$

But $[\partial R/\partial n]_s$ is a continuous function on the surface S and satisfies the Lyapunov conditions. Hence, as in Lecture 16, § 2, the function s will take the form of a potential of a single layer, whose density (by virtue of (16.4) and Theorem 6 of Lecture 15) will in its turn satisfy the Lyapunov conditions. It follows from Theorem 4 of Lecture 16 that the function s will have a regular normal derivative.

It has been proved (see Theorem 2, Lecture 21) that the solution of equation (25.33) with the conditions (25.34) can be presented in the form

$$v(P_0) = \lambda_i \iiint_\Omega G^*(P_0, P) \, V_i(P) \, dP. \tag{25.37}$$

Consequently the function $v(P_0)$ coincides with the function V_i. Hence V_i has a regular normal derivative, which vanishes on the boundary, and it satisfies the equation (25.30).

We shall now prove Property 1. We have still to show that there are no negative eigenvalues among those, λ_i, for a kernel representable by a Green's function. Using equation (23.7) we have

$$\int_\Omega U_i \, \nabla^2 U_i \, dx \, dy \, dz = -\lambda_i \int_\Omega U_i^2 \, dx \, dy \, dz = -\lambda_i. \qquad (25.38)$$

On the other hand, by Green's formula, we have, using the boundedness of U_i and the regularity of the normal derivatives,

$$\iiint_\Omega U_i \, \nabla^2 U_i \, dx \, dy \, dz$$

$$= -\iiint_\Omega \left\{ \left(\frac{\partial U_i}{\partial x}\right)^2 + \left(\frac{\partial U_i}{\partial y}\right)^2 + \left(\frac{\partial U_i}{\partial z}\right)^2 \right\} dx \, dy \, dz + \iint_s U_i \frac{\partial U_i}{\partial n} \, dS.$$

$$(25.39)$$

But if $[U_i]_S = 0$ or $[\partial U_i/\partial n]_S = 0$, then the last term, the surface integral, vanishes, and we get

$$\lambda_i = \iiint_\Omega \left\{ \left(\frac{\partial U_i}{\partial x}\right)^2 + \left(\frac{\partial U_i}{\partial y}\right)^2 + \left(\frac{\partial U_i}{\partial z}\right)^2 \right\} dx \, dy \, dz \geq 0, \quad (25.40)$$

as we had to prove.

We now prove the third property of a system of eigenfunctions—its completeness. Let φ be an arbitrary function which satisfies on the boundary of the domain the condition either $[\varphi]_S = 0$ or $[\partial\varphi/\partial n]_S = 0$ and which has continuous second-order derivatives everywhere. Then, if we put

$$\nabla^2 \varphi = 4\pi\psi,$$

the function ψ is continuous in the closed domain Ω. By the property of Green's function

$$\varphi = -\iiint_\Omega G(P_0, P) \, \psi(P) \, dP.$$

Thus the function φ is sourcewise represented by ψ with the aid of the kernel G and consequently may be expanded in a uniformly convergent series of eigenfunctions. Thus the system of eigenfunctions is complete relative to any function φ which is continuously twice-differentiable and which satisfies the boundary conditions. But with the aid of such functions it is obviously possible to represent approximately in the mean any function which is quadratically integrable in the domain.

As we have already proved (see Lemma 3, p. 362), the best approximation in the mean to a given function is a finite Fourier series. Hence a finite Fourier series also gives an approximation in the mean to any function which is quadratically integrable over the domain Ω. This implies that any function f which is quadratically integrable over the domain Ω can be expanded as a Fourier series of eigenfunctions converging in the mean. Hence

Property 3 has been proved. It follows, incidentally, that there are infinitely many eigenfunctions $\{U_i\}$, for otherwise they could not form a complete system.

We now prove Property 4. Let U_i and U_j be two eigenfunctions. Integrating by parts and using the boundary conditions, we have:

$$\iiint_\Omega \left(\frac{\partial U_i}{\partial x} \frac{\partial U_j}{\partial x} + \frac{\partial U_i}{\partial y} \frac{\partial U_j}{\partial y} + \frac{\partial U_i}{\partial z} \frac{\partial U_j}{\partial z} \right) \mathrm{d}x\, \mathrm{d}y\, \mathrm{d}z$$

$$= -\iiint_\Omega U_i \, \nabla^2 U_j \, \mathrm{d}x\, \mathrm{d}y\, \mathrm{d}z + \iint_S U_i \frac{\partial U_j}{\partial n} \, \mathrm{d}S = \lambda_j \iiint_\Omega U_i U_j \, \mathrm{d}x\, \mathrm{d}y\, \mathrm{d}z,$$

$$(25.41)$$

from which Property 4 follows.

It remains to prove Property 5. To do this, we consider the integral

$$J = \iiint_\Omega \left\{ \left[\frac{\partial}{\partial x} (\varphi - \varphi_N) \right]^2 + \left[\frac{\partial}{\partial y} (\varphi - \varphi_N) \right]^2 + \left[\frac{\partial}{\partial z} (\varphi - \varphi_N) \right]^2 \right\} \times$$

$$\times \ \mathrm{d}x\, \mathrm{d}y\, \mathrm{d}z, \qquad\qquad (25.42)$$

where $\varphi_N = \sum_{i=1}^{N} a_i U_i$, and the quantities a_i are the Fourier coefficients of the function φ in the system $\{U_i\}$.

Removing the brackets and integrating, in exactly the same way as in the derivation of Bessel's inequality, and making use of Property 4, then:

$$J = \iiint_\Omega \left\{ \left(\frac{\partial \varphi}{\partial x} \right)^2 + \left(\frac{\partial \varphi}{\partial y} \right)^2 + \left(\frac{\partial \varphi}{\partial z} \right)^2 \right\} \mathrm{d}x\, \mathrm{d}y\, \mathrm{d}z$$

$$-2 \sum_{i=1}^{N} a_i \iiint_\Omega \left\{ \frac{\partial \varphi}{\partial x} \frac{\partial U_i}{\partial x} + \frac{\partial \varphi}{\partial y} \frac{\partial U_i}{\partial y} + \frac{\partial \varphi}{\partial z} \frac{\partial U_i}{\partial z} \right\} \mathrm{d}x\, \mathrm{d}y\, \mathrm{d}z + \sum_{i=1}^{N} \lambda_i a_i^2 .$$

$$(25.43)$$

It is easily proved that

$$\iiint_\Omega \left\{ \frac{\partial \varphi}{\partial x} \frac{\partial U_i}{\partial x} + \frac{\partial \varphi}{\partial y} \frac{\partial U_i}{\partial y} + \frac{\partial \varphi}{\partial z} \frac{\partial U_i}{\partial z} \right\} \mathrm{d}x\, \mathrm{d}y\, \mathrm{d}z = \lambda_i a_i .$$

For, integrating by parts,

$$\iiint_\Omega \left(\frac{\partial \varphi}{\partial x} \frac{\partial U_i}{\partial x} + \frac{\partial \varphi}{\partial y} \frac{\partial U_i}{\partial y} + \frac{\partial \varphi}{\partial z} \frac{\partial U_i}{\partial z} \right) \mathrm{d}x\, \mathrm{d}y\, \mathrm{d}z$$

$$= -\iiint_\Omega \varphi \, \nabla^2 U_i \, \mathrm{d}x\, \mathrm{d}y\, \mathrm{d}z = \lambda_i \iiint_\Omega \varphi U_i \, \mathrm{d}x\, \mathrm{d}y\, \mathrm{d}z = \lambda_i a_i .$$

Thus, from (25.43) we have

$$\iiint_\Omega \left\{ \left[\frac{\partial}{\partial x} (\varphi - \varphi_N) \right]^2 + \left[\frac{\partial}{\partial y} (\varphi - \varphi_N) \right]^2 + \left[\frac{\partial}{\partial z} (\varphi - \varphi_N) \right]^2 \right\} dx \, dy \, dz$$

$$= \iiint_\Omega \left[\left(\frac{\partial \varphi}{\partial x} \right)^2 + \left(\frac{\partial \varphi}{\partial y} \right)^2 + \left(\frac{\partial \varphi}{\partial z} \right)^2 \right] dx \, dy \, dz - \sum_{i=1}^{N} \lambda_i a_i^2 .$$

The left-hand side of this last equation is non-negative, and we obtain an inequality similar to Bessel's inequality:

$$\sum_{i=1}^{N} \lambda_i a_i^2 \leq \iiint_\Omega \left[\left(\frac{\partial \varphi}{\partial x} \right)^2 + \left(\frac{\partial \varphi}{\partial y} \right)^2 + \left(\frac{\partial \varphi}{\partial z} \right)^2 \right] dx \, dy \, dz .$$

This inequality shows that the series $\sum \lambda_i a_i^2$ converges. And from this follows the convergence in the mean for the series

$$\sum_{i=1}^{\infty} a_i \frac{\partial U_i}{\partial x}, \quad \sum_{i=1}^{\infty} a_i \frac{\partial U_i}{\partial y}, \quad \sum_{i=1}^{\infty} a_i \frac{\partial U_i}{\partial z} .$$

For, consider the integral

$$\iiint_\Omega \left[\left(\sum_{i=n}^{n+p} a_i \frac{\partial U_i}{\partial x} \right)^2 + \left(\sum_{i=n}^{n+p} a_i \frac{\partial U_i}{\partial y} \right)^2 + \left(\sum_{i=n}^{n+p} a_i \frac{\partial U_i}{\partial z} \right)^2 \right] dx \, dy \, dz .$$

If we establish that this integral is as small as we please for a sufficiently large n and arbitrary p, then the convergence in which we are interested will follow. Using Property 4, we can evaluate this integral directly; we find

$$\iiint_\Omega \left[\left(\sum_{i=n}^{n+p} a_i \frac{\partial U_i}{\partial x} \right)^2 + \left(\sum_{i=n}^{n+p} a_i \frac{\partial U_i}{\partial y} \right)^2 + \left(\sum_{i=n}^{n+p} a_i \frac{\partial U_i}{\partial z} \right)^2 \right] dx \, dy \, dz$$

$$= \iiint_\Omega \left\{ \sum_{i=n}^{n+p} a_i^2 \left[\left(\frac{\partial U_i}{\partial x} \right)^2 + \left(\frac{\partial U_i}{\partial y} \right)^2 + \left(\frac{\partial U_i}{\partial z} \right)^2 \right] \right\} dx \, dy \, dz$$

$$+ 2 \sum_{\substack{i,j=n \\ i \neq j}}^{n+p} a_i a_j \iiint_\Omega \left(\frac{\partial U_i}{\partial x} \frac{\partial U_j}{\partial x} + \frac{\partial U_i}{\partial y} \frac{\partial U_j}{\partial y} + \frac{\partial U_i}{\partial z} \frac{\partial U_j}{\partial z} \right) dx \, dy \, dz$$

$$= \sum_{i=n}^{n+p} \lambda_i a_i^2 .$$

Since the series $\sum_{i=1}^{\infty} \lambda_i a_i^2$ converges, it follows by Cauchy's criterion that the last sum is arbitrarily small for sufficiently large n. We have proved that the series obtained by termwise differentiation with respect to x, y and z of the series

$$\sum_{i=1}^{\infty} a_i U_i(x, y, z)$$

converges in the mean.

To complete the proof of Property 5 we have to prove one more theorem.

THEOREM 8. *If the series* $\varphi = \sum_{i=1}^{\infty} v_i$ *of terms continuously differentiable in the domain* Ω *converges uniformly, if the sum of the series has a partial derivative with respect to* x, *and if the series* $\sum_{i=1}^{\infty} \partial v_i/\partial x$ *obtained by termwise differentiation converges in the mean, then the series* $\sum_{i=1}^{\infty} v_i$ *can be differentiated termwise, i.e.,*

$$\frac{\partial \varphi}{\partial x} = \sum_{i=n}^{\infty} \frac{\partial v_i}{\partial x}, \tag{25.44}$$

where, of course, by convergence is to be understood convergence in the mean.

Proof. Let ψ be an arbitrary, continuously differentiable function of the variables x, y, z, which is different from zero only in a certain domain Ω_ψ lying wholly within the domain. Then the following formula for integration by parts holds:

$$\iiint_{\Omega} \left(\psi \frac{\partial \chi}{\partial x} + \chi \frac{\partial \psi}{\partial x} \right) dx\, dy\, dz = 0, \tag{25.45}$$

where χ is an arbitrary function having a continuous derivative $\partial \chi/\partial x$. In particular,

$$\iiint_{\Omega} \left(\psi \frac{\partial \varphi}{\partial x} + \varphi \frac{\partial \psi}{\partial x} \right) dx\, dy\, dz = 0. \tag{25.46}$$

We now substitute in place of χ in (25.45) the function $\varphi_N = \sum_{i=1}^{N} v_i$. Then

$$\iiint_{\Omega} \left(\psi \frac{\partial \varphi_N}{\partial x} + \varphi_N \frac{\partial \psi}{\partial x} \right) dx\, dy\, dz = 0.$$

In this formula we can pass to the limit as $N \to \infty$. Let φ' denote the sum of the series $\sum_{i=1}^{\infty} \partial v_i/\partial x$. By the hypothesis, this sum exists. We then have

$$\iiint_{\Omega} \left(\psi \varphi' + \varphi \frac{\partial \psi}{\partial x} \right) dx\, dy\, dz = 0.$$

Subtracting this equality from (25.46) we get

$$\iiint_{\Omega} \psi \left(\frac{\partial \varphi}{\partial x} - \varphi' \right) dx\, dy\, dz = 0.$$

This equality holds for *any* function, subject only to the requirements stated above; hence

$$\frac{\partial \varphi}{\partial x} = \dot{\varphi}' = \sum_{i=1}^{\infty} \frac{\partial v_i}{\partial x},$$

as was to be proved.

§ 4. An Application of the Theory of Integral Equations with Symmetric Kernel

One of the most important applications of the theory of integral equations with symmetric kernel is to the theory of the so-called *Sturm–Liouville* equations, *i.e.*, of ordinary differential equations

$$(py')' + ry + \lambda\varrho(x)y = 0 \tag{25.47}$$

depending on a parameter λ, with certain boundary conditions.

We shall seek a solution of such an equation satisfying the homogeneous conditions:

$$[py' + \alpha y]_{x=0} = 0, \quad [py' + \beta y]_{x=1} = 0. \tag{25.48}$$

Assuming that, for $\lambda = 0$, the homogeneous problem has at least a trivial solution, and applying Green's formula, we have

$$y(x_0) = -\int_0^1 \lambda G(x, x_0)\, y(x)\, \varrho(x)\, dx. \tag{25.49}$$

The kernel $G(x, x_0)$ will be symmetrical, because the operator Ly and the boundary conditions are self-adjoint.

Consequently the eigenfunctions of the Sturm–Liouville problem, *i.e.*, functions satisfying (25.47) and (25.48), will serve as eigenfunctions of the integral equation (25.49) with weighted symmetric kernel and will have all the properties of eigenfunctions of such equations. The converse is easily established: any solution of equation (25.49) will satisfy the differential equation (25.47) and the boundary conditions (25.48). We leave the reader to verify this assertion.

As an example of a more general equation with a singularity we may consider an equation of the form

$$(xy')' - \frac{m^2}{x} y + \lambda xy = 0,$$

or

$$xy'' + y' + x\left(\lambda - \frac{m^2}{x}\right)y = 0.$$

Green's function for this equation was constructed earlier [see (20.52)]. All our theory applies to this equation, which is known as *Bessel's Equation*.

THE INHOMOGENEOUS INTEGRAL
EQUATION WITH A SYMMETRIC KERNEL

§ 1. Expansion of the Resolvent

In previous lectures we have examined in some detail the homogeneous integral equation of Fredholm type with a symmetric kernel. In this lecture we shall study the inhomogeneous equation, from the point of view of the theory already developed. We have explained already how the kernel of a symmetrical integral equation can be expressed when its eigenvalues and eigenfunctions are known. The solution of an inhomogeneous equation can be expressed in a similar way, using the same eigenvalues and eigenfunctions.

To prove this, we shall apply the Hilbert–Schmidt theorem to an inhomogeneous equation with a symmetric kernel. We have

$$\mu(P) = f(P) + \lambda \int_{\Omega} K(P, P_1)\, \mu(P_1)\, dP_1$$

(for simplicity we take $\varrho = 1$).

Suppose that λ is not an eigenvalue. Then a solution of the equation exists. We shall find an expression for this solution by means of eigenfunctions.

It is clear that the function $\mu(P) - f(P)$ is sourcewise representable by means of the kernel. Consequently

$$\mu(P) - f(P) = \lambda \sum_{i=1}^{\infty} \frac{\mu_i \varphi_i(P)}{\lambda_i} \tag{26.1}$$

where

$$\mu_i = \int_{\Omega} \mu(P)\, \varphi_i(P)\, dP.$$

On the other hand, multiplying both sides of (26.1) by $\varphi_i(P)$ and integrating, we get

$$\mu_i - f_i = \frac{\lambda \mu_i}{\lambda_i},$$

whence

$$\mu_i = \frac{f_i}{1 - \lambda/\lambda_i} = \frac{\lambda_i f_i}{\lambda_i - \lambda}.$$

Consequently the function μ, if it exists, can be represented in the form

$$\mu(P) = f(P) + \lambda \sum_{i=1}^{\infty} \frac{f_i}{\lambda_i - \lambda} \varphi_i(P). \tag{26.2}$$

If we now introduce the expression for f_i in terms of eigenfunctions and interchange the order in the limit of its partial sum, we obtain

$$\mu(P) = f(P) + \lambda \lim_{N \to \infty} \int_{\Omega} f(P_1) \sum_{i=1}^{N} \frac{\varphi_i(P) \varphi_i(P_1)}{\lambda_i - \lambda} dP_1.$$

We shall show that the formula just obtained really does give the solution of the problem. To do this, it is convenient to introduce a new function

$$\Gamma(P, P_1, \lambda)$$

called the *resolvent* of the integral equation under consideration and given by

$$\Gamma(P, P_1, \lambda) = \sum_{i=1}^{\infty} \frac{\varphi_i(P) \varphi_i(P_1)}{\lambda_i - \lambda}. \tag{26.3}$$

It is easy to show that the series on the right-hand side of (26.3) converges in the mean for all values of λ which do not coincide with any one of the λ_i. To do this, we notice that the series for the resolvent can be written in a different form. We have:

$$\Gamma(P, P_1, \lambda) = \sum_{i=1}^{\infty} \varphi_i(P) \varphi_i(P_1) \left[\frac{1}{\lambda_i} + \left(\frac{1}{\lambda_i - \lambda} - \frac{1}{\lambda_i} \right) \right]$$

$$= \sum_{i=1}^{\infty} \frac{\varphi_i(P) \varphi_i(P_1)}{\lambda_i} + \lambda \sum_{i=1}^{\infty} \frac{\varphi_i(P) \varphi_i(P_1)}{\lambda_i(\lambda_i - \lambda)},$$

or

$$\Gamma(P, P', \lambda) = K(P, P') + \lambda \sum_{i=1}^{\infty} \frac{\varphi_i(P) \varphi_i(P')}{\lambda_i(\lambda_i - \lambda)}$$

$$= \sum_{i=1}^{\infty} \frac{\varphi_i(P) \varphi_i(P')}{\lambda_i} + \lambda \sum_{i=1}^{\infty} \frac{\varphi_i(P) \varphi_i(P')}{\lambda_i(\lambda_i - \lambda)}, \tag{26.4}$$

where the series in the second term converges uniformly.

The necessary and sufficient condition for the uniform convergence of the series (26.3) is clearly the uniform convergence of the bilinear series for the kernel.

Note. We see that the resolvent of an integral equation with a symmetric kernel is a meromorphic function over the whole complex plane of the parameter λ. All the poles of this function are simple and are the eigenvalues of the kernel.

Using the demonstrated convergence of the series for $\Gamma(P, P_1, \lambda)$ we can interchange the order of summation and integration in the formula so that

$$\mu(P) = f(P) + \lambda \int_{\Omega} \Gamma(P, P_1, \lambda) f(P_1) \, dP_1,$$

this result being meaningful, by what has been proved. By applying to both sides the integral operation with kernel $K(P_0, P)$ we can verify that $\mu(P)$ does really satisfy the integral equation

$$\mu(P) = f(P) + \lambda \int_{\Omega} K(P, P_1) \, \mu(P_1) \, dP_1.$$

§ 2. Representation of the Solution by means of Analytical Functions

Expansion as a Fourier series, which we have investigated in previous lectures (see § 2, Lecture 23), was interpreted in a purely geometric way as a representation of a certain function in function space, in which the characteristic directions of the linear operator were taken as coordinates axes.

By using the resolvent, we can approach the question in another way. Consider the integral

$$\chi(P, \lambda) = -\int_{\Omega} \Gamma(P, P_1, \lambda) f(P_1) \, dP_1 = -\sum_{i=1}^{\infty} \frac{f_i \varphi_i(P)}{\lambda_i - \lambda}.$$

This integral is a meromorphic function of the parameter λ with simple poles at the points λ_i. The residues at these poles are equal to $f_i \varphi_i(P)$ and form the successive terms of a Fourier series. The Fourier series for the function f is the sum of these residues, and a part of this series is the sum of the residues at a certain number of the poles. Hence such a part may be represented in the form

$$f_N(P) = \frac{1}{2\pi i} \int_{C_N} \chi(P, \lambda) \, d\lambda$$

where C_N is a contour in the plane of the complex variable λ enclosing the first N singular points of the resolvent.

It is again convenient to use a symbolic notation.

From formula (26.2), the function $\chi(P, \lambda)$ can be expressed in the form

$$\chi(P, \lambda) = -\frac{\mu(P) - f(P)}{\lambda},$$

where $\mu(P)$ is a solution of the equation

$$\mu(P) - \lambda \iiint_{\Omega} K(P, P_1)\, \mu(P_1)\, dP_1 = f(P),$$

i.e., of the equation

$$(E - \lambda A)\, \mu = f.$$

As we have seen earlier, for small values of λ this solution can be written in series form

$$\mu = (E + \lambda A + \lambda^2 A^2 + \cdots + \lambda^m A^m + \cdots) f;$$

on the other hand

$$\mu = (E - \lambda A)^{-1} f,$$

whence, in symbolic notation,

$$\chi = - \frac{\mu - f}{\lambda} = -(E - \lambda A)^{-1} Af.$$

Hence

$$f_N(P) = -\frac{1}{2\pi i} \int_{C_N} \frac{A}{E - \lambda A} f\, d\lambda. \tag{26.5}$$

We have previously seen the analogy between certain symbolic formulae containing polynomials or power series in the operator A and corresponding formulae of algebra or analysis. Our formula (26.5) is also the analogue of a corresponding formula in the theory of functions. For, let a and f be any numbers. Consider the integral

$$\psi = -\frac{1}{2\pi i} \int_{C} \frac{a}{1 - \lambda a} f\, d\lambda = -\frac{1}{2\pi i} \int_{C} \frac{f}{1/a - \lambda}\, d\lambda. \tag{26.6}$$

It is clear that $\psi = f$ if the contour C encloses the point $\lambda_0 = 1/a$, and that otherwise $\psi = 0$. The formula (26.5) is a generalization of (26.6).

The number λ_0 satisfies the relation

$$\lambda_0 a \varphi_k = \varphi_k \tag{26.7}$$

where φ_k is any non-zero number.

The equation (26.7) goes over into (25.1) defining λ_k, if we replace a in it by the operator A. And the formula (26.5) can be used to obtain solutions of problems of mathematical physics expressed in the form of definite integrals.

Without going into the general theory of this question in detail, we shall analyse an example of heat conduction, *viz.*, the problem of integrating the equation

$$\nabla^2 u - \frac{\partial u}{\partial t} = 0 \tag{26.8}$$

with the conditions

$$[u]_S = 0, \quad [u]_{t=0} = u_0(P).$$

We recite briefly the usual argument in the Fourier method.

Applying Green's formula to the equation, we have

$$u = -\int_\Omega G(P, P_1) \frac{\partial u(P_1, t)}{\partial t} \, dP_1$$

or in symbolic notation

$$u + A \frac{\partial u}{\partial t} = 0.$$

Particular solutions of this equation will be given by

$$u = e^{-\lambda_i t} u_i(P)$$

where

$$u_i - \lambda_i A u_i = 0.$$

The sum of the particular solutions corresponding to different eigenvalues λ_i, i.e., the solution in the form of a Fourier series will take the form

$$u = \sum_{i=1}^{\infty} e^{-\lambda_i t} u_i(P). \tag{26.9}$$

It follows from the condition

$$[u]_{t=0} = u_0$$

that the $u_i(P)$ are the members of the Fourier series for the function u_0, and this implies that the functions u_i are the residues at the poles of the resolvent of the function

$$-\iiint_\Omega \Gamma(P, P_1, \lambda) \, u_0(P_1) \, dP_1 .$$

Further, $e^{-\lambda_i t} u_i(P)$ are the residues at the same poles of the function

$$-e^{-\lambda t} \iiint_\Omega \Gamma(P, P_1, \lambda) \, u_0(P_1) \, dP_1 ,$$

and a finite number of terms of the series (26.9) can be expressed in terms of the integral

$$u_N = -\frac{1}{2\pi i} \int_{C_N} e^{-\lambda t} \left(\iiint_\Omega \Gamma(P, P_1, \lambda) \, u_0(P_1) \, dP_1 \right) d\lambda, \tag{26.10}$$

thus giving the complete solution of the problem.

The solution may be written symbolically in the form

$$u_N = -\frac{1}{2\pi i} \int_{C_N} e^{-\lambda t} \frac{A}{E - \lambda A} u_0 \, d\lambda. \tag{26.11}$$

We compare our solution with the solution of the ordinary equation

$$a \frac{\partial u}{\partial t} + u = 0,$$

where a is a constant number. It is clear that here

$$u = e^{-t/a} u_0. \tag{26.12}$$

By the theorem on residues

$$u = -\frac{1}{2\pi i} \int_C e^{-\lambda t} \frac{a u_0}{1 - \lambda a} \, d\lambda.$$

This integral is equal to its residue at the point $\lambda_0 - 1/a$ if λ_0 lies within the contour C, and is zero if λ_0 is outside this contour.

On this basis it is convenient to denote

$$-\frac{1}{2\pi i} \int_C e^{-\lambda t} \frac{A}{E - \lambda a} f_0 \, dt$$

symbolically by $e^{-t/A} f_0$. The solution of equation (26.8) can then be written in the form

$$u = e^{-t/A} u_0. \tag{26.13}$$

The general theory of functions of operators (which we cannot develop here) enables a whole series of generalizations of this formula to be obtained for a considerably wider class of problems of mathematical physics. It enables similar methods to be applied to the solution of problems where the singularities of the resolvent are not merely isolated points in the plane of the variable λ.

In particular, these ideas enable the solution of problems of mathematical physics to be investigated in unbounded domains or in cases where the equations in question have singularities within the domain investigated. Often the occurrence of such singularities is bound up with a change in the analytical character of the resolvent, producing singularities of the resolvent which are not simple poles.

In the following chapters we shall not go more deeply into this question, but we shall study concrete applications of the Fourier method to particular problems of mathematical physics. The examples which we shall consider will be extremely instructive, since it was, in fact, from the investigation of these questions that the theory developed.

VIBRATIONS OF A RECTANGULAR
PARALLELEPIPED

AS AN EXAMPLE of the use of the Fourier method we take first the problem of the vibrations of a rectangular parallelepiped, which we shall formulate in the following way. We shall seek a solution of the equation

$$\nabla^2 u - \frac{\partial^2 u}{\partial t^2} = 0 \qquad (27.1)$$

in the domain

$$0 \leqq x \leqq a, \quad 0 \leqq y \leqq b, \quad 0 \leqq z \leqq c$$

with the boundary conditions

$$[u]_{x=0} = [u]_{x=a} = [u]_{y=0} = [u]_{y=b} = [u]_{z=0} = [u]_{z=c} = 0 \qquad (27.2)$$

and the initial conditions

$$[u]_{t=0} = u_0(x, y, z), \quad \left[\frac{\partial u}{\partial t}\right]_{t=0} u_1(x, y, z). \qquad (27.3)$$

By the general theory (see Lecture 23, § 1), the solution will have the form

$$u = \sum U_i(x, y, z) \, (a_i \cos \lambda_i t + b_i \sin \lambda_i t), \qquad (27.4)$$

where the U_i are solutions of the equation

$$\nabla^2 u_i + \lambda_i^2 U_i = 0, \qquad (27.5)$$

satisfying the conditions (27.2). In the present case these solutions can be expressed in finite form by means of elementary functions by using the device known as complete separation of the variables. With this object we shall write the solution of equation (27.5) in the form

$$U_i(x, y, z) = X_i(x) \, V_i(y, z),$$

where X_i depends only on x, and V_i is independent of x. To determine the functions X_i and V_i we have

$$V_i X_i'' + X_i \left(\frac{\partial^2 V_i}{\partial y^2} + \frac{\partial^2 V_i}{\partial z^2} \right) + \lambda_i^2 X_i V_i = 0,$$

or

$$\frac{X_i''}{X_i} + \frac{\dfrac{\partial^2 V_i}{\partial y^2} + \dfrac{\partial^2 V_i}{\partial z^2}}{V_i} + \lambda_i^2 = 0.$$

It is clear that

$$\frac{X_i''}{X_i} = -\varkappa_i^2, \quad \frac{\dfrac{\partial^2 V_i}{\partial y^2} + \dfrac{\partial^2 V_i}{\partial z^2}}{V_i} = -\tau_i^2,$$

where $\varkappa_i$ and τ_i are constants related by $\varkappa_i^2 + \tau_i^2 = \lambda_i^2$. The equation

$$X_i'' + \varkappa_1^2 X_i = 0 \tag{27.6}$$

under the conditions (27.2) has as eigenvalues the numbers $\varkappa_i = k_i \pi / a$ where the k_i are integers. The function

$$X_i = \sqrt{\left(\frac{2}{a} \right)} \left[\sin \frac{k_i \pi x}{a} \right]$$

satisfies the equation (27.6) with the boundary conditions at $x = 0$ and at $x = a$. The system of solutions $\{X_i\}$ is an orthogonal, normalized system of functions.

Considering the equation

$$\frac{\partial^2 V_i}{\partial y_i^2} + \frac{\partial^2 V_i}{\partial z_i^2} + \tau_i^2 V_i = 0,$$

again we seek a solution in the form

$$V_i = Y_i(y) \, Z_i(z),$$

where Y_i depends only on y, and Z_i only on z. We obtain

$$\frac{Y_i''}{Y_i} + \frac{Z_i''}{Z_i} + \tau_i^2 = 0,$$

whence it follows that

$$\frac{Y_i''}{Y_i} = -\sigma_i^2, \quad \frac{Z_i''}{Z_i} = -\mu_i^2,$$

where σ_i and μ_i are constants satisfying

$$\sigma_i^2 + \mu_i^2 = \tau_i^2.$$

The equation

$$Y_i'' + \sigma_i^2 Y_i = 0$$

has the eigenvalues

$$\sigma_i = \frac{l_i \pi}{b}$$

where the l_i are integers, and the eigenfunctions

$$Y_i = \sqrt{\left(\frac{2}{b}\right)}\left[\sin \frac{\pi l_i y}{b}\right].$$

The equation

$$Z_i'' + \mu_i^2 Z_i = 0$$

has the eigenvalues

$$\mu_i = \frac{m_i \pi}{c},$$

where the m_i are integers, and the eigenfunctions

$$Z_i = \sqrt{\left(\frac{2}{c}\right)}\left[\sin \frac{\pi m_i z}{c}\right].$$

Finally we obtain a set of eigenfunctions for the vibration of the parallel-epiped in the form

$$U = \sqrt{\left(\frac{8}{abc}\right)}\left[\sin \frac{\pi k_i x}{a} \sin \frac{\pi l_i y}{b} \sin \frac{\pi m_i z}{c}\right]. \tag{27.7}$$

These eigenfunctions correspond to the following eigenvalues of equation (27.5) under the boundary conditions (27.2):

$$\lambda_i^2 = \pi^2 \left(\frac{k_i^2}{a^2} + \frac{l_i^2}{b^2} + \frac{m_i^2}{c^2}\right), \tag{27.8}$$

where k_i, l_i, m_i run through all possible trios of integers.

We notice that the numbers λ_i may be repeated, i.e., it may happen that

$$\lambda_i = \lambda_{i+1} = \cdots = \lambda_{i+k}.$$

The number of eigenvalues which are equal to one another is equal to the number of solutions in integers of equation (27.8) as regards k_i, l_i, m_i.

It can be shown that equation (27.5) under the conditions (27.2) has no other eigenvalues or eigenfunctions. To do this, we prove the following lemma.

LEMMA 1. *Any function $\varphi(x, y, z)$ which satisfies the conditions (27.2) and has continuous second-order derivatives†, is expressible as a convergent series*

$$\varphi(x, y, z) = \sqrt{\left(\frac{8}{abc}\right)} \sum_{k=1}^{\infty} \sin \frac{\pi k x}{a} \left[\sum_{l=1}^{\infty} \sin \frac{\pi l y}{b} \left(\sum_{m=1}^{\infty} \varphi_{k,l,m} \sin \frac{\pi m z}{c} \right) \right]$$

where the series   (27.9)

$$\varphi_{k,l}(z) = \sum_{m=1}^{\infty} \varphi_{k,l,m} \sin \frac{\pi m z}{c}$$

$$\varphi_k(y, z) = \sum_{l=1}^{\infty} \varphi_{k,l}(z) \sin \frac{\pi l y}{b}$$

$$\varphi(x, y, z) = \sum_{m=1}^{\infty} \varphi_k(y, z) \sin \frac{\pi k x}{a}$$

converge uniformly with respect to the arguments which appear under the sine. Here

$$\varphi_{k,l,m} = \sqrt{\left(\frac{8}{abc}\right)} \int_0^a \int_0^b \int_0^c \varphi(x, y, z) \sin \frac{\pi k x}{a} \sin \frac{\pi l y}{b} \sin \frac{\pi m z}{c} \, dx \, dy \, dz.$$

(27.10)

We shall at first regard $\varphi(x, y, z)$ as a function of the variable x with y and z as parameters. In view of the fact that for the equation (27.6) and the given boundary conditions Green's function has been constructed (see Lecture 20, § 3), we can, using the Hilbert–Schmidt theorem, express the function $\varphi(x, y, z)$ as a series, uniformly convergent with respect to x, in the eigenfunctions of equation (27.6), *i.e.*,

$$\varphi(x, y, z) = \sqrt{\left(\frac{2}{a}\right)} \sum_{k=1}^{\infty} \varphi_k(y, z) \sin \frac{k \pi x}{a},$$

where

$$\varphi_k(y, z) = \sqrt{\left(\frac{2}{a}\right)} \int_0^a \varphi(x_1, y, z) \sin \frac{k \pi x_1}{a} \, dx_1. \qquad (27.11)$$

From formula (27.11) it is seen that $\varphi_k(y, z)$ is a continuous, twice-differentiable function of y and z, which satisfies in these variables the conditions (27.2). Consequently $\varphi_k(y, z)$ is expressible as a series uniformly convergent relative to y:

$$\varphi_k(y, z) = \sqrt{\left(\frac{2}{b}\right)} \sum_{l=1}^{\infty} \varphi_{k,l}(z) \sin \frac{\pi l y}{b},$$

† We require the second-order derivatives here only because we invoke the theorem on expansion in eigenfunctions.

where

$$\varphi_{k,l}(z) = \sqrt{\left(\frac{2}{b}\right)} \int_0^b \varphi_k(y_1, z) \sin \frac{\pi l y_1}{b} \, dy_1$$

Finally, $\varphi_{k,l}(z)$ can be expanded in a series uniformly convergent with respect to z:

$$\varphi_{k,l}(z) = \sqrt{\left(\frac{2}{c}\right)} \sum_{m=1}^{\infty} \varphi_{k,l,m} \sin \frac{\pi m z}{c}$$

where

$$\varphi_{k,l,m} = \sqrt{\left(\frac{2}{c}\right)} \int_0^c \varphi_{k,l}(z_1) \sin \frac{\pi m z_1}{c} \, dz_1 .$$

Combining the formulae for $\varphi_k(y, z)$, $\varphi_{k,l}(z)$, $\varphi_{k,l,m}$ and the series for $\varphi(x, y, z)$, $\varphi_k(y, z)$ and $\varphi_{k,l,m}$, we obtain at once the assertion in the lemma.

Suppose now, if possible, that U_0 is some eigenfunction of equation (27.5) with the conditions (27.12) which is different from all the functions (27.7). As was proved in Lecture 24, it can be regarded as orthogonal to all the functions (27.7). By the lemma, it can be expanded in a series (27.9). It follows from formula (27.10) that all the coefficients of such an expansion would be zero. Hence formula (27.9) gives

$$U_0 = 0.$$

Consequently our problem can have no eigenfunction different from (27.7).

Instead of the conditions (27.2) we could also consider conditions of a different type, for example

$$\left[\alpha \frac{\partial u}{\partial n} + \beta u\right]_S = 0 \qquad (27.12)$$

where α and β take constant values on each of the boundaries. On the basis of the general theory, we conclude that this problem too is completely solved by the Fourier method, since we can satisfy all the initial and boundary conditions by proper choice of the coefficients in the expansion of the solution as a series of eigenfunctions (27.7).

The vibrations of a rectangular membrane can be studied in the same way as the vibrations of a rectangular parallelepiped.

Attention may usefully be directed to the following circumstance. Suppose that the coefficients α and β in formula (27.12) are not constant. Then the decomposition of the eigenfunctions U_l into the product $X_l Y_l Z_l$ will not, in general, take place. The question arises: Would it be possible to carry out some transformation of the independent variables, introducing new coordinates t_1, t_2, t_3 in place of x, y, z, so as to achieve such a representation?

The coordinates for which it is possible in some problem of mathematical physics to carry out a separation of the variables are sometimes called *normal coordinates*. So we may formulate the question: Are there normal coordinates for a given problem, and, if so, how are they to be found?

In the next lecture we shall indicate the normal coordinates for certain particular problems of mathematical physics. But, as V. V. Stepanov has proved, normal coordinates exist only in a very restricted class of problems.

LECTURE 28

LAPLACE'S EQUATION IN CURVILINEAR COORDINATES. EXAMPLES OF THE USE OF FOURIER'S METHOD

§ 1. Laplace's Equation in Curvilinear Coordinates

Curvilinear coordinates of one sort or another, polar, cylindrical, *etc.*, are often used in mathematical physics. We shall investigate the form taken by Laplace's equation in such coordinates.

Let t_1, t_2, t_3 be the curvilinear coordinates, related to x, y, z by the formulae:

$$t_1 = t_1(x, y, z), \qquad t_2 = t_2(x, y, z), \qquad t_3 = t_3(x, y, z),$$

$$x = x(t_1, t_2, t_2), \quad y = y(t_1, t_2, t_3), \quad z = z(t_1, t_2, t_3).$$

Consider any two curves passing through one and the same point:

$$x_1(s_1), \quad y_1(s_1), \quad z_1(s_1);$$

$$x_2(s_2), \quad y_2(s_2), \quad z_2(s_2).$$

where s is the arc-length.

By a well-known formula of differential geometry, the cosine of the angle between these two curves is given by

$$\cos \varphi = \frac{dx_1}{ds_1} \frac{dx_2}{ds_2} + \frac{dy_1}{ds_1} \frac{dy_2}{ds_2} + \frac{dz_1}{ds_1} \frac{dz_2}{ds_2}.$$

In the new coordinates this formula will take the form

$$\cos \varphi = \sum_{i=1}^{3} \sum_{j=1}^{3} \left(\frac{\partial x}{\partial t_i} \frac{\partial x}{\partial t_j} + \frac{\partial y}{\partial t_i} \frac{\partial y}{\partial t_j} + \frac{\partial z}{\partial t_i} \frac{\partial z}{\partial t_j} \right) \frac{dt_i^{(1)}}{ds_1} \frac{dt_j^{(2)}}{ds_2}$$

$$= \sum_{i=1}^{3} \sum_{j=1}^{3} \alpha_{ij} \frac{dt_i^{(1)}}{ds_1} \frac{dt_j^{(2)}}{ds_2},$$

388

or

$$\cos \varphi \, ds_1 \, ds_2 = \sum_{i=1}^{3} \sum_{j=1}^{3} \alpha_{ij} \, dt_i^{(1)} \, dt_j^{(2)}, \qquad (28.1)$$

where $dt_i^{(1)}$ and $dt_j^{(2)}$ are differentials taken respectively along each of the two curves considered. We note that

$$\alpha_{ij} = \alpha_{ji} = \frac{\partial x}{\partial t_i} \frac{\partial x}{\partial t_j} + \frac{\partial y}{\partial t_i} \frac{\partial y}{\partial t_j} + \frac{\partial z}{\partial t_i} \frac{\partial z}{\partial t_j}.$$

If the two curves coincide, then (28.1) takes the form

$$ds^2 = \sum_{i=1}^{3} \sum_{j=1}^{3} \alpha_{ij} \, dt_i \, dt_j. \qquad (28.2)$$

In deriving Laplace's equation, as in many other questions involving curvilinear coordinates, it is convenient to express all formulae in terms of the coefficients α_{ij}.

If the system of coordinates t_1, t_2, t_3 is orthogonal, i.e., if the coordinate curves intersect at right angles, then all $\alpha_{ij} = 0$ when $i \neq j$. For, let us fix arbitrarily chosen values $i = i_0, j = j_0, i_0 \neq j_0$. Then on the one coordinate curve, only $dt_{i_0}^{(1)} \neq 0$, and on the other curve, only $dt_{j_0}^{(2)} \neq 0$. For these coordinate curves the left-hand side of (28.1) vanishes since the angle $\varphi = \pi/2$, and the right-hand side will be $2\alpha_{i_0 j_0}, dt_{i_0}^{(1)}, dt_{j_0}^{(2)}$, whence $\alpha_{i_0 j_0} = 0$. Restricting ourselves then to orthogonal systems, we put

$$\alpha_{ij} = \begin{cases} h_i^2, & i = j, \\ 0, & i \neq j. \end{cases}$$

Then

$$ds^2 = h_1^2 \, dt_1^2 + h_2^2 \, dt_2^2 + h_3^2 \, dt_3^2.$$

EXAMPLES.

1. For polar coordinates in 3 dimensions,

$$x = r \sin \theta \cos \varphi, \quad y = r \sin \theta \sin \varphi, \quad z = r \cos \theta,$$

whence

$$ds^2 = dx^2 + dy^2 + dz^2 = dr^2 + r^2 \, d\theta^2 + r^2 \sin^2 \theta \, d\varphi^2.$$

2. For cylindrical coordinates, we have

$$x = r \cos \varphi, \quad y = r \sin \varphi, \quad z = z,$$

$$ds^2 = dx^2 + dy^2 + dz^2 = dr^2 + r^2 \, d\varphi^2 + dz^2.$$

It is useful to evaluate the determinant of the transformation $D(x, y, z)/D(t_1, t_2, t_3)$ in terms of the same quantities h_1, h_2 and h_3. To do this we introduce an intermediate, rectangular coordinate system x_1, y_1, z_1, differing from x, y, z only in the direction of the axes. We choose the directions of x_1, y_1, z_1 so that at the given point x, y, z they coincide with the directions of t_1, t_2, t_3. In this case we have at the point considered:

$$\frac{\partial(x, y, z)}{\partial(t_1, t_2, t_3)} = \frac{\partial(x, y, z)}{\partial(x_1, y_1, z_1)} \frac{\partial(x_1, y_1, z_1)}{\partial(t_1, t_2, t_3)} = \pm \frac{\partial(x_1, y_1, z_1)}{\partial(t_1, t_2, t_3)},$$

for at this point $\partial(x, y, z)/\partial(x_1, y_2, z_1) = \pm 1$, the sign being determined by the orientation of the coordinate system x_1, y_1, z_1. Hence

$$\frac{\partial(x, y, z)}{\partial(t_1, t_2, t_3)} = \pm \begin{vmatrix} \dfrac{\partial x_1}{\partial t_1} & \dfrac{\partial x_1}{\partial t_2} & \dfrac{\partial x_1}{\partial t_3} \\[2mm] \dfrac{\partial y_1}{\partial t_1} & \dfrac{\partial y_1}{\partial t_2} & \dfrac{\partial y_1}{\partial t_3} \\[2mm] \dfrac{\partial z_1}{\partial t_1} & \dfrac{\partial z_1}{\partial t_2} & \dfrac{\partial z_1}{\partial t_3} \end{vmatrix}$$

But in the determinant $\partial(x_1, y_1, z_1)/\partial(t_1, t_2, t_3)$ all terms except those on the principal diagonal are zero and so

$$\frac{\partial(x_1, y_1, z_1)}{\partial(t_1, t_2, t_3)} = \frac{\partial x_1}{\partial t_1} \frac{\partial y_1}{\partial t_2} \frac{\partial z_1}{\partial t_3}.$$

Further, on the x_1-axis and on any curve tangent to x_1 at our point we have

$$dx_1^2 = ds^2 = h_1^2 dt_1^2;$$

similarly on the y_1-axis and on any curve touching it

$$dy_1^2 = ds^2 = h_2^2 dt_2^2;$$

and finally on the z_1-axis and on any curve touching it

$$dz_1^2 = ds^2 = h_3^2 dt_3^2;$$

whence

$$\left.\begin{array}{lll} \dfrac{\partial t_1}{\partial x_1} = \pm \dfrac{1}{h_1}, & \dfrac{\partial t_2}{\partial x_1} = 0, & \dfrac{\partial t_3}{\partial x_1} = 0, \\[3mm] \dfrac{\partial t_1}{\partial y_1} = 0, & \dfrac{\partial t_2}{\partial y_1} = \pm \dfrac{1}{h_2}, & \dfrac{\partial t_3}{\partial y_1} = 0, \\[3mm] \dfrac{\partial t_1}{\partial z_1} = 0, & \dfrac{\partial t_2}{\partial z_1} = 0, & \dfrac{\partial t_3}{\partial z_1} = \pm \dfrac{1}{h_3} \end{array}\right\} \quad (28.3)$$

and

$$
\frac{\partial x_1}{\partial t_1} = \pm h_1, \quad \frac{\partial x_2}{\partial t_2} = 0, \quad \frac{\partial x_3}{\partial t_3} = 0,
$$

$$
\frac{\partial y_1}{\partial t_1} = 0, \quad \frac{\partial y_2}{\partial t_2} = \pm h_2, \quad \frac{\partial y_3}{\partial t_3} = 0,
$$

$$
\frac{\partial z_1}{\partial t_1} = 0, \quad \frac{\partial z_2}{\partial t_2} = 0, \quad \frac{\partial z_3}{\partial t_3} = \pm h_3;
$$

finally we get

$$
\frac{\partial(x_1, y_1, z_1)}{\partial(t_1, t_2, t_3)} = \pm h_1 h_2 h_3,
$$

the sign being determined by the orientation of the coordinate systems (x_1, y_1, z_1) and (t_1, t_2, t_3).

As we have seen earlier, if L and M are mutually adjoint operators, then the following integral relation holds:

$$
\iiint_\Omega (uLv - vMu) \, d\Omega = \iint_S [uP(v) - vQ(u)] \, dS,
$$

where Ω is a certain volume bounded by a smooth surface S, and $P(v), Q(u)$ are linear conbinations of derivatives of the coefficients u, v with coefficients independent of v and u.

If one of the functions, for example v, vanishes outside a certain domain Ω_v which with its boundary lies wholly within the domain Ω, then the surface integral in the above identity vanishes, and we have

$$
\iiint_\Omega (uLv - vMu) \, d\Omega = 0.
$$

We can take this last relation as the definition of the operator M, if we specify that it shall hold for an arbitrary function u and for any function v which is different from zero only in a certain interior domain Ω_v.

Before we evaluate the Laplace operator in curvilinear coordinates we prove a lemma.

LEMMA 1. *If a second-order operator is self-adjoint, then it must be representable in the form*

$$
Lu = \sum_{i=1}^\infty \sum_{j=1}^\infty \frac{\partial}{\partial x_i} \left(A_{ij} \frac{\partial u}{\partial x_j} \right) + cu, \tag{28.4}
$$

where $A_{ij} = A_{ji}$.

We note that an operator which has the form (28.4) must be self-adjoint: this follows at once from the definition of self-adjointness of an operator (see § 2, Lecture 5). We now have to prove the converse proposition, expressed by the lemma.

We show first that there could be only one operator adjoint to the given one. For, if M_1 and M_2 were two operators adjoint to the operator L, and if v is a function which is zero everywhere except in a domain Ω_v interior to Ω, then

$$\iiint_\Omega (uLv - vM_1u)\, d\Omega = 0$$

and

$$\iiint_\Omega (uLv - vM_2u)\, d\Omega = 0$$

for any function u. Subtracting, we get

$$\iiint_\Omega v(M_1 - M_2)u\, d\Omega = 0.$$

But v is an arbitrary function within the domain Ω_v. Hence $(M_1 - M_2)\,u = 0$, and consequently the operators M_1 and M_2 are identical.

If L has the form

$$\sum_{i=1}^{n} \sum_{j=1}^{n} \frac{\partial}{\partial t_i}\left(A_{ij}\frac{\partial u}{\partial t_j}\right) + \sum_{i=1}^{n} B_i \frac{\partial u}{\partial t_i} + cu,$$

then the adjoint operator will have the form

$$\sum_{i=1}^{n} \sum_{j=1}^{n} \frac{\partial}{\partial t_i}\left(A_{ji}\frac{\partial u}{\partial t_j}\right) - \sum_{i=1}^{n} B_i \frac{\partial u}{\partial t_i} + cu,$$

and the two operators can be equal only if $B_i = 0$ and $A_{ij} = A_{ji}$. Hence the lemma.

The Laplace operator ∇^2 will not be self-adjoint in the variables t_1, t_2, t_3, but it can be readily expressed in terms of such an operator. For, choosing any pair of functions u, v as in the proof of Lemma 1, we have

$$\iiint_\Omega (u\nabla^2 v - v\nabla^2 u)\, dx\, dy\, dz = 0,$$

or, going over to the new variables,

$$\iiint_\Omega (u\nabla^2 v - v\nabla^2 u)\, h_1 h_2 h_3\, dt_1\, dt_2\, dt_3 = 0,$$

since, as we have already proved, the Jacobian $\partial(x, y, z)/\partial(t_1, t_2, t_3)$ can differ from $h_1 h_2 h_3$ only in sign.

This implies that the operator L defined by $Lu = h_1h_2h_3\nabla^2u$ is already self-adjoint.

If

$$Lu = \sum_{i=1}^{3} \sum_{j=1}^{3} \frac{\partial}{\partial t_i} \left(A_{ij} \frac{\partial u}{\partial t_j} \right)$$

(the term cu obviously cannot appear either in ∇^2 or in L), then

$$\Delta u = \frac{1}{h_1h_2h_3} \sum_{i=1}^{3} \frac{\partial}{\partial t_i} \left(\sum_{j=1}^{3} A_{ij} \frac{\partial u}{\partial t_j} \right). \tag{28.5}$$

The coefficients β_{ij} of the second-order derivatives in the expression (28.5) will clearly be $A_{ij}/h_1h_2h_3$.

On the other hand, these coefficients β_{ij} can be determined directly. We first change the variables to x_1, y_1, z_1: this does not change the form of the operator ∇^2u, since, as may be immediately verified, the Laplace operator is invariant for a linear, orthogonal transformation of the variables. Now, using the formulae (28.3), we find

$$\beta_{ij} = \frac{\partial t_i}{\partial x_1} \frac{\partial t_j}{\partial x_1} + \frac{\partial t_i}{\partial y_1} \frac{\partial t_j}{\partial y_1} + \frac{\partial t_i}{\partial z_1} \frac{\partial t_j}{\partial z_1} = \begin{cases} \dfrac{1}{h_i^2}, & i = j, \\ 0, & i \neq j. \end{cases}$$

Hence we have

$$A_{ij} = \begin{cases} \dfrac{h_1h_2h_3}{h_i^2}, & i = j, \\ 0, & i \neq j, \end{cases}$$

i.e.,

$$\nabla^2u = \frac{1}{h_1h_2h_3} \left[\frac{\partial}{\partial t_1} \left(\frac{h_2h_3}{h_1} \frac{\partial u}{\partial t_1} \right) + \frac{\partial}{\partial t_2} \left(\frac{h_3h_1}{h_2} \frac{\partial u}{\partial t_2} \right) + \frac{\partial}{\partial t_3} \left(\frac{h_1h_2}{h_3} \frac{\partial u}{\partial t_3} \right) \right]. \tag{28.6}$$

In polar coordinates, $h_1 = 1$, $h_2 = r$, $h_3 = r\sin\theta$, and

$$\nabla^2u = \frac{1}{r^2\sin\theta} \left[\frac{\partial}{\partial r} \left(r^2\sin\theta \frac{\partial u}{\partial r} \right) + \frac{\partial}{\partial \theta} \left(\sin\theta \frac{\partial u}{\partial \theta} \right) + \frac{\partial}{\partial \varphi} \left(\frac{1}{\sin\theta} \frac{\partial u}{\partial \varphi} \right) \right]$$

$$= \frac{1}{r^2} \frac{\partial}{\partial r} \left(r^2 \frac{\partial u}{\partial r} \right) + \frac{1}{r^2\sin\theta} \frac{\partial}{\partial \theta} \left(\sin\theta \frac{\partial u}{\partial \theta} \right) + \frac{1}{r^2\sin^2\theta} \frac{\partial^2 u}{\partial \varphi^2}. \tag{28.7}$$

In cylindrical coordinates, $h_1 = 1$, $h_2 = r$, $h_3 = 1$,

$$\nabla^2 u = \frac{1}{r}\left[\frac{\partial}{\partial r}\left(r\,\frac{\partial u}{\partial r}\right) + \frac{\partial}{\partial \varphi}\left(\frac{1}{r}\,\frac{\partial u}{\partial \varphi}\right) + \frac{\partial}{\partial z}\left(r\,\frac{\partial u}{\partial z}\right)\right]$$

$$= \frac{1}{r}\,\frac{\partial}{\partial r}\left(r\,\frac{\partial u}{\partial r}\right) + \frac{1}{r^2}\,\frac{\partial^2 u}{\partial \varphi^2} + \frac{\partial^2 u}{\partial z^2} \tag{28.8}$$

§ 2. Bessel Functions

Consider the ordinary differential equation

$$y'' + \frac{1}{x}\,y' + \left(1 - \frac{\nu^2}{x^2}\right)y = 0, \tag{28.9}$$

which is known as Bessel's equation. In books on the analytical theory of differential equations and the theory of special functions, a number of important properties of the solutions of this equation are derived. Here we shall merely give the results without proof; readers wishing to learn the proofs of these properties may consult:

V.I. Smirnov: "Course of Higher Mathematics", Vol. 3, Chapter 2, or
R.O. Kuz'min: "Bessel Functions".
For a detailed exposition of the theory of Bessel functions, see
G.N. Watson: "A Treatise on the Theory of Bessel Functions".

We shall examine equation (28.9) with integral values of ν and with non-integral values separately.

The following propositions hold:
1. For non-integral values of ν, equation (28.9) has two linearly independent integrals,

$$J_\nu(x) \quad \text{and} \quad J_{-\nu}(x),$$

which may be expanded in series uniformly convergent over the whole plane of the complex variable x:

$$J_\nu(x) = \sum_{s=0}^{\infty} \frac{(-1)^s \left(\dfrac{x}{2}\right)^{\nu+2s}}{\Gamma(s+1)\cdot\Gamma(\nu+s+1)},$$

$$J_{-\nu}(x) = \sum_{s=0}^{\infty} \frac{(-1)^s \left(\dfrac{x}{2}\right)^{-\nu+2s}}{\Gamma(s+1)\cdot\Gamma(-\nu+s+1)}.$$

In other words, $x^{-\nu}J_\nu(x)$ and $x^\nu J_{-\nu}(x)$ are integral functions of x. The functions $J_\nu(x)$ and $J_{-\nu}(x)$ are called respectively *Bessel functions of order ν and*

$-\nu$; the linear combination of them,

$$N_\nu(x) = \frac{J_\nu(x) \cos (\nu\pi) - J_{-\nu}(x)}{\sin (\nu\pi)}$$

is called *Neumann's function*.

2. For integral values of $\nu = m$, Bessel functions of order m and $-m$ as defined above are no longer independent:

$$J_{-m}(x) = (-1)^m J_m(x).$$

In this case we may take as the linearly independent solutions of equation (28.7) the two functions:

$$J_m(x) = \sum_{s=0}^{\infty} \frac{(-1)^s \left(\dfrac{x}{2}\right)^{m+2s}}{\Gamma(s + 1) \cdot \Gamma(m + s + 1)},$$

and

$$N_m(x) = \frac{2}{\pi} J_m(x) \left(\log_e \frac{x}{2} + C\right) - \frac{1}{\pi} \sum_{s=0}^{m-1} \frac{(m - s - 1)!}{\Gamma(s + 1)} \left(\frac{x}{2}\right)^{-m+2s}$$

$$- \frac{1}{\pi} \left(\frac{x}{2}\right)^m \frac{1}{m!} \left(\frac{1}{m} + \frac{1}{m - 1} + \cdots + 1\right)$$

$$- \frac{1}{\pi} \sum_{s=1}^{\infty} \left[\frac{(-1)^s}{\Gamma(s + 1) \cdot \Gamma(m + s + 1)} \left(\frac{x}{2}\right)^{2s+m} \times\right.$$

$$\left. \times \left(\frac{1}{m + s} + \frac{1}{m + s - 1} + \cdots + 1 + \frac{1}{s} + \frac{1}{s - 1} + \cdots + 1\right)\right]$$

(28.10)

where $C = 0 \cdot 577215 \ldots$ is Euler's constant.

As may be seen from (28.10), Bessel's function $J_m(x)$ is an integral function of its argument and so is $x^{-m} J_m(x)$. Neumann's function $N_m(x)$, which is the limit of the function $N_\nu(x)$ as $\nu \to m$, has a singularity at the origin in the form of a branch-point coincident with the pole.

3. For pure imaginary values of the argument ($x = it$) the function

$$i^{-m} J_m(x) = i^{-m} J_m(it) = I_m(t),$$

which is the solution of Bessel's equation in the variable x, will be real.

The other solution of equation (28.9) which is real for imaginary values of x has the form

$$i^{-m} \left[N_m(x) - \frac{i\pi}{2} J_m(x)\right] = i^{-m} \left[N_m(it) - \frac{i\pi}{2} J_m(it)\right].$$

4. The linear combinations

$$H_\nu^{(1)}(x) = J_\nu(x) + iN_\nu(x), \qquad H_\nu^{(2)}(x) = J_\nu(x) - iN_\nu(x)$$

are known as *Hankel functions of the first and second order* respectively. These functions, which are clearly solutions of Bessel's equation (although they are complex for real values of x), have the following asymptotic expressions for large, real values of x (valid both for integral and fractional values of ν):

$$\left. \begin{array}{l} H_\nu^{(1)}(x) = \sqrt{\left(\dfrac{2}{\pi x}\right)} \, e^{i(x - \nu\pi/2 - \pi/4)} \, [1 + O(x^{-1})] \\[4mm] H_\nu^{(2)}(x) = \sqrt{\left(\dfrac{2}{\pi x}\right)} \, e^{-i(x - \nu\pi/2 - \pi/4)}[1 + O(x^{-1})]. \end{array} \right\} \quad (28.11)\dagger$$

For $H_\nu^{(1)}(x)$ this representation is valid for large x satisfying the condition

$$\left| \arg x - \frac{\pi}{2} \right| < \frac{3\pi}{2},$$

and for $H_\nu^{(2)}(x)$ for large x satisfying the condition $|\arg x + \pi/2| < 3\pi/2$.

The following asymptotic expressions for the Bessel and Neumann functions are obtained from (28.11):

$$\left. \begin{array}{l} J_\nu(x) = \sqrt{\left(\dfrac{2}{\pi x}\right)} \cos\left(x - \dfrac{\nu\pi}{2} - \dfrac{\pi}{4}\right) + O\!\left(x^{-\frac{3}{2}}\right), \\[4mm] N_\nu(x) = \sqrt{\left(\dfrac{2}{\pi x}\right)} \sin\left(x - \dfrac{\nu\pi}{2} - \dfrac{\pi}{4}\right) + O\!\left(x^{-\frac{3}{2}}\right). \end{array} \right\} \quad (28.12)$$

The formulae (28.12) show how the Bessel and Neumann functions behave as the argument increases. They are oscillatory functions, passing through zero infinitely many times. The amplitude of their oscillation gradually dies away. The roots of the Bessel and Neumann functions of the same suffix alternate, *i.e.*, the functions interlace.

For large, real values of z, the function $I_n(z)$ has the representation

$$I_n(z) \sim \frac{e^z}{\sqrt{2\pi z}} \left\{ 1 + O\left(\frac{1}{z}\right) \right\}.$$

The equation satisfied by $I_n(z)$ can be obtained by elementary means:

$$\frac{d^2 I_n(z)}{dz^2} + \frac{1}{z}\frac{dI_n(z)}{dz} - \left(1 + \frac{n^2}{z^2}\right) I_n(z) = 0.$$

This function is often encountered in applications.

$\dagger\ f(x) = O(\varphi(x))$ means that the ratio $f(x)/|\varphi(x)|$ remains bounded as $x \to \infty$.

Usually in the tables only the values of the functions $J_0(z)$ and $J_1(z)$ are given. There is a relation between the functions $J_n(z)$ for different n, which enables $J_n(z)$ and its derivative to be expressed, for integral n, in terms of these two basic functions:

$$J_{n-1}(z) + J_{n+1}(z) = \frac{2n}{z} J_n(z), \quad J_{n-1}(z) - J_{n+1}(z) = 2J'_n(z).$$

§ 3. Complete Separation of the Variables in the Equation $\nabla^2 u = 0$ in Polar Coordinates

Consider the wave equation in a plane:

$$\frac{\partial^2 u}{\partial x^2} + \frac{\partial^2 u}{\partial y^2} - \frac{\partial^2 u}{\partial t^2} = 0.$$

Suppose we wish to investigate the vibration of a circular membrane $r \leqq 1$ with the boundary condition

$$[u]_{r=1} = 0 \tag{28.13}$$

and the initial conditions

$$[u]_{t=0} = u_0, \quad \left[\frac{\partial u}{\partial t} \right]_{t=0} = u_1.$$

Going over to the variables r and φ by means of the formulae $x = r \cos \varphi$, $y = r \sin \varphi$, we transform the wave equation into the form [see (28.8)]

$$\frac{1}{r} \frac{\partial}{\partial r} \left(r \frac{\partial u}{\partial r} \right) + \frac{\partial^2 u}{\partial \varphi^2} - \frac{\partial^2 u}{\partial t^2} = 0. \tag{28.14}$$

We shall seek a solution of (28.14) in the form

$$u = \sum_{i=1}^{\infty} v_i(r, \varphi) (a_i \cos \lambda_i t + b_i \sin \lambda_i t),$$

where v_i is a solution of the equation

$$\frac{1}{r} \frac{\partial}{\partial r} \left(r \frac{\partial v_i}{\partial r} \right) + \frac{1}{r^2} \frac{\partial^2 v_i}{\partial \varphi^2} + \lambda_i^2 v_i = 0. \tag{28.15}$$

We shall show that r and φ are normal coordinates for this problem.

We seek a solution v_i in the form of a product

$$v_i = \Phi_i(\varphi) R_i(r). \tag{28.16}$$

We shall prove later that such a solution exists. Substituting (28.16) into (28.15) we get

$$\frac{\Phi_i}{r} \frac{d}{dr} \left(r \frac{dR_i}{dr} \right) + \frac{1}{r^2} R_i \frac{d^2\Phi_i}{d\varphi^2} + \lambda_i^2 R_i \Phi_i = 0,$$

or, dividing by $R_i \Phi_i$ and multiplying by r^2,

$$\frac{r \dfrac{\mathrm{d}}{\mathrm{d}r}\left(r \dfrac{\mathrm{d}R_i}{\mathrm{d}r}\right)}{R_i} + \frac{\dfrac{\mathrm{d}^2\Phi_i}{\mathrm{d}\varphi^2}}{\Phi_i} + \lambda_i^2 r^2 = 0,$$

i.e.,

$$-\frac{\dfrac{\mathrm{d}^2\Phi_i}{\mathrm{d}\varphi^2}}{\Phi_i} = \frac{r \dfrac{\mathrm{d}}{\mathrm{d}r}\left(r \dfrac{\mathrm{d}R_i}{\mathrm{d}r}\right)}{R_i} + r^2 \lambda_i^2.$$

Equating both sides to a constant m_i^2, we obtain the two equations

$$\Phi_i'' + m_i^2 \Phi_i = 0, \tag{28.17}$$

$$r(rR_i')' - (m_i^2 - r^2\lambda_i^2)\, R_i = 0. \tag{28.18}$$

The solution of equation (28.17) must have the period 2π if it is to have a definite physical meaning; so that m_i must be real and integral. Then

$$\Phi_i^{(1)} = \cos m_i\varphi, \quad \Phi_i^{(2)} = \sin m_i\varphi.$$

Equation (28.18) can be transformed to the self-adjoint form

$$(rR_i')' - \left(\frac{m_i^2}{r} - \lambda_i^2 r\right) R_i = 0. \tag{28.19}$$

In this form the equation for R_i is a self-adjoint equation of the second order. We shall seek a solution of it which vanishes for $r = 1$ and which remains bounded for $r = 0$. By the general theory (Lecture 20 and Lecture 24, § 4) such a solution exists for a certain set of values λ_i^2.

LEMMA 1.† *The fundamental functions of equation* (28.19) *which satisfy the specified boundary conditions are orthogonal with weight* r:

$$\int_0^1 R_i(r)\, R_j(r)\, r\, \mathrm{d}r = \begin{cases} 0, & i \neq j, \\ 1, & i = j. \end{cases}$$

For, let

$$LR_i \equiv (rR_i')' - \frac{m_i^2}{r}\, R_i = -\lambda_i^2 r R_i;$$

then

$$\int_0^1 [R_j L R_i - R_i L R_j]\, \mathrm{d}r = (\lambda_j^2 - \lambda_i^2) \int_0^1 r R_j R_i\, \mathrm{d}r.$$

† This lemma follows from the general theory, but we repeat the proof. The lemma is valid for a more general condition at $r = 1$, *viz.*, $[\alpha R_i + \beta R_i']_{r=1} = 0$.

But the integral on the left-hand side is identically zero: hence the integral on the right is zero, as was to be proved.

We make the substitution $\lambda_i r = \varrho$, so that

$$\frac{\mathrm{d}}{\mathrm{d}r} = \lambda_i \frac{\mathrm{d}}{\mathrm{d}\varrho}$$

and equation (28.19) becomes

$$\lambda_i \frac{\mathrm{d}}{\mathrm{d}\varrho}\left(\varrho \cdot \frac{\mathrm{d}R_i}{\mathrm{d}\varrho}\right) - \left(\frac{m_i^2 \lambda_i}{\varrho} - \lambda_i \varrho\right) R_i = 0,$$

or

$$\frac{\mathrm{d}}{\mathrm{d}\varrho}\left(\varrho \frac{\mathrm{d}R_i}{\mathrm{d}\varrho}\right) - \left(\frac{m_i^2}{\varrho} - \varrho\right) R_i = 0. \qquad (28.20)$$

Denoting R_i by y, we have

$$y'' + \frac{1}{\varrho} y' + \left(1 - \frac{m_i^2}{\varrho}\right) y = 0,$$

so that the problem reduces to Bessel's equation, which we have discussed earlier.

It is natural to take for R the regular solution of equation (28.20), *i.e.*,

$$R_i(r) = J_{m_i}(\lambda_i r).$$

The value of λ_i is obtained from the boundary condition (28.13). We have

$$R_i(1) = J_{m_i}(\lambda_i) = 0. \qquad (28.21)$$

As was to be expected, this equation has an infinite set of solutions; this is obvious from the asymptotic representation of Bessel's function (28.12).

We thus obtain a system of solutions of our boundary-value problem in the form

$$c_i \cos (m_i \varphi) J_{m_i}(\lambda_i r), \quad d_i \sin (m_i \varphi) J_{m_i}(\lambda_i r) \qquad (28.22)$$

with eigenvalue λ_i^2, where c_i and d_i are certain constants which can be found from the normalization condition. All these solutions are mutually orthogonal. It is easily proved, as in the case of vibrations of a rectangular parallelepiped, that the functions (28.22) in which the pairs of numbers m_i and λ_i are all possible pairs of numbers satisfying equation (28.21) include *all* the eigenfunctions of the problem. To prove this we first establish a lemma.

LEMMA 2. *An arbitrary function $\omega(r, \varphi)$ having continuous derivatives up to the second order inclusive and satisfying the boundary condition (28.13) can be represented in the form of a convergent series*

$$\omega(r, \varphi) = \sum_{m=0}^{\infty} \sum_{k=1}^{\infty} (\alpha_{mk} \cos m\varphi + \beta_{mk} \sin m\varphi) J_m(\lambda_k^{(m)} r), \qquad (28.23)$$

where
$$\alpha_{mk} = \frac{1}{\pi} \int_0^{2\pi} \int_0^1 \omega(r, \varphi) \cos m\varphi \, J_m(\lambda_k^{(m)} r) \, r \, dr \, d\varphi,$$

$$\beta_{mk} = \frac{1}{\pi} \int_0^{2\pi} \int_0^1 \omega(r, \varphi) \sin m\varphi \, J_m(\lambda_k^{(m)} r) \, r \, dr \, d\varphi.$$

First, since $\omega(r, \varphi)$ is a periodic function it can be expressed in the form

$$\omega(r, \varphi) = \sum_{m=0}^{\infty} [a_m(r) \cdot \cos m\varphi + b_m(r) \sin m\varphi], \qquad (28.24)$$

where

$$a_m(r) = \frac{1}{\pi} \int_0^{2\pi} \omega(r, \varphi) \cos m\varphi \, d\varphi, \quad b_m(r) = \frac{1}{\pi} \int_0^{2\pi} \omega(r, \varphi) \sin m\varphi \, d\varphi.$$

The coefficients $a_m(r)$ have continuous second-order derivatives, satisfy the boundary condition (28.13), and must remain bounded for $r = 0$; hence they can be expanded in uniformly convergent series in the eigenfunctions of equation (28.20), *i.e.*, in the functions $J_m(\lambda_k^{(m)}r)$:

$$a_m(r) = \sum_{k=1}^{\infty} \alpha_{mk} J_m(\lambda_k^{(m)}r), \qquad (28.25)$$

where

$$\alpha_{mk} = \frac{\displaystyle\int_0^1 a_m(r) \, J_m(\lambda_k^{(m)}r) \, r \, dr}{\left\{\displaystyle\int_0^1 [J_m(\lambda_k^{(m)}r)]^2 \, r \, dr\right\}^{\frac{1}{2}}} \qquad (28.26)$$

Similarly

$$b_m(r) = \sum_{k=1}^{\infty} \beta_{mk} J_m(\lambda_k^{(m)}r), \qquad (28.27)$$

where

$$\beta_{mk} = \frac{\displaystyle\int_0^1 b_m(r) \, J_m(\lambda_k^{(m)}r) \, r \, dr}{\left\{\displaystyle\int_0^1 [J_m(\lambda_k^{(m)}r)]^2 \, r \, dr\right\}^{\frac{1}{2}}}. \qquad (28.28)$$

Combining (28.25), (28.26), (28.27) and (28.28), we obtain the assertion of the lemma, and the series (28.24) converges uniformly relative to φ, and the series (28.25) and (28.27) converge uniformly relative to r.

It follows at once from Lemma 2 that our problem cannot have any other eigenfunctions than those already found, for any such eigenfunction would be expressible as a series (28.23) with zero coefficients, and consequently would be zero.

The problem of the vibrating membrane is thus completely solved.

HARMONIC POLYNOMIALS AND SPHERICAL FUNCTIONS

§ 1. Definition of Spherical Functions

Before proceeding to further applications of the Fourier method, we examine another important topic. Consider Laplace's equation in three variables:

$$\frac{\partial^2 u}{\partial x^2} + \frac{\partial^2 u}{\partial y^2} + \frac{\partial^2 u}{\partial z^2} = 0,$$

and suppose we wish to find solutions of this equation having the form of homogeneous polynomials of degree n in x, y and z.

In the plane case, as may easily be verified, such homogeneous solutions of Laplace's equation will have the form $(x + iy)^n$, $(x - iy)^n$.

We shall seek solutions of the problem in the form†

$$v = (x + iy)^m \sum_{j=0}^{\frac{n-m}{2}} a_j \varrho^{2j} z^{n-m-2j}$$

$$= (x + iy)^m f(\varrho^2, z), \quad (m = 0, 1, 2, ..., n) \quad (29.1)$$

where

$$\varrho = \sqrt{x^2 + y^2}\,.$$

Calculating the Laplacian of v, we obtain:

$$\frac{\partial^2 v}{\partial x^2} = m(m - 1)\,(x + iy)^{m-2} f(\varrho^2, z) + 4m(x + iy)^{m-1} \times \frac{\partial f(\varrho^2, z)}{\partial(\varrho^2)}$$

$$+ 2(x + iy)^m \frac{\partial f(\varrho^2, z)}{\partial(\varrho^2)} + 4x^2(x + iy)^m \frac{\partial^2 f(\varrho^2, z)}{\partial(\varrho^2)^2},$$

---

† The summation is from $j = 0$ to $j = E(n - m)/2$ where $E(n - m)/2$ is the integral part of the number $(n - m)/2$.

$$\frac{\partial^2 v}{\partial y^2} = -m(m-1)(x+iy)^{m-2} f(\varrho^2, z) + 4miy(x+iy)^{m-1} \frac{\partial f(\varrho^2, z)}{\partial(\varrho^2)}$$

$$+ 2(x+iy)^m \frac{\partial f(\varrho^2, z)}{\partial(\varrho^2)} + 4y^2(x+iy)^m \frac{\partial^2 f(\varrho^2, z)}{\partial(\varrho^2)^2},$$

$$\frac{\partial^2 v}{\partial z^2} = (x+iy)^m \frac{\partial^2 f(\varrho^2, z)}{\partial z^2}.$$

whence

$$\nabla^2 v = (x+iy)^m \left[\frac{\partial^2 f(\varrho^2, z)}{\partial z^2} + 4(m+1)\frac{\partial f(\varrho^2, z)}{\partial(\varrho^2)} + 4\varrho^2 \frac{\partial^2 f(\varrho^2, z)}{\partial(\varrho^2)^2} \right]$$

$$= (x+iy)^m \left\{ \frac{\partial^2 f}{\partial z^2} + \frac{4}{(\varrho^2)^m} \frac{\partial}{\partial(\varrho^2)} \left[\varrho^{2(m+1)} \frac{\partial f}{\partial(\varrho^2)} \right] \right\}.$$

Substituting the value of $f(\varrho^2, z)$ gives

$$\nabla^2 v = (x+iy)^m \left[\sum_{j=0}^{(n-m)/2} (n-m-2j)(n-m-2j-1) a_j \varrho^{2j} z^{n-m-2j-2} \right.$$

$$+ \left. \sum_{j=1}^{(n-m)/2} j(j+m) a_j \varrho^{2j-2} z^{n-m-2j} \right].$$

Replacing j by $j+1$ in the second summation, we get

$$\nabla^2 v = (x+iy)^m \left\{ \sum_{j=0}^{(n-m)/2} [(n-m-2j)(n-m-2j-1) a_j \right.$$

$$+ \left. (j+1)(j+m+1) a_{j+1}] \varrho^{2j} z^{n-m-2j-2} \right\}.$$

Equating the coefficients of $\varrho^{2j} z^{n-m-2j-2}$ to zero, we obtain a system of equations from which all the a_j can be expressed one after the other in terms of any one of them. Thus we have:

$$a_1 = (-1)\frac{(n-m)(n-m-1)}{1.(m+1)} a_0,$$

$$a_2 = (-1)^2 \frac{(n-m)(n-m-1)(n-m-2)(n-m-3)}{1.2.(m+1)(m+2)} a_0,$$

. . . .

$$a_{j+1} = (-1)^{j+1} \frac{(n-m)(n-m-1)\dots(n-m-2j+1)}{(j+1)!(m+1)(m+2)\dots(m+j+1)} a_0.$$

If a_0 is real and positive, then all the a_{2j} will be positive, and all the a_{2j+1} negative. For definiteness, let

$$a_0 = \frac{(n + m)!}{2^m \, m!(n - m)!}.$$

We have thus established the existence of one and, with accuracy up to a constant factor, only one harmonic polynomial of the form

$$v = (x + iy)^m f_{n,m}(\varrho^2, z),$$

for any n and $m = 0, 1, \ldots, n$. Associating with these a further n polynomials of the form

$$(x - iy)^m f_{n,m}(\varrho^2, z) \quad m = 1, \ldots, n \tag{29.2}$$

we obtain *a system of* $2n + 1$ *harmonic polynomials.*

It is not difficult to see that these polynomials are linearly independent. Taking $\zeta_1 = x + iy$ and $\zeta_2 = x - iy$ as new independent variables, we obtain in each of these polynomials such highest terms in ζ_1 or ζ_2 as cannot be encountered in any linear combination of polynomials having a lower value of the index m.

We shall prove that there are no other homogeneous, harmonic polynomials of degree n which are independent of those given.

Any homogeneous polynomial $R_n(x, y, z)$ will, if we introduce in it new independent variables z, $\zeta_1 = x + iy$, $\zeta_2 = x - iy$, be a polynomial in z, ζ_1 and ζ_2, and therefore can be uniquely expressed in the form

$$R_n = \varphi_{n,0}(\zeta_1\zeta_2, z) + \sum_{m=1}^{n} [\zeta_1^m \varphi_{n,m}(\zeta_1\zeta_2, z) + \zeta_2^m \varphi_{n,m}(\zeta_1\zeta_2, z)]. \tag{29.3}$$

To do this we must group together the terms in the full expression for the polynomial which have the form $\zeta_1^j \zeta_2^k z^l$ where $j - k = \pm m$. Hence it follows that *any homogeneous polynomial of degree* n *is expressed uniquely in the form of a sum of polynomials by the formula* (29.3).

If we apply the Laplace operator to the polynomial R_n, we obtain a homogeneous polynomial of degree $(n - 2)$ which is also uniquely expressible as a sum of polynomials by the formula (29.3) with a change of n into $(n - 2)$. If R_n is a harmonic polynomial, then all terms in such an expansion of $\nabla^2 R_n$ must consequently vanish (it follows from the uniqueness that there is no other expansion besides that in which all the terms vanish). On the other hand, applying Laplace's operator to formula (29.3) and noting that when this operator is applied to polynomials of the form (29.1) and (29.2) polynomials of the same form but with n replaced by $(n - 2)$ are obtained, we see that the result from each term of the sum (29.3) in turn is zero. This implies that $\varphi_{n,\,m} = cf_{n,\,m}$, as was to be shown.

Continuing to operate by analogy with the plane case, we change to polar coordinates putting $x = r \sin \theta \cos \varphi$, $y = r \sin \theta \sin \varphi$, $z = r \cos \theta$, where

$r = \sqrt{x^2 + y^2 + z^2}$. Then any harmonic polynomial of the type considered will take the form

$$r^n e^{\pm im\phi} f_{n,m}(\sin^2 \theta, \cos \theta) \sin^m \theta. \qquad (29.4)$$

Let

$$f_{n,m}(\sin^2 \theta, \cos \theta) = \psi_{n,m}(\cos \theta);$$

then it is not difficult to see that $\psi_{n,m}(\cos \theta)$ is a polynomial in $\cos \theta$ of degree precisely $(n - m)$. For, each term of the form $a_j z^{n-m-2j} \varrho^{2j}$ is replaced by a polynomial of the form

$$a_j(\cos \theta)^{n-m-2j} \sin^{2j} \theta = a_j(1 - \cos^2 \theta)^j (\cos \theta)^{n-m-2j},$$

in which the sign in front of terms containing $(\cos \theta)^{n-m-2j}$ is $(-1)^k$. In the sum of such polynomials not a single term can cancel out. Moreover, it is clear that

$$\psi_{n,m}(1) = a_0 = \frac{(n + m)!}{m! \, 2^m \, (n - m)!}.$$

Putting

$$\sin^m \theta \, \psi_{n,m}(\cos \theta) = P_n^{(m)}(\cos \theta), \qquad (29.5)$$

and using (29.4), we can write the complete system of homogeneous harmonic polynomials of x, y, z of degree n in the form

$$r^n P_n^{(0)}(\cos \theta), \quad r^n \sin \varphi \, P_n^{(1)}(\cos \theta), \quad r^n \sin 2\varphi \, P_n^{(2)}(\cos \theta), \dots ,$$

$$r^n \sin n\varphi \, P_n^{(n)}(\cos \theta); \qquad (29.6)$$

$$r^n \cos \varphi \, P_n^{(1)}(\cos \theta), \quad r^n \cos 2\varphi \, P_n^{(2)}(\cos \theta), \dots , \quad r^n \cos n\varphi \, P_n^{(n)}(\cos \theta).$$

The functions (29.6) serve as three-dimensional analogues of the functions

$$\varrho^n \cos n\varphi, \quad \varrho^n \sin n\varphi,$$

which appear with the solutions of the harmonic polynomial problem, and the functions

$$\sin m\varphi \, P_n^{(m)}(\cos \theta), \quad \cos m\varphi \, P_n^{(m)}(\cos \theta) \qquad (29.7)$$

are the analogues of the trigonometrical functions with multiple arguments, $\cos n\varphi$ and $\sin n\varphi$. These functions, or linear combinations of them, are usually known as *spherical harmonics of order n*.

§ 2. Approximation by means of Spherical Harmonics

THEOREM 1. *An arbitrary continuous function can be represented (in the sense of uniform convergence) on a sphere of unit radius by a linear combination of the functions (29.7) of orders 0, 1, ..., N with any desired accuracy provided N is sufficiently large.*

To prove this theorem we first establish one or two auxiliary propositions.

LEMMA 1. *An arbitrary function $F(\theta, \varphi)$ which is continuous on the sphere of unit radius can be represented thereon with any desired accuracy in the form of a polynomial*

$$Q(x, y, z) = \sum_{k=1}^{N} \sum_{l=1}^{N} \sum_{m=1}^{N} a_{klm} x^k y^l z^m.$$

Proof. We form the expression

$$Q_N = \frac{N+1}{4\pi} \int_0^{2\pi} \int_0^{\pi} \left(\frac{1 + \cos \gamma}{2} \right)^N F(\theta_1, \varphi_1) \sin \theta_1 \, d\theta_1 \, d\varphi_1$$

where $\cos \gamma$ denotes the cosine of the angle between the vector

$$x = \sin \theta \cos \varphi, \quad y = \sin \theta \sin \varphi, \quad z = \cos \theta$$

and the vector

$$x_1 = \sin \theta_1 \cos \varphi_1, \quad y_1 = \sin \theta_1 \sin \varphi_1, \quad z_1 = \cos \theta_1.$$

We have $\cos \gamma = xx_1 + yy_1 + zz_1$.

It is easily established that Q_N is a polynomial of degree N in x, y, z. We shall prove, on the other hand, that $Q_N - F = \varepsilon(N)$ tends to zero uniformly as N tends to infinity. Consider the integral

$$\frac{N+1}{4\pi} \int_0^{2\pi} \int_0^{\pi} \left(\frac{1 + \cos \gamma}{2} \right)^N F(\theta, \varphi) \sin \theta_1 \, d\theta_1 \, d\varphi_1$$

$$= \frac{N+1}{4\pi} \int_0^{2\pi} \int_0^{\pi} \left(\frac{1 + \cos \gamma}{2} \right)^N \sin \theta_1 \, d\theta_1 \, d\varphi_1.$$

Let S be the surface of the sphere of unit radius. The magnitude of

$$\frac{N+1}{4\pi} \int_0^{2\pi} \int_0^{\pi} \left(\frac{1 + \cos \gamma}{2} \right)^N \sin \theta_1 \, d\theta_1 \, d\varphi_1 = \frac{N+1}{4\pi} \iint_S \left(\frac{1 + \cos \gamma}{2} \right)^N dS$$

does not depend on the point (θ, φ). Consequently, in calculating it, we can put $\varphi = 0$. In this case, $\cos \gamma = \cos \theta_1$, and the integral becomes

$$\frac{N+1}{4\pi} \int_0^{2\pi} \int_0^{\pi} \left(\frac{1 + \cos \theta_1}{2} \right)^N \sin \theta_1 \, d\theta_1 \, d\varphi_1$$

$$= -\frac{N+1}{4\pi} 4\pi \int_0^{\pi} \left(\frac{1 + \cos \theta_1}{2} \right)^N d \left(\frac{1 + \cos \theta_1}{2} \right)$$

$$= -\left[\left(\frac{1 + \cos \theta_1}{2} \right)^{N+1} \right]_0^{\pi} = 1.$$

This implies

$$\frac{N+1}{4\pi} \int_0^{2\pi} \int_0^{\pi} \left(\frac{1+\cos\gamma}{2}\right)^N F(\theta, \varphi) \sin\theta_1 \, d\theta_1 \, d\varphi_1 = F(\theta, \varphi).$$

Moreover,

$$Q_N - F(\theta, \varphi)$$

$$= \frac{N+1}{4\pi} \int_0^{2\pi} \int_0^{\pi} \left(\frac{1+\cos\gamma}{2}\right)^N [F(\theta_1, \varphi_1) - F(\theta, \varphi)] \sin\theta_1 \, d\theta_1 \, d\varphi_1.$$

We shall divide the last integral into two terms, by separating off around the point (θ, φ) a domain $\gamma < \delta$ on the sphere so that

$$\left|F(\theta, \varphi) - F(\theta_1, \varphi_1)\right| < \frac{\varepsilon}{2}$$

for points (θ_1, φ_1) belonging to the domain $\gamma < \delta$.

Since the function $F(\theta, \varphi)$ is continuous, it is bounded, $i.e.$, $|F(\theta, \varphi)| \leq M$. We now choose N so large that the inequality

$$(N+1)\left(\frac{1+\cos\delta}{2}\right)^N < \frac{\varepsilon}{4M}$$

holds. Then obviously

$$(N+1)\left(\frac{1+\cos\gamma}{2}\right)^N < \frac{\varepsilon}{4M} \quad \text{for} \quad \gamma \geq \delta.$$

For such N,

$$|Q_N - F|$$

$$\leq \frac{N+1}{4\pi} \iint_{\gamma \geq \delta} \left(\frac{1+\cos\gamma}{2}\right)^N \left|F(\theta, \varphi) - F(\theta_1, \varphi_1)\right| \sin\theta_1 \, d\theta_1 \, d\varphi_1$$

$$+ \frac{N+1}{4\pi} \iint_{\gamma < \delta} \left(\frac{1+\cos\gamma}{2}\right)^N \left|F(\theta, \varphi) - F(\theta_1, \varphi_1)\right| \sin\theta_1 \, d\theta_1 \, d\varphi_1$$

$$\leq \frac{N+1}{4\pi} 2M \iint_{\gamma \geq \delta} \left(\frac{1+\cos\gamma}{2}\right)^N dS$$

$$+ \frac{\varepsilon}{2} \frac{N+1}{4\pi} \iint_{\gamma < \delta} \left(\frac{1+\cos\gamma}{2}\right)^N dS \leq \frac{\varepsilon}{4M} 2M + \frac{\varepsilon}{2} = \varepsilon.$$

Hence the lemma.

LEMMA 2. Let $Q_k(x, y, z)$ be a completely arbitrary polynomial in the variables x, y, z of degree k. Then there is a harmonic polynomial $P_k(x, y, z)$

of degree not higher than k which takes the same values as $Q_k(x, y, z)$ on the sphere $x^2 + y^2 + z^2 = 1$.

Proof. From what has already been proved, the polynomial $Q_k(x, y, z)$ can be represented in the form

$$Q_k = q_0(\varrho^2, z) + \sum_{m=1}^{k} [(x + iy)^m q_{1,m}(\varrho^2, z) + (x - iy)^m q_{2,m}(\varrho^2, z)]$$

[see (29.3)].

Hence on the sphere $x^2 + y^2 + z^2 = 1$ we shall have

$$Q_k = \tau_0(\cos \theta) + \sum_{m=1}^{k} \sin^m \theta \, [\cos m\varphi \, \tau_{1,m}(\cos \theta) + \sin m\varphi \, \tau_{2,m}(\cos \theta)].$$

(29.8)

where $\tau_0(\cos \theta)$ is a polynomial of degree not higher than k in $\cos \theta$, and $\tau_{1,m}$ and $\tau_{2,m}$ are polynomials of degree not higher than $k - m$ in $\cos \theta$. But $\tau_0(\cos \theta)$ can obviously be represented in the form

$$\tau_0(\cos \theta) = \sum_{j=0}^{k} c_j P_j^{(0)}(\cos \theta)$$

(29.9)

and each of the quantities $\sin^m \theta \, \tau_{1,m}$ and $\sin^m \theta \, \tau_{2,m}$ can be represented in the form

$$\left.\begin{array}{l}
\sin^m \theta \, \tau_{1,m}(\cos \theta) = \sum_{m=n}^{k} c_{n,m}^{(1)} P_n^{(m)}(\cos \theta) \\[2mm]
\sin^m \theta \, \tau_{2,m}(\cos \theta) = \sum_{n=m}^{k} c_{n,m}^{(2)} P_n^{(m)}(\cos \theta).
\end{array}\right\}$$

(29.10)

The coefficients of these expansions can be found one after the other, beginning with the highest, for the degree of each $P_n^{(m)}(\xi)$ increases by unity as m does.

Substituting (29.9) and (29.10) in (29.8) we have the proof of Lemma 2. Theorem 1 immediately follows from Lemmas 1 and 2. For, by Lemma 1, an arbitrary continuous function may be replaced on the sphere with any desired accuracy by a polynomial. And by Lemma 2 this polynomial can be replaced directly by a harmonic polynomial.

§ 3. The Dirichlet Problem for a Sphere

The theorem just proved gives us a new method of solving the Dirichlet problem for a sphere. We know from the earlier lectures that this problem always has a continuous solution and that if we change the boundary values of the function

$$[u]_S = \varphi$$

to approximate values φ' such that

$$|\varphi - \varphi'| < \varepsilon,$$

then the solution of the equation $\nabla^2 u = 0$ subject to the condition

$$[u]_S = \varphi'$$

will differ by no more than ε from the solution of the same equation subject to the condition

$$[u]_S = \varphi.$$

By taking as φ' a linear combination of spherical functions, we at once find a solution of the problem in the form of a sum of the corresponding polynomials. This will be the approximate solution of the Dirichlet problem. At the end of this lecture we shall give an explicit expression for the exact solution by means of spherical functions.

§ 4. The Differential Equations for Spherical Functions

Let $Y_n(\theta, \varphi)$ be any spherical harmonic of order n. The product $r^n Y_n(\theta, \varphi)$ is a harmonic polynomial. Consequently

$$\nabla^2 r^n Y_n(\theta, \varphi) = 0.$$

Using the expression for the Laplace operator in polar coordinates [see (28.7)], we obtain

$$\frac{1}{r^2} \frac{\partial}{\partial r} \left[r^2 \frac{\partial}{\partial r} (r^n Y_n(\theta, \varphi)) \right] + \frac{1}{r^2 \sin \theta} \frac{\partial}{\partial \theta} \left[\sin \theta \frac{\partial}{\partial \theta} (r^n Y_n(\theta, \varphi)) \right]$$

$$+ \frac{1}{r^2 \sin \theta} \frac{\partial^2}{\partial \varphi^2} [r^n Y_n(\theta, \varphi)] = 0,$$

whence

$$\frac{1}{\sin \theta} \frac{\partial}{\partial \theta} \left[\sin \theta \frac{\partial Y_n}{\partial \theta} \right] + \frac{1}{\sin^2 \theta} \frac{\partial^2 Y_n}{\partial \varphi^2} + n(n + 1)Y_n = 0. \quad (29.11)$$

Equation (29.11) is obtained from the equation

$$\frac{1}{\sin \theta} \frac{\partial}{\partial \theta} \left[\sin \theta \frac{\partial Y}{\partial \theta} \right] + \frac{1}{\sin^2 \theta} \frac{\partial^2 Y}{\partial \varphi^2} + \lambda Y = 0 \quad (29.12)$$

with

$$\lambda = n(n + 1). \quad (29.13)$$

We let D denote the operator

$$\frac{1}{\sin \theta} \frac{\partial}{\partial \theta} \left(\sin \theta \frac{\partial}{\partial \theta} \right) + \frac{1}{\sin^2 \theta} \frac{\partial^2}{\partial \varphi^2}$$

operating on a function $\chi(\theta, \varphi)$ defined on the surface of the unit sphere. Then equation (29.12) may be written in the form

$$DY + \lambda Y = 0. \tag{29.14}$$

At first sight the operator D seems to depend on the choice of pole on the sphere. However, it does in fact remain unchanged for any possible choice of the coordinates θ and φ on this sphere. For, suppose the function $\chi(\theta, \varphi)$ is specified in some manner on the sphere. We shall regard it as specified throughout the whole space r, θ, φ but independent of r. Then

$$D(\chi(\theta, \varphi)) = r^2 \nabla^2 \chi.$$

The Laplace operator appearing on the right-hand side does not depend on the choice of the coordinate axes, and hence the operator D will not depend on this choice, as was to be shown.

The arc-length ds of a curve on the sphere is given by

$$ds^2 = d\theta^2 + \sin^2\theta \, d\varphi^2,$$

or

$$ds^2 = h_1^2 \, d\theta^2 + h_2^2 \, d\varphi^2$$

where

$$h_1 = 1, \quad h_2 = \sin\theta.$$

The operator D is expressed in terms of h_1 and h_2 by the formula

$$D\chi = \frac{1}{h_1 h_2} \frac{\partial}{\partial \theta}\left(\frac{h_2}{h_1} \frac{\partial \chi}{\partial \theta}\right) + \frac{1}{h_1 h_2} \frac{\partial}{\partial \varphi}\left(\frac{h_1}{h_2} \frac{\partial \chi}{\partial \varphi}\right)$$

which is analogous to formula (28.6) expressing the Laplace operator in terms of curvilinear coordinates. This formula may therefore be called the Laplace operator on the surface of the sphere.

The equation (29.12) can have a non-trivial solution which is continuous over the whole surface of the sphere, but not for every value of λ, only for particular values. The problem of finding such values of λ and the solutions themselves is similar to that of finding solutions of the equations $y'' + \lambda y = 0$ which are continuous in a plane. We dealt with the latter problem in Lecture 20. The present problem can be reduced to the theory of integral equations by using Green's function. Let us examine this in more detail.

We shall seek a solution of the equation

$$Dv = \psi(\theta, \varphi) \tag{29.15}$$

on a sphere. We notice that the corresponding homogeneous equation

$$Du = 0 \tag{29.16}$$

has a non-trivial solution $u = 1$. Hence the problem of finding solutions of (29.15) is not always soluble.

We shall take as the Green's function for equation (29.15) the function of two variable points on the sphere

$$G = \frac{1}{2\pi} \log_e \sin \gamma/2,$$

where γ denotes the angular distance between the two points considered. We know from spherical trigonometry that

$$\cos \gamma = \cos \theta_1 \cos \theta_2 - \sin \theta_1 \sin \theta_2 \cos (\varphi_1 - \varphi_2),$$

where (θ_1, φ_1), (θ_2, φ_2) are the coordinates of the two points.

We shall prove that if the function ψ satisfies the equation

$$\iint_S \psi \, dS = \int_0^\pi \int_0^{2\pi} \psi \sin \theta \, d\varphi \, d\theta = 0, \tag{29.17}$$

then the solution of equation (29.15) which satisfies the same condition

$$\int_0^\pi \int_0^{2\pi} v \sin \theta \, d\varphi \, d\theta = 0 \tag{29.17'}$$

is given by the formula

$$v(\theta_0, \varphi_0) = \int_0^\pi \int_0^{2\pi} G(\theta, \varphi, \theta_0, \varphi_0) \, \psi(\theta, \varphi) \sin \theta \, d\varphi \, d\theta = \iint_S G\psi \, dS, \tag{29.18}$$

where dS denotes an element of the surface of the sphere. This will justify our calling G the Green's function for the equation.

Suppose a solution of equation (29.15) exists. We choose the pole of the sphere as the point (θ_0, φ_0). We then have

$$\iint_S G\psi \, dS = \int_0^\pi \int_0^{2\pi} GDv \sin \theta \, d\varphi \, d\theta = \int_0^\pi \left(\int_0^{2\pi} GDv \, d\varphi \right) \sin \theta \, d\theta$$

$$= \int_0^\pi \log_e \sin \frac{\theta}{2} \left\{ \frac{\partial}{\partial \theta} \sin \theta \frac{\partial}{\partial \theta} \left[\frac{1}{2\pi} \int_0^{2\pi} v \, d\varphi \right] + \frac{1}{\sin \theta} \frac{1}{2\pi} \int_0^{2\pi} \frac{\partial^2 v}{\partial \varphi^2} \, d\varphi \right\} d\theta$$

$$= -\int_0^\pi \sin \theta \frac{\partial}{\partial \theta} \left[\frac{1}{2\pi} \int_0^{2\pi} v \, d\varphi \right] \frac{1}{2} \frac{\cos \theta/2}{\sin \theta/2} \, d\theta$$

$$+ \left[\sin \theta \frac{\partial}{\partial \theta} \left(\frac{1}{2\pi} \int_0^{2\pi} v \, d\varphi \right) \log_e \sin \frac{\theta}{2} \right]_0^\pi$$

$$= -\frac{1}{2} \int_0^\pi \sin \theta \left[\frac{1}{2\pi} \int_0^{2\pi} v \, d\varphi \right] d\theta - \left[\cos^2 \frac{\theta}{2} \frac{1}{2\pi} \int_0^{2\pi} v \, d\varphi \right]_0^\pi.$$

By virtue of the condition (29.17),

$$\iint_S G\psi \, dS = [v]_{\theta=0}. \tag{29.19}$$

It follows from (29.19) by symmetry considerations that formula (29.18) holds for any point on the sphere. We can also prove that if ψ is an arbitrary function satisfying the condition (29.17), then the function $v(\theta_0, \varphi_0)$ definable by equation (29.18) gives the solution of the equation

$$Dv = \psi.$$

The proof of this assertion is similar to that which we have given more than once, and we shall omit it.

Since the Green's function which we have constructed is symmetrical, all the theory of integral equations with symmetric kernels is applicable to the equation

$$u = \lambda \iint_S Gu \, dS$$

which is equivalent to (29.14).

If we set ourselves the task of finding all regular solutions of equation (29.12) on a sphere, then from what has been said it follows that the values (29.13) will be the eigenvalues. Each eigenvalue has $(2n + 1)$ eigenfunctions. We have already obtained, in (29.7), the set of solutions of equation (29.12). It follows from the general theory of integral equations that those of the functions (29.7) which correspond to different values of n are orthogonal. The functions (29.7) which correspond to the same value of n are also orthogonal, since sines and cosines of multiples of angles in the interval $0 \leqq \varphi \leqq 2\pi$ are orthogonal. Hence all the functions (29.7) are orthogonal.

THEOREM 2. *The functions* (29.7) *exhaust the whole set of eigenfunctions of equation* (29.12).

Proof. Let $Y^*(\theta, \varphi)$ be an eigenfunction of equation (29.12) which is distinct from the functions (29.7), which we suppose to have been ordered in some way to give the set $u_i(\theta, \varphi)$.

By Theorem 1, the function $Y^*(\theta, \varphi)$ can be represented in the form

$$Y^*(\theta, \varphi) = \sum_{i=1}^{N} c_i^{(N)} u_i(\theta, \varphi) + \eta_N,$$

where η_N is arbitrarily small. This implies

$$\iint_S \left\{ Y^*(\theta, \varphi) - \sum_{i=1}^{N} c_i^{(N)} u_i(\theta, \varphi) \right\}^2 dS = \iint_S \eta_N^2 \, dS < \varepsilon(N).$$

Let

$$y_i = \iint_S Y^*(\theta, \varphi) \, u_i(\theta, \varphi) \, dS.$$

Then

$$\iint_S \left\{ Y^*(\theta, \varphi) - \sum_{i=1}^N y_i u_i(\theta, \varphi) \right\}^2 \mathrm{d}S \leqq \iint_S \left\{ Y^*(\theta, \varphi) - \sum_{i=1}^N c_i^{(N)} u_i(\theta, \varphi) \right\}^2 \mathrm{d}S.$$

Taking into account the fact that, from the general theory of integral equations, $Y^*(\theta, \varphi)$ is orthogonal to all $u_i(\theta, \varphi)$ and hence that $y_i = 0$, we get

$$\iint_S [Y^*(\theta, \varphi)]^2 \, \mathrm{d}S < \varepsilon$$

for any positive ε. This implies

$$Y^*(\theta, \varphi) = 0,$$

as was to be shown.

It follows at once from Theorem 2 that any function $\varphi(S)$ which is quadratically integrable on the sphere, *i.e.*, any function satisfying the condition

$$\iint_S \varphi^2(S) \, \mathrm{d}S < \infty,$$

can be represented in the form of a series convergent in the mean:

$$\sum_{i=1}^\infty \varphi_i u_i$$

where

$$\varphi_i = \iint_S \varphi u_i \, \mathrm{d}S.$$

The convergence of the series will be uniform, by the Hilbert–Schmidt theorem, if φ has continuous second-order derivatives.

If we substitute one of the functions (29.7) in the equation (29.12) and use the fact that

$$\frac{\partial^2}{\partial \varphi^2} [\sin m\varphi \, P_n^{(m)}(\cos \theta)] = -m^2 \sin m\varphi \, P_n^{(m)}(\cos \theta),$$

we get

$$\frac{1}{\sin \theta} \frac{\partial}{\partial \theta} \left[\sin \theta \frac{\partial P_n^{(m)}(\cos \theta)}{\partial \theta} \right] + \left[n(n+1) - \frac{m^2}{\sin^2 \theta} \right] P_n^{(m)}(\cos \theta) = 0.$$

$$(29.20)$$

Equation (29.20) is a differential equation for the function $P_n^{(m)}(\cos \theta)$.

If we put $\cos \theta = \mu$, then

$$\frac{\partial}{\partial \theta} = -\sin \theta \frac{\partial}{\partial \mu},$$

and equation (29.20) can be written

$$\frac{d}{d\mu}\left[\sin^2\theta\,\frac{dP_n^{(m)}(\mu)}{d\mu}\right] + \left[n(n+1) - \frac{m^2}{\sin^2\theta}\right]P_n^{(m)}(\mu) = 0,$$

or

$$\frac{d}{d\mu}\left[(1-\mu^2)\,\frac{dP_n^{(m)}(\mu)}{d\mu}\right] + \left[n(n+1) - \frac{m^2}{1-\mu^2}\right]P_n^{(m)}(\mu) = 0.$$

(29.21)

The differential equation just obtained is a linear, second-order equation. It is obtained from the equation

$$\frac{d}{d\mu}\left[(1-\mu^2)\,\frac{dy}{d\mu}\right] + \left[\lambda_1 - \frac{\lambda_2}{1-\mu^2}\right]y = 0$$

by putting $\lambda_1 = n(n+1)$ and $\lambda_2 = m^2$.

If we seek values of λ_1 and λ_2 for which the last equation has a bounded solution, then it can be asserted that, if $\lambda_2 = m^2$, the equation will have no other eigenvalues than $\lambda_1 = n(n+1)$. This follows from the fact that, otherwise, we should have an eigenfunction y^* of equation (29.21) and a corresponding eigenfunction for equation (29.12) in the form $P^*(\cos\theta)\sin m\varphi$ which would be different from all the functions (29.7), and this has been proved impossible.

The relations which we have obtained enable us to study the problem of solving Laplace's equation for the sphere by the method of separation of the variables. Indeed, we have already practically carried out this method, though starting from other premises.

If we separate the variables in Laplace's equation

$$\nabla^2 u = \frac{1}{r^2}\left(\frac{\partial}{\partial r}\,r^2\,\frac{\partial u}{\partial r}\right) + \frac{1}{r^2\sin\theta}\frac{\partial}{\partial\theta}\left(\sin\theta\,\frac{\partial u}{\partial\theta}\right) + \frac{1}{r^2\sin^2\theta}\frac{\partial^2 u}{\partial\varphi^2} = 0,$$

we obtain after some simplification, which we leave to the reader,

$$u = \sum_{\substack{n,m \\ m\leq n}} r^n P_n^{(m)}(\cos\theta)\,[a_n^{(m)}\cos m\varphi + b_n^{(m)}\sin m\varphi],$$

where $\cos m\varphi$ and $\sin m\varphi$ are solutions of the equation

$$\frac{d^2\Phi}{d\varphi^2} + m^2\Phi = 0,$$

$P_n^{(m)}$ is a solution of equation (29.20) or (29.21),

and r^n is a solution of the equation

$$\frac{1}{r^2}\frac{d}{dr}\left(r^2\,\frac{dR}{dr}\right) - n(n+1)R = 0.$$

SOME ELEMENTARY PROPERTIES
OF SPHERICAL FUNCTIONS

§ 1. Legendre Polynomials

We shall examine one or two properties of spherical functions.

THEOREM 1. *A solution of the equation*

$$\frac{d}{d\mu}\left[(1 - \mu^2)\frac{dy}{d\mu}\right] + \left[n(n + 1) - \frac{m^2}{1 - \mu^2}\right]y = 0$$

is given by the function

$$P_n^{(m)}(\mu) = \frac{(-1)^n}{2^n \cdot n!}(1 - \mu^2)^{\frac{m}{2}}\frac{d^{n+m}(1 - \mu^2)^n}{d\mu^{n+m}} \qquad (30.1)$$

Proof. Let $(1 - \mu^2)^n = \Phi$. Then the relation

$$(1 - \mu^2)\frac{d^2\Phi}{d\mu^2} + 2(n - 1)\mu\frac{d\Phi}{d\mu} + 2n\Phi = 0$$

can be verified immediately. Differentiating it $(m + n)$ times with respect to μ and using Leibniz's formula, we get

$$(1 - \mu^2)\frac{d^{m+n+2}\Phi}{d\mu^{m+n+2}} - 2(m + 1)\mu\frac{d^{m+n+1}\Phi}{d\mu^{m+n+1}}$$

$$+ (n - m)(n + m + 1)\frac{d^{n+m}\Phi}{d\mu^{n+m}} = 0,$$

or, putting

$$\frac{d^{m+n}\Phi}{d\mu^{m+n}} = \psi, \quad (1 - \mu^2)\psi'' - 2(m + 1)\mu\psi' + (n - m)(n + m + 1)\psi = 0.$$

Further, let $P_n^{(m)}(\mu) = C(1 - \mu^2)^{\frac{m}{2}}\psi$, then by direct substitution we get

$$(1 - \mu^2)\frac{d^2 P_n^{(m)}}{d\mu^2} - 2\mu\frac{dP_n^{(m)}}{d\mu} + \left[n(n + 1) - \frac{m^2}{1 - \mu^2}\right]P_n^{(m)}$$

$$= C(1 - \mu^2)^{\frac{m}{2}}[(1 - \mu^2)\psi'' - 2(m + 1)\mu\psi' + (n - m)(n + m + 1)\psi] = 0,$$

as was to be shown.

414

Formula (30.1) obviously gives us a new explicit expression for $P_n^{(m)}(\cos\theta)$, since $r^n \sin m\varphi\, P_n^{(m)}$ is a harmonic polynomial, and there cannot be two such polynomials. A number of useful recurrence formulae can be obtained from this formula which facilitate the calculation of $P_n^{(m)} (\cos\theta)$.

The polynomials

$$P_n^{(0)}(\mu) = P_n(\mu) = \frac{(-1)^n}{2^n \cdot n!} \frac{d^n(1 - \mu^2)^n}{d\mu^n}$$

are known as *Legendre polynomials*, and play a special role among the spherical functions.

Strictly speaking, we have not yet the right to denote the function (30.1) by $P_n^{(m)} (\mu)$, as previously defined, since the two might differ by a constant multiplier. It is, however, easily established that such a multiplier could only be unity. For, firstly, it follows from (30.1) that

$$P_n(1) = 1. \tag{30.2}$$

This may be seen by putting $\mu - 1 = z$; then

$$(1 - \mu^2)^n = (-1)^n z^n (2 + z)^n = (-1)^n (2^n z^n + \cdots),$$

from which (30.2) follows immediately.

Also, if

$$\frac{P_n^{(m)}(\cos\theta)}{\sin^m\theta} = \psi_{n,m}$$

then similarly

$$\psi_{n,m}(1) = \frac{(n + m)!}{2^m \cdot n!(n - m)!},$$

which proves our assertion.

§ 2. The Generating Function

THEOREM 2. *The expansion of the function* $\dfrac{1}{r_1} = \dfrac{1}{\sqrt{1 - 2r\cos\theta + r^2}}$ *as a power series in r has the form*

$$\frac{1}{r_1} = \sum_{n=0}^{\infty} r^n P_n(\cos\theta). \tag{30.3}$$

The function $1/r_1$ is therefore called the generating function of the Legendre polynomials.

This theorem could be proved directly by calculating the coefficients of the power-series, but it can also be proved by a different method, as follows.

Considering the function $1/r_1$ in a certain sphere $r \leq \varrho \leq 1$, we may expand it as a power series in r, the coefficients of which will evidently be polynomials in $\cos \theta$, *i.e.*,

$$\frac{1}{r_1} = \sum_{n=0}^{\infty} r^n Q_n(\cos \theta).$$

For,

$$1 - 2r \cos \theta + r^2 = (1 - re^{i\theta})(1 - re^{i\theta}), \tag{30.4}$$

and hence the expansion will have a radius of convergence equal to unity. This implies, that, for $r = \varrho$,

$$\frac{1}{r_1} = \sum_{n=0}^{\infty} \varrho^n Q_n(\cos \theta). \tag{30.5}$$

As we proved in the last lecture, a harmonic function which takes the specified values (30.5) on the surface $r = \varrho$ can be expanded in a uniformly convergent series

$$\frac{1}{r_1} = \sum_{n=0}^{\infty} a_n r^n P_n(\cos \theta).$$

Since the expansion of an arbitrary function as a power series in r must be unique, it follows that

$$Q_n(\cos \theta) = a_n P_n(\cos \theta).$$

To determine the value of a_n, we note that, for $\cos \theta = 1$, we have

$$(1 - 2r \cos \theta + r^2)_{\theta=0} = (1 - r)^2$$

and

$$\frac{1}{r_1} = \frac{1}{1-r} = \sum_{n=0}^{\infty} r^n.$$

Comparing this with the expression

$$\frac{1}{r_1} = \sum_{n=0}^{\infty} r^n a_n P_n(1),$$

we have $a_n = 1$. Hence Theorem 2.

COROLLARY 1.

$$\frac{1}{\sqrt{R^2 - 2Rr \cos \theta + r^2}} = \begin{cases} \sum_{n=0}^{\infty} \dfrac{r^n}{R^{n+1}} P_n(\cos \theta), & r < R \\[2ex] \sum_{n=0}^{\infty} \dfrac{R^n}{r^{n+1}} P_n(\cos \theta), & r > R. \end{cases} \tag{30.6}$$

For,

$$\frac{1}{\sqrt{R^2 - 2Rr\cos\theta + r^2}} = \frac{1}{R}\frac{1}{\sqrt{1 - 2\dfrac{r}{R}\cos\theta + \left(\dfrac{r}{R}\right)^2}}$$

$$= \frac{1}{r}\frac{1}{\sqrt{1 - 2\dfrac{R}{r}\cos\theta + \left(\dfrac{R}{r}\right)^2}}$$

from which (30.6) follows.

COROLLARY 2.

$$\frac{\partial}{\partial\varrho}\frac{1}{\sqrt{\varrho^2 - 2\varrho r\cos\theta + r^2}} = -\sum_{n=0}^{\infty}(n+1)\frac{r^n}{\varrho^{n+2}}P_n(\cos\theta),\ r < \varrho.$$

$$(30.7)$$

We may further remark that the convergence of the two series (30.6) and of (30.7) will be uniform relative to the variable θ provided that, respectively,

$$\left.\begin{array}{c} R - r > \varepsilon, \\ r - R > \varepsilon, \\ \varrho - r > \varepsilon, \end{array}\right\} \text{ for a fixed value of } R.$$

or

To prove this assertion, consider the series with positive terms

$$\frac{1}{\sqrt{1-r}} = 1 + \frac{1}{2}r + \frac{1.3}{2.4}r^2 + \cdots. \tag{30.8}$$

All the terms of this series are equal in absolute value to the corresponding terms of the series

$$\frac{1}{\sqrt{1 - re^{i\theta}}} = \sum_{n=0}^{\infty}a_n r^n, \quad \frac{1}{\sqrt{1 - re^{-i\theta}}} = \sum_{n=0}^{\infty}\bar{a}_n r^n. \tag{30.9}$$

Consequently, the terms of the series

$$\frac{1}{1-r} = \frac{1}{\sqrt{1-r}}\frac{1}{\sqrt{1-r}} = 1 + r + r^2 + \cdots,$$

which can be obtained by the termwise multiplication of the series (39.8) by itself, will be not less in absolute value than the corresponding terms of the series

$$\frac{1}{\sqrt{1 - 2r\cos\theta + r^2}} = \frac{1}{\sqrt{1 - re^{i\theta}}}\frac{1}{\sqrt{1 - re^{-i\theta}}} = \sum_{n=0}^{\infty}r^n P_n(\cos\theta),$$

which is obtained by the termwise multiplication of the series (30.9) one by the other. Hence we obtain the inequality $P_n(\cos \theta) \leq 1$, in which the equality sign can hold only for $\theta = 0$, *i.e.*, if $\cos \theta = 1$, as may easily be seen from the proof. From this follows our assertion that the convergence is uniform.

§ 3. Laplace's Formula

Now let $r^n Y_n(\theta, \varphi)$ be a certain harmonic polynomial. Applying Green's formula to it, we obtain

$$r^n Y_n(\theta, \varphi)$$

$$= \frac{1}{4\pi} \iint_{\rho=1} \left\{ \frac{\partial}{\partial n} \left(\frac{1}{r_1} \right) \varrho^n Y_n(\theta_1, \varphi_1) - \frac{1}{r_1} \frac{\partial}{\partial n} (\varrho^n Y_n(\theta_1, \varphi_1)) \right\} \, dS_1 \quad (30.10)$$

where

$$r_1 = \sqrt{r^2 - 2r\varrho \cos \gamma + \varrho^2}.$$

dS_1 is an element of the surface of the sphere with coordinates θ_1, φ_1,

$r_1 < \varrho$ is the distance between the points $M(r_1, \theta, \varphi)$ and $M(\varrho, \theta_1, \varphi_1)$,

and $\quad \gamma \quad$ is the angle between the vectors from the origin to the points M_1, M_2.

Clearly,

$$r\varrho \cos \gamma$$

$$= r\varrho \sin \theta \cos \varphi \sin \theta_1 \cos \varphi_1 + r\varrho \sin \theta \sin \varphi \sin \theta_1 \sin \varphi_1 + r\varrho \cos \theta \cos \theta_1$$

$$= r\varrho \left[\cos \theta \cos \theta_1 + \sin \theta \sin \theta_1 \cos (\varphi - \varphi_1) \right].$$

Substituting in formula (30.10) the expressions (30.6) and (30.7) for $1/r_1$ and

$$\frac{\partial}{\partial n} \left(\frac{1}{r_1} \right) = - \frac{\partial}{\partial \varrho} \left(\frac{1}{r_1} \right),$$

we have

$$r^n Y_n(\theta, \varphi)$$

$$= \frac{1}{4\pi} \iint_{\rho=1} Y_n(\theta_1, \varphi_1) \left\{ \varrho^n (n + 1) \sum_{k=0}^{\infty} \frac{r^k}{\varrho^{k+2}} P_k(\cos \gamma) \right.$$

$$\left. + n\varrho^{n-1} \sum_{k=0}^{\infty} \frac{r^k}{\varrho^{k+1}} P_k(\cos \gamma) \right\} \sin \theta_1 \, d\theta_1 \, d\varphi_1$$

$$= \frac{1}{4\pi} \int_0^{2\pi} \int_0^{\pi} Y_n(\theta_1, \varphi_1) \, (2n + 1) \left\{ \sum_{k=0}^{\infty} r^k P_k(\cos \gamma) \right\} \sin \theta_1 \, d\theta_1 \, d\varphi_1.$$

It is clear that the function $P_j(\cos \gamma)$ for $j \neq n$ is orthogonal to the function $Y_n(\theta, \varphi)$, since they are harmonics of different orders. The series under the integral sign is convergent in the mean and may therefore be integrated termwise. But then all terms save one disappear, because of the orthogonality. Hence

$$r^n Y^n(\theta, \varphi) = \frac{2n + 1}{4\pi} \int_0^{2\pi} \int_0^{\pi} r^n Y_n(\theta_1, \varphi_1) P_n(\cos \gamma) \sin \theta_1 \, d\theta_1 \, d\varphi_1$$

i.e.,

$$Y_n(\theta, \varphi) = \frac{2n + 1}{4\pi} \int_0^{2\pi} \int_0^{\pi} Y_n(\theta_1, \varphi_1) P_n(\cos \gamma) \, d\theta_1 \, d\varphi_1. \quad (30.11)$$

Formula (30.11) enables the coefficients of the expansion of a given function $F(\theta, \varphi)$ in spherical harmonics to be obtained immediately. For, suppose

$$F(\theta, \varphi) = \sum_{n=0}^{\infty} Y_n(\theta, \varphi) \quad (30.12)$$

Multiplying both sides of (30.12) by $P_k(\cos \gamma)$ and integrating over a sphere we get

$$\frac{2n + 1}{4\pi} \int_0^{2\pi} \int_0^{\pi} F(\theta_1, \varphi_1) P_k(\cos \gamma) \sin \theta_1 \, d\theta_1 \, d\varphi_1 = Y_k(\theta, \varphi). \quad (30.13)$$

Formula (30.13) is known as *Laplace's Formula*.

Laplace's formula enables the solution of the Dirichlet problem in a sphere to be written explicitly in the form of a series of harmonic polynomials. For, by multiplying each harmonic of order k by r^k and adding, we can obtain a harmonic function which takes the given values $F(\theta, \varphi)$ on the boundary of the sphere. We obtain the solution in the following form:

$$u(r, \theta, \varphi) = \frac{1}{4\pi} \sum_{k=0}^{\infty} (2k + 1) r^k \int_0^{2\pi} \int_0^{\pi} F(\theta_1, \varphi_1) P_k(\cos \gamma) \sin \theta_1 \, d\theta_1 \, d\varphi_1.$$

To conclude this exposition of the theory of spherical functions, we shall give without proof a few formulae which are often useful:

$$\left. \begin{array}{l} \displaystyle\int_{-1}^{+1} [P_n(\mu)]^2 \, d\mu = \frac{2}{2n + 1}, \\[2ex] \displaystyle\int_{-1}^{+1} [P_n^{(m)}(\mu)]^2 \, d\mu = \frac{2}{2n + 1} \frac{(n + m)!}{(n - m)!} \\[2ex] \displaystyle\int_{-1}^{+1} P_n(\mu) P_{n'}(\mu) \, d\mu = 0, \quad \int_{-1}^{+1} P_n^{(m)}(\mu) P_{n'}^{(m)}(\mu) \, d\mu = 0, \\[2ex] \hspace{6cm} n \neq n'. \end{array} \right\} \quad (30.14)$$

The formulae (30.14) enable us to calculate directly the coefficients of the expansion of a function in a series of spherical functions of the form (29.7).

Finally, we may mention the asymptotic representation of Legendre polynomials for large values of n:

$$P_n(\cos \theta) = \sqrt{\frac{2}{\pi n \sin \theta}} \left\{\cos\left[\left(n + \frac{1}{2}\right)\theta - \frac{\pi}{4}\right] + \varepsilon_n\right\},$$

where $\varepsilon_n \to 0$ as $n \to \infty$ uniformly relative to θ, for $\varepsilon < \theta < \pi - \varepsilon$.

At this point we bring to an end this elementary course dealing with the principal characteristic properties of the equations of mathematical physics and their classical methods of solution. In this we have, of course, far from exhausted the content of contemporary mathematical physics. Thus, beyond the scope of our course there remain the following highly important topics, which would furnish material for another book of at least the same size:

1. Problems of mathematical physics for an unbounded medium and the applications of the Fourier integral transformation.

2. Special problems for a semi-bounded medium, diffraction of waves, *etc.*

3. Variational methods in mathematical physics.

4. Approximate solution of problems of mathematical physics by the method of finite differences.

And among these most important topics we ought also to include the theory of non-linear equations.

The scope of this course has not allowed us to go into any of the topics mentioned. We hope, however, that an attentive reader who has familiarized himself in these lectures with the basic ideas in the theory of the equations of mathematical physics will be able, on his own, to explore profitably the contemporary literature dealing with these questions.

INDEX

A CATALOG OF SELECTED
DOVER BOOKS
IN SCIENCE AND MATHEMATICS

A CATALOG OF SELECTED

DOVER BOOKS

IN SCIENCE AND MATHEMATICS

QUALITATIVE THEORY OF DIFFERENTIAL EQUATIONS, V.V. Nemytskii and V.V. Stepanov. Classic graduate-level text by two prominent Soviet mathematicians covers classical differential equations as well as topological dynamics and erqodic theory. Bibliographies. 523pp. 5⅜ × 8½. 65954-2 Pa. $10.95

MATRICES AND LINEAR ALGEBRA, Hans Schneider and George Phillip Barker. Basic textbook covers theory of matrices and its applications to systems of linear equations and related topics such as determinants, eigenvalues and differential equations. Numerous exercises. 432pp. 5⅜ × 8½. 66014-1 Pa. $8.95

QUANTUM THEORY, David Bohm. This advanced undergraduate-level text presents the quantum theory in terms of qualitative and imaginative concepts, followed by specific applications worked out in mathematical detail. Preface. Index. 655pp. 5⅜ × 8½. 65969-0 Pa. $10.95

ATOMIC PHYSICS (8th edition), Max Born. Nobel laureate's lucid treatment of kinetic theory of gases, elementary particles, nuclear atom, wave-corpuscles, atomic structure and spectral lines, much more. Over 40 appendices, bibliography. 495pp. 5⅜ × 8½. 65984-4 Pa. $11.95

ELECTRONIC STRUCTURE AND THE PROPERTIES OF SOLIDS: The Physics of the Chemical Bond, Walter A. Harrison. Innovative text offers basic understanding of the electronic structure of covalent and ionic solids, simple metals, transition metals and their compounds. Problems. 1980 edition. 582pp. 6⅛ × 9¼. 66021-4 Pa. $14.95

BOUNDARY VALUE PROBLEMS OF HEAT CONDUCTION, M. Necati Özisik. Systematic, comprehensive treatment of modern mathematical methods of solving problems in heat conduction and diffusion. Numerous examples and problems. Selected references. Appendices. 505pp. 5⅜ × 8½. 65990-9 Pa. $11.95

A SHORT HISTORY OF CHEMISTRY (3rd edition), J.R. Partington. Classic exposition explores origins of chemistry, alchemy, early medical chemistry, nature of atmosphere, theory of valency, laws and structure of atomic theory, much more. 428pp. 5⅜ × 8½. (Available in U.S. only) 65977-1 Pa. $10.95

A HISTORY OF ASTRONOMY, A. Pannekoek. Well-balanced, carefully reasoned study covers such topics as Ptolemaic theory, work of Copernicus, Kepler, Newton, Eddington's work on stars, much more. Illustrated. References. 521pp. 5⅜ × 8½. 65994-1 Pa. $11.95

PRINCIPLES OF METEOROLOGICAL ANALYSIS, Walter J. Saucier. Highly respected, abundantly illustrated classic reviews atmospheric variables, hydrostatics, static stability, various analyses (scalar, cross-section, isobaric, isentropic, more). For intermediate meteorology students. 454pp. 6⅛ × 9¼. 65979-8 Pa. $12.95

THE FOUR-COLOR PROBLEM: Assaults and Conquest, Thomas L. Saaty and Paul G. Kainen. Engrossing, comprehensive account of the century-old combinatorial topological problem, its history and solution. Bibliographies. Index. 110 figures. 228pp. 5⅜ × 8½.   65092-8 Pa. $6.00

CATALYSIS IN CHEMISTRY AND ENZYMOLOGY, William P. Jencks. Exceptionally clear coverage of mechanisms for catalysis, forces in aqueous solution, carbonyl- and acyl-group reactions, practical kinetics, more. 864pp. 5⅜ × 8½.   65460-5 Pa. $18.95

PROBABILITY: An Introduction, Samuel Goldberg. Excellent basic text covers set theory, probability theory for finite sample spaces, binomial theorem, much more. 360 problems. Bibliographies. 322pp. 5⅜ × 8½.   65252-1 Pa. $7.95

LIGHTNING, Martin A. Uman. Revised, updated edition of classic work on the physics of lightning. Phenomena, terminology, measurement, photography, spectroscopy, thunder, more. Reviews recent research. Bibliography. Indices. 320pp. 5⅜ × 8¼.   64575-4 Pa. $7.95

PROBABILITY THEORY: A Concise Course, Y.A. Rozanov. Highly readable, self-contained introduction covers combination of events, dependent events, Bernoulli trials, etc. Translation by Richard Silverman. 148pp. 5⅜ × 8¼.   63544-9 Pa. $4.50

THE CEASELESS WIND: An Introduction to the Theory of Atmospheric Motion, John A. Dutton. Acclaimed text integrates disciplines of mathematics and physics for full understanding of dynamics of atmospheric motion. Over 400 problems. Index. 97 illustrations. 640pp. 6 × 9.   65096-0 Pa. $16.95

STATISTICS MANUAL, Edwin L. Crow, et al. Comprehensive, practical collection of classical and modern methods prepared by U.S. Naval Ordnance Test Station. Stress on use. Basics of statistics assumed. 288pp. 5⅜ × 8½.
    60599-X Pa. $6.00

WIND WAVES: Their Generation and Propagation on the Ocean Surface, Blair Kinsman. Classic of oceanography offers detailed discussion of stochastic processes and power spectral analysis that revolutionized ocean wave theory. Rigorous, lucid. 676pp. 5⅜ × 8½.   64652-1 Pa. $14.95

STATISTICAL METHOD FROM THE VIEWPOINT OF QUALITY CONTROL, Walter A. Shewhart. Important text explains regulation of variables, uses of statistical control to achieve quality control in industry, agriculture, other areas. 192pp. 5⅜ × 8½.   65232-7 Pa. $6.00

THE INTERPRETATION OF GEOLOGICAL PHASE DIAGRAMS, Ernest G. Ehlers. Clear, concise text emphasizes diagrams of systems under fluid or containing pressure; also coverage of complex binary systems, hydrothermal melting, more. 288pp. 6½ × 9¼.   65389-7 Pa. $8.95

STATISTICAL ADJUSTMENT OF DATA, W. Edwards Deming. Introduction to basic concepts of statistics, curve fitting, least squares solution, conditions without parameter, conditions containing parameters. 26 exercises worked out. 271pp. 5⅜ × 8½.   64685-8 Pa. $7.95

DE RE METALLICA, Georgius Agricola. The famous Hoover translation of greatest treatise on technological chemistry, engineering, geology, mining of early modern times (1556). All 289 original woodcuts. 638pp. 6¾ × 11.
60006-8 Clothbd. $15.95

SOME THEORY OF SAMPLING, William Edwards Deming. Analysis of the problems, theory and design of sampling techniques for social scientists, industrial managers and others who find statistics increasingly important in their work. 61 tables. 90 figures. xvii + 602pp. 5⅜ × 8½.
64684-X Pa. $14.95

THE VARIOUS AND INGENIOUS MACHINES OF AGOSTINO RAMELLI: A Classic Sixteenth-Century Illustrated Treatise on Technology, Agostino Ramelli. One of the most widely known and copied works on machinery in the 16th century. 194 detailed plates of water pumps, grain mills, cranes, more. 608pp. 9 × 12.
25497-6 Clothbd. $34.95

LINEAR PROGRAMMING AND ECONOMIC ANALYSIS, Robert Dorfman, Paul A. Samuelson and Robert M. Solow. First comprehensive treatment of linear programming in standard economic analysis. Game theory, modern welfare economics, Leontief input-output, more. 525pp. 5⅜ × 8½.
65491-5 Pa. $12.95

ELEMENTARY DECISION THEORY, Herman Chernoff and Lincoln E. Moses. Clear introduction to statistics and statistical theory covers data processing, probability and random variables, testing hypotheses, much more. Exercises. 364pp. 5⅜ × 8½.
65218-1 Pa. $8.95

THE COMPLEAT STRATEGYST: Being a Primer on the Theory of Games of Strategy, J.D. Williams. Highly entertaining classic describes, with many illustrated examples, how to select best strategies in conflict situations. Prefaces. Appendices. 268pp. 5⅜ × 8½.
25101-2 Pa. $5.95

MATHEMATICAL METHODS OF OPERATIONS RESEARCH, Thomas L. Saaty. Classic graduate-level text covers historical background, classical methods of forming models, optimization, game theory, probability, queueing theory, much more. Exercises. Bibliography. 448pp. 5⅜ × 8¼.
65703-5 Pa. $12.95

CONSTRUCTIONS AND COMBINATORIAL PROBLEMS IN DESIGN OF EXPERIMENTS, Damaraju Raghavarao. In-depth reference work examines orthogonal Latin squares, incomplete block designs, tactical configuration, partial geometry, much more. Abundant explanations, examples. 416pp. 5⅜ × 8¼.
65685-3 Pa. $10.95

THE ABSOLUTE DIFFERENTIAL CALCULUS (CALCULUS OF TENSORS), Tullio Levi-Civita. Great 20th-century mathematician's classic work on material necessary for mathematical grasp of theory of relativity. 452pp. 5⅜ × 8½.
63401-9 Pa. $9.95

VECTOR AND TENSOR ANALYSIS WITH APPLICATIONS, A.I. Borisenko and I.E. Tarapov. Concise introduction. Worked-out problems, solutions, exercises. 257pp. 5⅜ × 8¼.
63833-2 Pa. $6.95

CHALLENGING MATHEMATICAL PROBLEMS WITH ELEMENTARY SOLUTIONS, A.M. Yaglom and I.M. Yaglom. Over 170 challenging problems on probability theory, combinatorial analysis, points and lines, topology, convex polygons, many other topics. Solutions. Total of 445pp. 5⅜ × 8½. Two-vol. set.
Vol. I 65536-9 Pa. $5.95
Vol. II 65537-7 Pa. $5.95

FIFTY CHALLENGING PROBLEMS IN PROBABILITY WITH SOLUTIONS, Frederick Mosteller. Remarkable puzzlers, graded in difficulty, illustrate elementary and advanced aspects of probability. Detailed solutions. 88pp. 5⅜ × 8½.
65355-2 Pa. $3.95

EXPERIMENTS IN TOPOLOGY, Stephen Barr. Classic, lively explanation of one of the byways of mathematics. Klein bottles, Moebius strips, projective planes, map coloring, problem of the Koenigsberg bridges, much more, described with clarity and wit. 43 figures. 210pp. 5⅜ × 8½. 25933-1 Pa. $4.95

RELATIVITY IN ILLUSTRATIONS, Jacob T. Schwartz. Clear non-technical treatment makes relativity more accessible than ever before. Over 60 drawings illustrate concepts more clearly than text alone. Only high school geometry needed. Bibliography. 128pp. 6⅛ × 9¼. 25965-X Pa. $5.95

AN INTRODUCTION TO ORDINARY DIFFERENTIAL EQUATIONS, Earl A. Coddington. A thorough and systematic first course in elementary differential equations for undergraduates in mathematics and science, with many exercises and problems (with answers). Index. 304pp. 5⅜ × 8¼. 65942-9 Pa. $7.95

FOURIER SERIES AND ORTHOGONAL FUNCTIONS, Harry F. Davis. An incisive text combining theory and practical example to introduce Fourier series, orthogonal functions and applications of the Fourier method to boundary-value problems. 570 exercises. Answers and notes. 416pp. 5⅜ × 8½. 65973-9 Pa. $8.95

THE THOERY OF BRANCHING PROCESSES, Theodore E. Harris. First systematic, comprehensive treatment of branching (i.e. multiplicative) processes and their applications. Galton-Watson model, Markov branching processes, electron-photon cascade, many other topics. Rigorous proofs. Bibliography. 240pp. 5⅜ × 8½. 65952-6 Pa. $6.95

AN INTRODUCTION TO ALGEBRAIC STRUCTURES, Joseph Landin. Superb self-contained text covers "abstract algebra": sets and numbers, theory of groups, theory of rings, much more. Numerous well-chosen examples, exercises. 247pp. 5⅜ × 8½. 65940-2 Pa. $6.95

GAMES AND DECISIONS: Introduction and Critical Survey, R. Duncan Luce and Howard Raiffa. Superb non-technical introduction to game theory, primarily applied to social sciences. Utility theory, zero-sum games, n-person games, decision-making, much more. Bibliography. 509pp. 5⅜ × 8½. 65943-7 Pa. $10.95

Prices subject to change without notice.

Available at your book dealer or write for free Mathematics and Science Catalog to Dept. GI, Dover Publications, Inc., 31 East 2nd St., Mineola, N.Y. 11501. Dover publishes more than 175 books each year on science, elementary and advanced mathematics, biology, music, art, literary history, social sciences and other areas.